Lecture Notes in Physics

The Editorial Policy for Proceedings

The series Lecture Notes in Physics reports new developments in physical research and teaching – quickly, informally, and at a high level. The proceedings to be considered for publication in this series should be limited to only a few areas of research, and these should be closely related to each other. The contributions should be of a high standard and should avoid lengthy redraftings of papers already published or about to be published elsewhere. As a whole, the proceedings should aim for a balanced presentation of the theme of the conference including a description of the techniques used and enough motivation for a broad readership. It should not be assumed that the published proceedings must reflect the conference in its entirety. (A listing or abstracts of papers presented at the meeting but not included in the proceedings could be added as an appendix.)

When applying for publication in the series Lecture Notes in Physics the volume's editor(s) should submit sufficient material to enable the series editors and their referees to make a fairly accurate evaluation (e.g. a complete list of speakers and titles of papers to be presented and abstracts). If, based on this information, the proceedings are (tentatively) accepted, the volume's editor(s), whose name(s) will appear on the title pages, should select the papers suitable for publication and have them refereed (as for a journal) when appropriate. As a rule discussions will not be accepted. The series editors and Springer-Verlag will normally not interfere with the detailed editing except in fairly obvious cases or on technical matters.

Final acceptance is expressed by the series editor in charge, in consultation with Springer-Verlag only after receiving the complete manuscript. It might help to send a copy of the authors' manuscripts in advance to the editor in charge to discuss possible revisions with him. As a general rule, the series editor will confirm his tentative acceptance if the final manuscript corresponds to the original concept discussed, if the quality of the contribution meets the requirements of the series, and if the final size of the manuscript does not greatly exceed the number of pages originally agreed upon. The manuscript should be forwarded to Springer-Verlag shortly after the meeting. In cases of extreme delay (more than six months after the conference) the series editors will check once more the timeliness of the papers. Therefore, the volume's editor(s) should establish strict deadlines, or collect the articles during the conference and have them revised on the spot. If a delay is unavoidable, one should encourage the authors to update their contributions if appropriate. The editors of proceedings are strongly advised to inform contributors about these points at an early stage.

The final manuscript should contain a table of contents and an informative introduction accessible also to readers not particularly familiar with the topic of the conference. The contributions should be in English. The volume's editor(s) should check the contributions for the correct use of language. At Springer-Verlag only the prefaces will be checked by a copy-editor for language and style. Grave linguistic or technical shortcomings may lead to the rejection of contributions by the series editors. A conference report should not exceed a total of 500 pages. Keeping the size within this bound should be achieved by a stricter selection of articles and not by imposing an upper limit to the length of the individual papers. Editors receive jointly 30 complimentary copies of their book. They are entitled to purchase further copies of their book at a reduced rate. As a rule no reprints of individual contributions can be supplied. No royalty is paid on Lecture Notes in Physics volumes. Commitment to publish is made by letter of interest rather than by signing a formal contract. Springer-Verlag secures the copyright for each volume.

The Production Process

The books are hardbound, and the publisher will select quality paper appropriate to the needs of the author(s). Publication time is about ten weeks. More than twenty years of experience guarantee authors the best possible service. To reach the goal of rapid publication at a low price the technique of photographic reproduction from a camera-ready manuscript was chosen. This process shifts the main responsibility for the technical quality considerably from the publisher to the authors. We therefore urge all authors and editors of proceedings to observe very carefully the essentials for the preparation of camera-ready manuscripts, which we will supply on request. This applies especially to the quality of figures and halftones submitted for publication. In addition, it might be useful to look at some of the volumes already published. As a special service, we offer free of charge LATEX and TEX macro packages to format the text according to Springer-Verlag's quality requirements. We strongly recommend that you make use of this offer, since the result will be a book of considerably improved technical quality. To avoid mistakes and time-consuming correspondence during the production period the conference editors should request special instructions from the publisher well before the beginning of the conference. Manuscripts not meeting the technical standard of the series will have to be returned for improvement.

For further information please contact Springer-Verlag, Physics Editorial Department II, Tiergartenstrasse 17, D-69121 Heidelberg, FRG

H. Latal   W. Schweiger (Eds.)

Matter Under Extreme Conditions

Proceedings of the
33. Internationale Universitätswochen
für Kern- und Teilchenphysik
Schladming, Austria, 27 February - 5 March 1994

Springer-Verlag
Berlin Heidelberg GmbH

Editors

Heimo Latal
Wolfgang Schweiger
Institut für Theoretische Physik
Karl-Franzens-Universität Graz
Universitätsplatz 5, A-8010 Graz, Austria

Supported by the Bundesministerium für Wissenschaft und Forschung, Vienna, Austria

ISBN 978-3-662-13964-6   ISBN 978-3-540-49042-5 (eBook)
DOI 10.1007/978-3-540-49042-5

CIP data applied for

This book was processed by the authors/editors using the T$_E$X/L^{A}T$_E$X macro package from Springer-Verlag Berlin Heidelberg GmbH.
SPIN: 10127090   55/3140-543210 - Printed on acid-free paper

Preface

This volume contains the written versions of the invited lectures presented at the "33. Internationale Universitätswochen für Kern- und Teilchenphysik" in Schladming, Austria, which took place from February 27th to March 5th, 1994. The title of the School was "Matter under Extreme Conditions". Situations where this is the case can be found at high-energy accelerators producing extremely fast particles or heavy ions here on Earth, or out in the Cosmos at places of violent activity, as in supernova explosions, near white dwarfs or neutron stars, or finally back in time at the very beginning of our Universe. Some of the processes happening there and their theoretical implications have been presented by the lecturers of the Winter School. Of course, only a limited selection of the great manifold of problems available could be made. In the following they are characterized in a kind of topical order.

One of the basic questions in any model of the Universe is, "How did matter arise in the form as we see it today?" This first occurrence of matter under the most extreme conditions imaginable is the topics treated by Keith A. Olive in his lectures.

Another source of matter, especially of the chemical elements, are nuclear reactions at the center of stars. Immediate information on the processes happening within the Sun is provided by neutrinos generated there. Corresponding experiments at a number of laboratories have raised one of the most puzzling enigmas in astrophysics: why are we seeing only about one half of the neutrinos that are expected from the currently accepted model of the Sun? Some ideas about the solution to this problem were presented by Michel Spiro.

At the end of its life, a sufficiently massive star ejects a large fraction of its material in a supernova explosion and collapses into a neutron star. The physical processes during this stage of enormous densities and temperatures were treated in the lectures by Robert Mochkovitch.

Near the remnants of collapsed stars, that is, white dwarfs or neutron stars, one expects and has measured magnetic fields of immense strengths. In this environment the properties of atoms will be drastically altered. On Earth, a similar situation arises for highly excited atoms, so-called Rydberg atoms, in moderate magnetic fields. With this situation, interestingly enough, the question as to the existence or non-existence of quantum chaos is closely linked. In his lectures Günter Wunner gave a review of the current state of knowledge on these topics.

Another type of fields, namely strong electric ones, and their influence on the properties of matter, and even the vacuum, were the subject of the lecture series by Walter Greiner. Experimentally, such situations can be realized in heavy ion collisions. But not only the electronic shells of the ions are drastically altered in these encounters, also the nuclei may be compressed to unimaginable densities such that their components, protons and neutrons, dissolve into their constituents and form a so-called quark-gluon plasma. Herbert Ströbele gave an account of the possible experimental evidence for this proposed exotic state of matter.

A kind of completion and conclusion of the program was provided in the report by Michel della Negra on the motivation for and the results expected from the world's largest accelerator, the "Large Hadron Collider" at CERN.

In addition many excellent seminars were presented at the School, but due to space limitations could not be included in these proceedings. However, a list of seminar contributions is given at the end of this volume, so that interested readers may request information or pertinent material directly from the authors.

Finally, we would like to express our thanks to the lecturers for all their efforts, to the sponsors of the School, above all the Austrian Ministry of Science and Research and the Government of Styria, for providing generous support, to our colleagues in the organizing committee for their assistance, and to Mrs. E. Neuhold for her help in bringing the files prepared by the authors in TEX or LATEX into their final form.

Graz, July 1994

> *H. Latal (Director of the School)*
> *W. Schweiger (Scientific Secretary)*

Contents

Theoretical Aspects of Quantum Electrodynamics in Strong Fields

By J. Reinhardt and W. Greiner

The Search for the Quark-Gluon Plasma

By H. Ströbele

Physics at LHC

By M. Della Negra

Contributors

Della Negra, M.
PPE-Division, CERN
CH-1211 Genève 23, Switzerland

Greiner, W.
Institut für Theoretische Physik, Johann Wolfgang Goethe-Universität
D-60054 Frankfurt am Main, Robert-Mayer-Str. 8-10, Germany

Mochkovitch, R.
Institut d'Astrophysique de Paris
F-75014 Paris, 98 bis Boulevard Arago, France

Olive, K.A.
School of Physics and Astronomy, University of Minnesota
Minneapolis MN 55455, USA

Reinhardt, J.
Institut für Theoretische Physik, Johann Wolfgang Goethe-Universität
D-60054 Frankfurt am Main, Robert-Mayer-Str. 8-10, Germany

Spiro, M.
DAPNIA, Centre d'Etudes Saclay
F-91191 Gif-sur-Yvette, France

Ströbele, H.
Fachbereich Physik, Johann Wolfgang Goethe-Universität
D-60486 Frankfurt am Main, August-Euler-Str. 6, Germany

Wunner, G.
Theoretische Physik I, Ruhr-Universität
D-44780 Bochum, Universitätsstr. 150, Germany

Big Bang Baryogenesis

Keith A. Olive

School of Physics and Astronomy, University of Minnesota,
Minneapolis MN 55455, USA

1 The Standard Model

One of the most basic questions we can ask about the Universe is: What is the origin of matter? There are of course many ways in which to interpret this question, and there are varying depths to which it can be answered. Essentially all of the mass in the Universe that is *observed* is in the form of baryons, today consisting of protons and neutrons in the nuclei of atoms. While baryons may not be the dominant component of the Universe, they are without a doubt present and essential to our existence. However, the fact that a significant part of the mass of the Universe is baryonic is not in and of itself surprising. The lightest baryons are relatively long lived by particle physics standards and massive. Protons have extremely long lifetimes (or, in a boring world may be stable) and neutrons live long enough to become incorporated primarily into helium nuclei in the early Universe (see below). While electrons are stable (so long as electric charge is conserved), and they are present in numbers equal to that of protons, they are too light to make a significant contribution to the mass density of the Universe. Other stable particles which may yet be found to be massive, such as neutrinos, or still to be discovered such as the lightest supersymmetric particle, may in fact dominate the overall mass density of the Universe. There are however, two known particles which on the basis of mass and lifetime could be expected to contribute to the mass of the Universe: the anti-proton and the anti-neutron. $\bar{p}$ and $\bar{n}$ have, of course, exactly the same mass and lifetime as p and n. Yet these antibaryons are not observed in any abundance in nature. The creation of this asymmetry between baryons and anti-baryons or between matter and anti-matter is the subject of these lectures.

To deal with the specific problem of the baryon asymmetry, it will be useful to briefly review the standard cosmological model as a framework towards a solution. To put the problem in perspective, it is useful to have an idea of the general sequence of events which are believed to have occurred since the big bang. The earliest times (after the big bang) that we are able to discuss are after $t \simeq 10^{-44}$s, or at temperatures of about 10^{18} GeV. This period is the Planck epoch and a description of events at or prior to this time would require a more complete theory of quantum gravity which may yet be found in string theory. The Grand Unified (GUT) scale is typically at $T \sim 10^{15}$ GeV at times of about 10^{-35}s. Standard models of baryogenesis and inflation may have played important roles at this time. Barring new interactions at an intermediate scale, electroweak symmetry breaking then occurred at times of order 10^{-10}s at the electroweak scale of 100 GeV. Quark-gluon confinement should have taken place at $t \sim 10^{-5}$s at $T \sim \Lambda_{QCD} \sim 100$ MeV. Big bang nucleosynthesis and the formation of the light element isotopes of D, ^{3}He, ^{4}He and ^{7}Li took place between 1 and 100 s, at temperatures below 1 MeV. It wasn't until $t \sim 10^{12}$ s or $T \sim 1$ eV that recombination of neutral hydrogen occurred and the formation of galaxies began. Finally to put things in perspective, the age of the Universe today is $\sim 10^{17}$ s and the temperature is the well known 2.726 K as measured by COBE [1]

The standard big bang model assumes homogeneity and isotropy, so that space-time can be described by the Friedmann-Robertson-Walker metric which in co-moving coordinates is given by

$$ds^2 = -dt^2 + R^2(t) \left[\frac{dr^2}{(1 - kr^2)} + r^2 \left(d\theta^2 + \sin^2 \theta d\phi^2 \right) \right] , \qquad (1)$$

where $R(t)$ is the cosmological scale factor and k is the three-space curvature constant ($k = 0, +1, -1$ for a spatially flat, closed or open Universe). k and R are the only two quantities in the metric which distinguish it from flat Minkowski space. It is also common to assume the perfect fluid form for the energy-momentum tensor

$$T_{\mu\nu} = pg^{\mu\nu} + (p + \rho)u^\mu u^\nu , \qquad (2)$$

where $g_{\mu\nu}$ is the space-time metric described by (1), p is the isotropic pressure, ρ is the energy density and $u^\mu = (1,0,0,0)$ is the velocity vector for the isotropic fluid. Einstein's equation yield the Friedmann equation

$$H^2 \equiv \left(\frac{\dot{R}}{R} \right)^2 = \frac{1}{3} 8\pi G_N \rho - \frac{k}{R^2} + \frac{1}{3}\Lambda \qquad (3)$$

and

$$\left(\frac{\ddot{R}}{R} \right) = \frac{1}{3}\Lambda - \frac{1}{6} 8\pi G_N(\rho + 3p) , \qquad (4)$$

where Λ is the cosmological constant, or equivalently from $T^{\mu\nu}{}_{;\nu} = 0$

$$\dot{\rho} = -3H(\rho + p) . \qquad (5)$$

These equations form the basis of the standard big bang model.

At early times ($t < 10^5$ yrs) the Universe is thought to have been dominated by radiation so that the equation of state can be given by $p = \rho/3$. If we neglect the contributions to H from k and Λ (this is always a good approximation for small enough R) then we find that

$$R(t) \sim t^{1/2} \qquad (6)$$

and $\rho \sim R^{-4}$ so that $t \sim (3/32\pi G_N\rho)^{1/2}$. Similarly for a matter or dust dominated Universe with $p = 0$,

$$R(t) \sim t^{2/3} \tag{7}$$

and $\rho \sim R^{-3}$. The Universe makes the transition between radiation and matter domination when $\rho_{\rm rad} = \rho_{\rm matter}$ or when $T \simeq$ few $\times 10^3$ K.

In the absence of a cosmological constant, one can define a critical energy density ρ_c such that $\rho = \rho_c$ for $k = 0$

$$\rho_c = 3H^2/8\pi G_N . \tag{8}$$

In terms of the present value of the Hubble parameter this is

$$\rho_c = 1.88 \times 10^{-29} h_0{}^2 {\rm gcm}^{-3} , \tag{9}$$

where

$$h_0 = H_0/(100\,{\rm kmMpc}^{-1}{\rm s}^{-1}) . \tag{10}$$

The cosmological density parameter is then defined by

$$\Omega \equiv \frac{\rho}{\rho_c} \tag{11}$$

in terms of which the Friedmann equation (3) can be rewritten as

$$(\Omega - 1)H^2 = \frac{k}{R^2} \tag{12}$$

so that $k = 0, +1, -1$ corresponds to $\Omega = 1, \Omega > 1$ and $\Omega < 1$. Observational limits on h_0 and Ω are [2]

$$0.4 \leq h_0 \leq 1.0 , \qquad 0.1 \leq \Omega \leq 2 . \tag{13}$$

It is important to note that Ω is a function of time or of the scale factor. The evolution of Ω is shown in Fig. 1 for $\Lambda = 0$. For a spatially flat Universe, $\Omega = 1$ always. When $k = +1$, there is a maximum value for the scale factor R. At early times (small values of R), Ω always tends to one. Note that the fact that we do not yet know the sign of k, or equivalently whether Ω is larger than or smaller than unity, implies that we are at present still at the very left in the figure. What makes this peculiar is that one would normally expect that the sign of k to become apparent after a Planck time of 10^{-43} s. It is extremely puzzling that some 10^{60} Planck times later, we still do not know the sign of k.

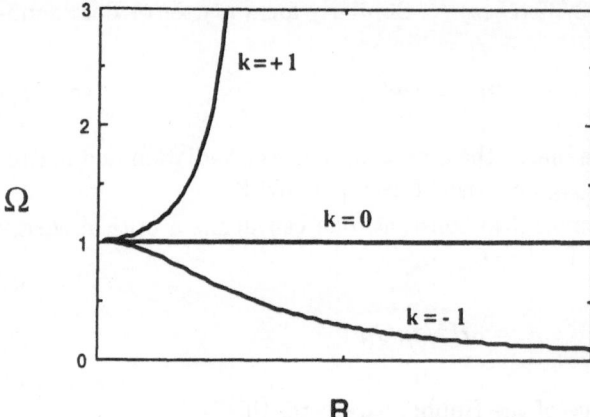

Fig. 1. The evolution of the cosmological density parameter, Ω, as a function of the scale factor for a closed, open and spatially flat Universe.

1.1 The Hot Thermal Universe

The epoch of recombination occurs when electrons and protons form neutral hydrogen through $e^- + p \rightarrow H + \gamma$ at a temperature $T_R \sim$ few $\times 10^3$ K ~ 1 eV. For $T < T_R$, photons are decoupled while for $T > T_R$, photons are in thermal equilibrium and the Universe is usually taken to be radiation dominated so that the content of the radiation plays a very important role. Today, the content of the microwave background consists of photons with $T_0 = 2.726 \pm .01$ K [1]. We can calculate the energy density of photons from

$$\rho_\gamma = \int E_\gamma dn_\gamma , \tag{14}$$

where the density of states is given by

$$dn_\gamma = \frac{g_\gamma}{2\pi^2}[\exp(E_\gamma/T) - 1]^{-1}q^2 dq , \tag{15}$$

and $g_\gamma = 2$ simply counts the number of degrees of freedom for photons, $E_\gamma = q$ is just the photon energy (momentum). (I am using units such that $\hbar = c = k_B = 1$ and will do so through the remainder of these lectures.) Integrating (14) gives

$$\rho_\gamma = \frac{\pi^2}{15}T^4 \tag{16}$$

which is the familiar blackbody result.

In general, at very early times, at very high temperatures, other particle degrees of freedom join the radiation background when $T \sim m_i$ for each particle type i if that type is brought into thermal equilibrium through interactions. In equilibrium the energy density of a particle type i is given by

$$\rho_i = \int E_i dn_{q_i} , \tag{17}$$

and

$$dn_{q_i} = \frac{g_i}{2\pi^2}[\exp[(E_{q_i} - \mu_i)/T] \pm 1]^{-1}q^2 dq , \qquad (18)$$

where again g_i counts the total number of degrees of freedom for type i,

$$E_{q_i} = \left(m_i^2 + q_i^2\right)^{1/2} . \qquad (19)$$

μ_i is the chemical potential if present and $\pm$ corresponds to either Fermi or Bose statistics.

In the limit that $T \gg m_i$ the total energy density can be conveniently expressed by

$$\rho = \left(\sum_B g_B + \frac{7}{8}\sum_F g_F\right)\frac{\pi^2}{30}T^4 \equiv \frac{\pi^2}{30}N(T)T^4 , \qquad (20)$$

where $g_{B(F)}$ are the total number of boson (fermion) degrees of freedom and the sum runs over all boson (fermion) states with $m \ll T$. The factor of 7/8 is due to the difference between the Fermi and Bose integrals. Equation (20) defines $N(T)$ by taking into account new particle degrees of freedom as the temperature is raised.

In the radiation dominated epoch, eq. (5) can be integrated (neglecting the T-dependence of N) giving us a relationship between the age of the Universe and its temperature

$$t = \left(\frac{90}{32\pi^3 G_N N(T)}\right)^{1/2} T^{-2} . \qquad (21)$$

Put into a more convenient form

$$tT_{\mathrm{MeV}}^2 = 2.4[N(T)]^{-1/2} , \qquad (22)$$

where t is measured in seconds and T_{MeV} in units of MeV.

The value of $N(T)$ at any given temperature depends on the particle physics model. In the standard $SU(3) \times SU(2) \times U(1)$ model, we can specify $N(T)$ up to temperatures of $\mathcal{O}(100)$ GeV. The change in N can be seen in table 1.

At higher temperatures, $N(T)$ will be model dependent. For example, in the minimal $SU(5)$ model, one needs to add to $N(T)$, 24 states for the X and Y gauge bosons, another 24 from the adjoint Higgs, and another 6 (in addition to the 4 already counted in $W^{\pm}$, Z and H) from the 5 of Higgs. Hence for $T > M_X$ in minimal $SU(5)$, $N(T) = 160.75$. In a supersymmetric model this would at least double, with some changes possibly necessary in the table if the lightest supersymmetric particle has a mass below M_H.

The notion of equilibrium also plays an important role in the standard big bang model. If, for example, the Universe were not expanding, then given enough time, every particle state would come into equilibrium with each other. Because of the expansion of the Universe, certain rates might be too slow indicating, for example, in a scattering process that the two incoming states might never find each other to bring about an interaction. Depending on their rates, certain interactions may pass in and out of thermal equilibrium during the course of the

Table 1. Effective numbers of degrees of freedom in the standard model.

Temperature	New Particles	$4N(T)$
$T < m_e$	γ's $+ \nu$'s	29
$m_e < T < m_\mu$	$e^\pm$	43
$m_\mu < T < m_\pi$	$\mu^\pm$	57
$m_\pi < T < T_c^*$	π's	69
$T_c < T < m_{strange}$	$- \pi$'s $+$ u,ū,d,d̄ $+$ gluons	205
$m_s < T < m_{charm}$	s,s̄	247
$m_c < T < m_\tau$	c,c̄	289
$m_\tau < T < m_{bottom}$	$\tau^\pm$	303
$m_b < T < m_{W,Z}$	b,b̄	345
$m_{W,Z} < T < m_{top}$	$W^\pm$, Z	381
$m_t < T < m_{Higgs}$	t,t̄	423
$M_H < T$	H^0	427

*T_c corresponds to the confinement-deconfinement transition between quarks and hadrons. $g(T) = N(T)$ is shown in Fig. 2 for $T_c = 150$ and 400 MeV. It has been assumed that $m_{Higgs} > m_{top}$.

Universal expansion. Quantitatively, for each particle i, we will require that some rate Γ_i involving that type be larger than the expansion rate of the Universe or

$$\Gamma_i > H \tag{23}$$

in order to be in thermal equilibrium.

A good example for a process in equilibrium at some stage and out of equilibrium at others is that of neutrinos. If we consider the standard neutral or charged-current interactions such as $e^+ + e^- \leftrightarrow \nu + \bar{\nu}$ or $e + \nu \leftrightarrow e + \nu$ etc., very roughly the rates for these processes will be

$$\Gamma = n\langle\sigma v\rangle \,, \tag{24}$$

where $\langle\sigma v\rangle$ is the thermally averaged weak interaction cross section

$$\langle\sigma v\rangle \sim \mathcal{O}(10^{-2})T^2/M_W^4 \,, \tag{25}$$

and n is the number density of leptons. Hence the rate for these interactions is

$$\Gamma_{wk} \sim \mathcal{O}(10^{-2})T^5/M_W^4 \,. \tag{26}$$

The expansion rate, on the other hand, is just

$$H = \left(\frac{8\pi G_N \rho}{3}\right)^{1/2} = \left(\frac{8\pi^3}{90}N(T)\right)^{1/2} T^2/M_P \sim 1.66 N(T)^{1/2}T^2/M_P \,. \tag{27}$$

The Planck mass $M_P = G_N^{-1/2} = 1.22 \times 10^{19}$ GeV.

Neutrinos will be in equilibrium when $\Gamma_{wk} > H$ or

$$T > (500 M_W^4)/M_P)^{1/3} \sim 1\text{MeV} \,. \tag{28}$$

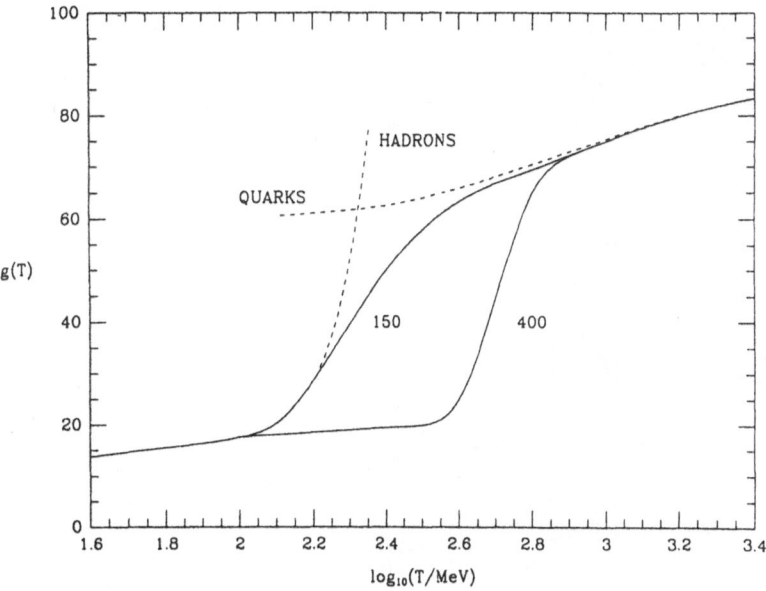

Fig. 2. The effective numbers of relativistic degrees of freedom as a function of temperature. The dashed lines correspond to free quarks and hadrons.

The temperature at which these rates are equal is commonly referred to as the decoupling or freeze-out temperature and is defined by

$$\Gamma(T_{\rm d}) = H(T_{\rm d}) \ . \tag{29}$$

For temperatures $T > T_{\rm d}$, neutrinos will be in equilibrium, while for $T < T_{\rm d}$ they will not. Basically, in terms of their interactions, the expansion rate is just too fast and they never *"see"* the rest of the matter in the Universe (nor themselves). Their momenta will simply redshift and their effective temperature (the shape of their momenta distribution is not changed from that of a blackbody) will simply fall with $T \sim 1/R$.

1.2 Big Bang Nucleosynthesis

An essential element of the standard cosmological model is big bang nucleosynthesis, the theory which predicts the abundances of the light element isotopes D, ^{3}He, ^{4}He, and ^{7}Li. As was mentioned earlier, nucleosynthesis takes place at a temperature scale of order 1 MeV. At temperatures above 1 MeV, the weak interactions, being in equilibrium, determined the ratio of neutrons to protons. Near 1 MeV, these interactions: $n + e^+ \leftrightarrow p + \bar{\nu}$; $n + \nu \leftrightarrow p + e^-$; and $n \leftrightarrow p + e^- + \bar{\nu}$; as the e,$\nu$ interactions discussed above, drop out of equilibrium. Although the binding energy of deuterium is 2.2 MeV, due to the high photon to baryon ratio (10^{10}), nucleosynthesis is delayed until about $T \sim 2.2/\ln 10^{10} \sim 0.1$ MeV, when deuterium can be formed without significant dissociation. Afterwhich, nucleosynthesis proceeds rapidly with the build-up of the light elements.

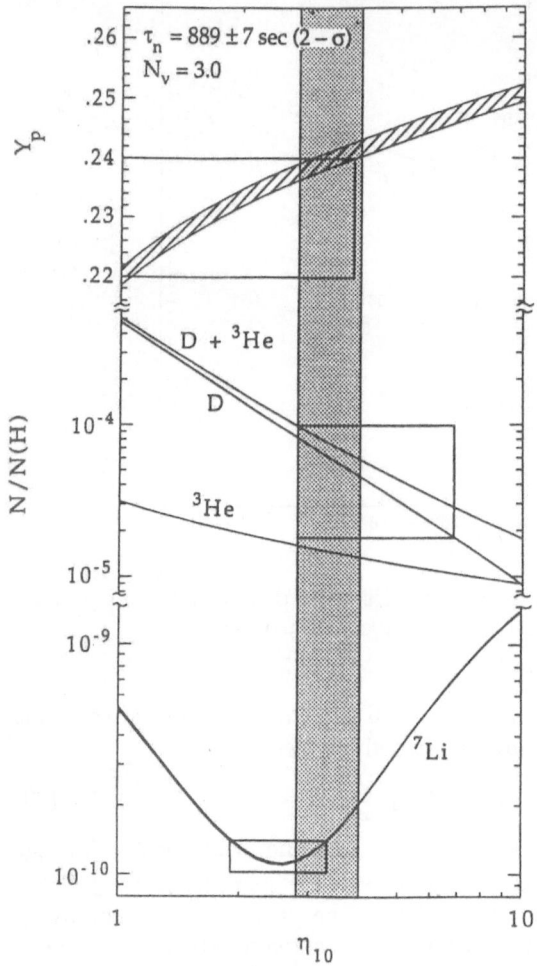

Fig. 3. The abundances of the light elements as a function of the baryon-to-photon ratio.

The nuclear processes lead primarily to ^{4}He, which is produced at about 24% by mass. Lesser amounts of the other light elements are produced: about 10^{-5} of D and ^{3}He and about 10^{-10} of ^{7}Li by number relative to H. The abundances of the light elements depend almost solely on one key parameter, the baryon-to-photon ratio, η. In Fig. 3, (taken from Ref. [3]) the predicted abundances of the light elements are shown as a function of $\eta_{10} = 10^{10}\eta$. In Fig. 3, the boxes correspond to acceptable values for the abundances as determined from the observations. The band for the ^{4}He curve shows the sensitivity to the neutron half life. The vertical band shows the overall range of η in which agreement is achieved between theory and observation for all of the light elements. From the figure we see that consistency is found for

$$2.8 \times 10^{-10} < \eta < 3.2(4) \times 10^{-10} , \tag{30}$$

where the bound can be as high as $\eta_{10} < 4$ when the uncertainties in ^{7}Li cross-sections are accounted for.

It is important to note that η is related to the fraction of Ω contained in baryons, Ω_B

$$\Omega_B = 3.66 \times 10^7 \eta h_0^{-2} (T_0/2.726)^3 . \tag{31}$$

Using the limits on η and h_0, one finds that Ω_B is restricted to a range

$$0.01 < \Omega_B < 0.08 \tag{32}$$

and one can conclude that the Universe is not closed by baryons. This value of η is the one that we try to explain by big bang baryogenesis.

1.3 Problems with the (Non-Inflationary) Standard Model

Despite the successes of the standard big bang model, there are a number of unanswered questions that appear difficult to explain without imposing what may be called unnatural initial conditions. The resolution of these problems may lie in a unified theory of gauge interactions or possibly in a theory which includes gravity. For example, prior to the advent of grand unified theories (GUTs), the baryon-to-photon ratio, could have been viewed as being embarrassingly small. Although, we still do not know the precise mechanism for generating the baryon asymmetry, many quite acceptable models are available as will be discussed in some detail for the better part of these lectures. In a similar fashion, it is hoped that a field theoretic description of inflation may resolve the problems outlined below.

The Curvature Problem. The bound on Ω in Eq. (13) is curious in the fact that at the present time we do not know even the sign of the curvature term in the Friedmann equation (3), i.e., we do not know if the Universe is open, closed or spatially flat.

The curvature problem (or flatness problem or age problem) can manifest itself in several ways. For a radiation dominated gas, the entropy density $s \sim T^3$ and $R \sim T^{-1}$. Thus assuming an adiabatically expanding Universe, the quantity $\hat{k} = k/R^2T^2$ is a dimensionless constant. If we now apply the limit in (13) to (12) we find

$$\hat{k} = \frac{k}{R^2T^2} = \frac{(\Omega_0 - 1)H_0^2}{T_0^2} < 2 \times 10^{-58} . \tag{33}$$

This limit on k represents an initial condition on the cosmological model. The problem then becomes what physical processes in the early Universe produced a value of $\hat{k}$ so extraordinarily close to zero (or Ω close to one). A more natural initial condition might have been $\hat{k} \sim \mathcal{O}(1)$. In this case the Universe would have become curvature dominated at $T \sim 10^{-1} M_P$. For $k = +1$, this would signify the onset of recollapse. As already noted earlier, one would naturally expect the effects of curvature (seen in Fig. 1 by the separation of the three curves) to manifest themselves at times on the order of the Planck time as gravity should provide the only dimensionful scale in this era. If we view the evolution of Ω in Fig. 1 as a function of time, then it would appear that the time $t_0 = 13$ Gyr $= 7.6 \times 10^{60} M_P^{-1}$ ($\sim$ the current age of the Universe) appears at the far left of x-axis, i.e. before the curves separate. Why then has the Universe lasted so long before revealing the true sign of k?

Even for $\hat{k}$ as small as $\mathcal{O}(10^{-40})$ the Universe would have become curvature dominated when $T \sim 10$ MeV, i.e., before the onset of big bang nucleosynthesis. Of course, it is also possible that $k = 0$ and the Universe is actually spatially flat. In fact, today we would really expect only one of two possible values for Ω: 0 or 1.

The Horizon Problem. Because of the cosmological principle, all physical length scales grow as the scale factor $R(t) \sim t^{2/3\gamma}$, with γ defined by $p = (\gamma-1)\rho$. However, causality implies the existence of a particle (or causal) horizon $d_H(t) \sim t$, which is the maximal physical distance light can travel from the co-moving position of an observer at some initial time ($t = 0$) to time t. For $\gamma > \frac{2}{3}$, scales originating outside of the horizon will eventually become part of our observable Universe. Hence we would expect to see anisotropies on large scales [4].

In particular, let us consider the microwave background today. The photons we observe have been decoupled since recombination at $T_d \sim 4000K$. At that time, the horizon volume was simply $V_d \propto t_d{}^3$, where t_d is the age of the Universe at $T = T_d$. Then $t_d = t_0(T_0/T_d)^{3/2} \sim 2 \times 10^5$ yrs, where $T_0 = 2.726K$ [1] is the present temperature of the microwave background. Our present horizon volume $V_0 \propto t_0{}^3$ can be scaled back to t_d (corresponding to that part of the Universe which expanded to our present visible Universe) $V_0(t_d) \propto V_0(T_0/T_d)^3$. We can now compare $V_0(t_d)$ and V_d. The ratio

$$\frac{V_0(t_d)}{V_d} \propto \frac{V_0 T_0{}^3}{V_d T_d{}^3} \propto \frac{t_0{}^3 T_0{}^3}{t_d{}^3 T_d{}^3} \sim 5 \times 10^4 \tag{34}$$

corresponds to the number of horizon volumes or casually distinct regions at decoupling which are encompassed in our present visible horizon.

In this context, it is astonishing that the microwave background appears highly isotropic on large scales with $\Delta T/T = 1.1 \pm 0.2 \times 10^{-5}$ at angular separations of $10°$ [5]. The horizon problem, therefore, is the lack of an explanation as to why nearly 10^5 causally disconnected regions at recombination all had the same temperature to within one part in 10^{-5}.

Density Perturbations. Although it appears that the Universe is extremely isotropic and homogeneous on very large scales (in fact the standard model assumes complete isotropy and homogeneity) it is very inhomogeneous on small scales. In other words, there are planets, stars, galaxies, clusters, etc. On these small scales there are large density perturbations namely $\delta\rho/\rho \gg 1$. At the same time, we know from the isotropy of the microwave background that on the largest scales, $\delta\rho/\rho \sim 3\Delta T/T \sim \mathcal{O}(10^{-5})$ [5] and these perturbations must have grown to $\delta\rho/\rho \sim 1$ on smaller scales.

In an expanding Universe, density perturbations evolve with time [6]. The evolution of the Fourier transformed quantity $\frac{\delta\rho}{\rho}(k,t)$ depends on the relative size of the wavelength $\lambda \sim k^{-1}$ and the horizon scale H^{-1}. For $k \ll H$, (always true at sufficiently early times) $\delta\rho/\rho \propto t$ while for $k \gg H$, $\delta\rho/\rho$ is $\simeq$ constant (or grows moderately as $\ln t$) assuming a radiation dominated Universe. In a matter dominated Universe, on scales larger than the Jean's length scale (determined by $k_{\mathrm{J}} = 4\pi G_{\mathrm{N}} \rho_{\mathrm{matter}}/v_{\mathrm{s}}^2$, $v_{\mathrm{s}} =$ sound speed) perturbations grow with the scale factor R. Because of the growth in $\delta\rho/\rho$, the microwave background limits force $\delta\rho/\rho$ to be extremely small at early times.

Consider a perturbation with wavelength on the order of a galactic scale. Between the Planck time and recombination, such a perturbation would have grown by a factor of $\mathcal{O}(10^{57})$ and the anisotropy limit of $\delta\rho/\rho \lesssim 10^{-5}$ implies that $\delta\rho/\rho < 10^{-61}$ on the scale of a galaxy at the Planck time. One should compare this value with that predicted from purely random (or Poisson) fluctuations of $\delta\rho/\rho \sim 10^{-40}$ (assuming 10^{80} particles (photons) in a galaxy) [7]. The extent of this limit is of course related to the fact that the present age of the Universe is so great.

An additional problem is related to the formation time of the perturbations. A perturbation with a wavelength large enough to correspond to a galaxy today must have formed with wavelength modes much greater than the horizon size if the perturbations are primordial as is generally assumed. This is due to the fact that the wavelengths red shift as $\lambda \sim R \sim t^{1/2}$ while the horizon size grows linearly. It appears that a mechanism for generating perturbations with acausal wavelengths is required.

The Magnetic Monopole Problem. In addition to the much desired baryon asymmetry produced by grand unified theories, a less favorable aspect is also present; GUTs predict the existence of magnetic monopoles. Monopoles will be produced [8] whenever any simple group [such as SU(5)] is broken down to a gauge group which contains a U(1) factor [such as SU(3) $\times$ SU(2) $\times$ U(1)]. The mass of such a monopole would be

$$M_{\mathrm{m}} \sim M_{\mathrm{GUT}}/\alpha_{\mathrm{GUT}} \sim 10^{16} GeV. \qquad (35)$$

The basic reason monopoles are produced is that in the breaking of SU(5) the Higgs adjoint needed to break SU(5) cannot align itself over all space [9]. On scales larger than the horizon, for example, there is no reason to expect the direction of the Higgs field to be aligned. Because of this randomness, topological knots are expected to occur and these are the magnetic monopoles. We can then estimate that the minimum number of monopoles produced [10] would be roughly one per horizon volume or causally connected region at the time of the SU(5) phase transition t_c

$$n_m \sim (2t_c)^{-3} \tag{36}$$

resulting in a monopole-to-photon ratio expressed in terms of the transition temperature of

$$\frac{n_m}{n_\gamma} \sim \left(\frac{10T_c}{M_P}\right)^3 . \tag{37}$$

The overall mass density of the Universe can be used to place a constraint on the density of monopoles. For $M_m \sim 10^{16}$ GeV and $\Omega_m h_0^2 \lesssim 1$ we have that

$$\frac{n_m}{n_\gamma} \lesssim \mathcal{O}(10^{-25}) . \tag{38}$$

The predicted density, however, from (37) for $T_c \sim M_{\mathrm{GUT}}$

$$\frac{n_m}{n_\gamma} \sim 10^{-9} . \tag{39}$$

Hence, we see that standard GUTs and cosmology have a monopole problem.

1.4 Inflation

All of the problems discussed above can be neatly resolved if the Universe underwent a period of cosmological inflation [11]. During a phase transition, our assumptions of an adiabatically expanding Universe may not be valid. If we look at a scalar potential describing a phase transition from a symmetric false vacuum state $\langle \phi \rangle = 0$ for some scalar field ϕ to the broken true vacuum at $\langle \phi \rangle = v$ as in Fig. 4, and suppose we find that upon solving the equations of motion for the scalar field that the field evolves slowly from the symmetric state to the global minimum (this will depend on the details of the potential). If the evolution is slow enough, the Universe may become dominated by the vacuum energy density associated with the potential near $\eta \approx 0$. The energy density of the symmetric vacuum, $V(0)$ acts as a cosmological constant with

$$\Lambda = 8\pi V(0) M_P^2 . \tag{40}$$

During this period of slow evolution, the energy density due, to say, radiation will fall below the vacuum energy density, $\rho \ll V(0)$. When this happens, the expansion rate will be dominated by the constant $V(0)$ and from (3) we find an exponentially expanding solution

$$R(t) \sim e^{\sqrt{\Lambda/3}\, t} . \tag{41}$$

When the field evolves towards the global minimum it will begin to oscillate about the minimum, energy will be released during its decay and a hot thermal Universe will be restored. If released fast enough, it will produce radiation at a temperature $T_R^4 \lesssim V(0)$. In this reheating process entropy has been created and $(RT)_f > (RT)_i$. Thus we see that during a phase transition the relation

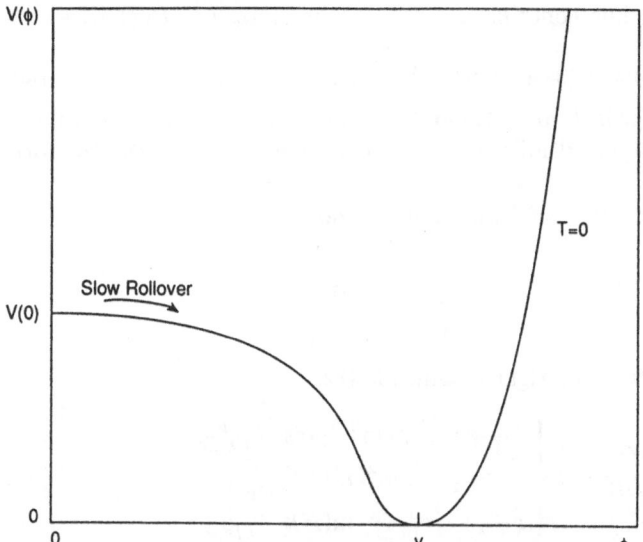

Fig. 4. A typical potential suitable for the inflationary Universe scenario.

$RT \sim$ constant, need not hold true and thus our dimensionless constant $\hat{k}$ may actually not have been constant.

If during the phase transition, the value of RT changed by a factor of $\mathcal{O}(10^{29})$, the cosmological problems would be solved. The isotropy would in a sense be generated by the immense expansion; one small causal region could get blown up and hence our entire visible Universe would have been at one time in thermal contact. In addition, the parameter $\hat{k}$ could have started out $\mathcal{O}(1)$ and have been driven small by the expansion. Density perturbations will be stretched by the expansion, $\lambda \sim R$. Thus it will appear that $\lambda \gg H^{-1}$ or that the perturbations have left the horizon. It is actually just that the size of the causally connected region is now no longer simply H^{-1}. However, not only does inflation offer an explanation for large scale perturbations, it also offers a source for the perturbations themselves [12]. Monopoles would also be diluted away.

The cosmological problems could be solved if

$$H\tau > 65 \,, \tag{42}$$

where τ is the duration of the phase transition, density perturbations are produced and do not exceed the limits imposed by the microwave background anisotropy, the vacuum energy density was converted to radiation so that the reheated temperature is sufficiently high, and baryogenesis is realized.

For the purposes of discussing baryogenesis, it will be sufficient to consider only a generic model of inflation whose potential is of the form

$$V(\eta) = \mu^4 P(\eta) \,, \tag{43}$$

where η is the scalar field driving inflation, the inflaton, μ is an as yet unspecified mass parameter, and $P(\eta)$ is a function of η which possesses the features necessary for inflation, but contains no small parameters. I.e. $P(\eta)$ takes the form

$$P(\eta) = P(0) + m^2\eta^2 + \lambda_3\eta^3 + \lambda_4\eta^4 + \dots , \tag{44}$$

where all of the couplings in P are $\mathcal{O}(1)$ and ... refers to possible non-renormaliz-able terms. Most of the useful inflationary potentials can be put into the form of (43).

The requirements for successful inflation boil down to:

1) enough inflation

$$\frac{\partial^2 V}{\partial\eta^2}\Big|_{\eta\sim\eta_i\pm H} < \frac{3H^2}{65} = \frac{8\pi V(0)}{65M_P{}^2} ; \tag{45}$$

2) density perturbations of the right magnitude [12]

$$\frac{\delta\rho}{\rho} \simeq \frac{H^2}{10\pi^{3/2}\dot\eta} \simeq \begin{cases} \left(\frac{32\lambda_4}{3\pi^3}\right)^{1/2}\frac{1}{10}\ln^{3/2}(Hk^{-1})\frac{\mu^2}{M_P{}^2} \\ \left(\frac{\lambda_3}{H\pi^{3/2}}\right)\frac{1}{10}\ln^2(Hk^{-1})\frac{\mu^4}{M_P{}^4} \\ \left(\frac{8}{3\pi^2}\right)^{1/2}\frac{1}{10}\frac{m}{M_P}\ln(Hk^{-1})\frac{\mu^2}{M_P{}^2} \end{cases} , \tag{46}$$

given here for scales which "re-enter" the horizon during the matter dominated era. These reduce approximately to

$$\frac{\delta\rho}{\rho} \sim \mathcal{O}(100)\frac{\mu^2}{M_P{}^2} ; \tag{47}$$

3) baryogenesis; the subject of the remaining lectures.

For large scale fluctuations of the type measured by COBE [5], we can use (47) to fix the inflationary scale μ. The magnitude of the density fluctuations can be related to the observed quadrupole [13] moment:

$$\langle a_2^2\rangle = \frac{5}{6}2\pi^2(\frac{\delta\rho}{\rho})^2 . \tag{48}$$

The observed quadrupole moment gives [5]:

$$\langle a_2^2\rangle = (4.7 \pm 2) \times 10^{-10} , \tag{49}$$

or

$$\frac{\delta\rho}{\rho} = (5.4 \pm 1.6) \times 10^{-6} , \tag{50}$$

which in turn fixes the coefficient μ of the inflaton potential [14]:

$$\frac{\mu^2}{M_P^2} = \text{few} \times 10^{-8} . \tag{51}$$

Fixing (μ^2/M_P^2) has immediate general consequences for inflation [15]. For example, the Hubble parameter during inflation, $H^2 \simeq (8\pi/3)(\mu^4/M_P^2)$ so that $H \sim 10^{-7}M_P$. The duration of inflation is $\tau \simeq M_P^3/\mu^4$, and the number of e-foldings of expansion is $H\tau \sim 8\pi(M_P^2/\mu^2) \sim 10^9$. If the inflaton decay rate goes as $\Gamma \sim m_\eta^3/M_P^2 \sim \mu^6/M_P^5$, the Universe recovers at a temperature $T_R \sim (\Gamma M_P)^{1/2} \sim \mu^3/M_P^2 \sim 10^{-11}M_P \sim 10^8\text{GeV}$. Recall that before COBE all that could be set was an upper limit on μ.

2 Big Bang Baryogenesis

It appears that there is apparently very little antimatter in the Universe and that the number of photons greatly exceeds the number of baryons. In the standard model, the entropy density today is related to n_γ by

$$s \sim 7 n_\gamma \tag{52}$$

so that (30) implies $n_B/s \sim 4 \times 10^{-11}$. In the absence of baryon number violation or entropy production this ratio is conserved however and hence represents a potentially undesirable initial condition.

Let us for the moment, assume that in fact $\eta = 0$. We can compute the final number density of nucleons left over after annihilations of baryons and antibaryons have frozen out. At very high temperatures (neglecting a quark-hadron transition) $T > 1$ GeV, nucleons were in thermal equilibrium with the photon background and $n_B = n_{\bar{B}} = (3/2) n_\gamma$ (a factor of 2 accounts for neutrons and protons and the factor $3/4$ for the difference between Fermi and Bose statistics). As the temperature fell below m_N annihilations kept the nucleon density at its equilibrium value $(n_B/n_\gamma) = (m_N/T)^{3/2} \exp(-m_N/T)$ until the annihilation rate $\Gamma_A \simeq n_B m_\pi^{-2}$ fell below the expansion rate. This occurred at $T \simeq 20$ MeV. However, at this time the nucleon number density had already dropped to

$$n_B/n_\gamma = n_{\bar{B}}/n_\gamma \simeq 10^{-18} \tag{53}$$

which is eight orders of magnitude too small [16] aside from the problem of having to separate the baryons from the antibaryons. If any separation did occur at higher temperatures (so that annihilations were as yet incomplete) the maximum distance scale on which separation could occur is the causal scale related to the age of the Universe at that time. At $T = 20$ McV, the age of the Universe was only $t = 2 \times 10^{-3}$ sec. At that time, a causal region (with distance scale defined by $2ct$) could only have contained $10^{-5} M_\odot$ which is very far from the galactic mass scales which we are asking for separations to occur, $10^{12} M_\odot$. In spite of all of these problems, $\eta = 0$, implies that the Universe as a whole is baryon symmetric, thus unless baryons are separated on extremely large (inflationary) domains, in which case we might just as well worry again about $\eta \neq 0$, there should be antimatter elsewhere in the Universe. To date, the only antimatter observed is the result of a high energy collision, either in an accelerator or in a cosmic-ray collision in the atmosphere. There has been no sign to date of any primary antimatter, such as an anti-helium nucleus $\bar{\alpha}$ found in cosmic-rays.

2.1 The Out-of-Equilibrium Decay Scenario

The production of a net baryon asymmetry requires baryon number violating interactions, C and CP violation and a departure from thermal equilibrium [17]. The first two of these ingredients are contained in GUTs, the third can be realized in an expanding Universe where as we have seen, it is not uncommon that interactions come in and out of equilibrium. In SU(5), the fact that quarks and leptons are in the same multiplets allows for baryon non-conserving interactions such as $e^- + d \leftrightarrow \bar{u} + \bar{u}$, etc., or decays of the supermassive gauge bosons X and Y such as $X \rightarrow e^- + d$, $\bar{u} + \bar{u}$. Although today these interactions are very ineffective because of the very large masses of the X and Y bosons, in the early Universe when $T \sim M_X \sim 10^{15}$ GeV these types of interactions should have been very important. C and CP violation is very model dependent. In the minimal SU(5) model, as we will see, the magnitude of C and CP violation is too small to yield a useful value of η. The C and CP violation in general comes from the interference between tree level and first loop corrections.

The departure from equilibrium is very common in the early Universe when interaction rates cannot keep up with the expansion rate. In fact, the simplest (and most useful) scenario for baryon production makes use of the fact that a single decay rate goes out of equilibrium. It is commonly referred to as the out of equilibrium decay scenario [18]. The basic idea is that the gauge bosons X and Y (or Higgs bosons) may have a lifetime long enough to insure that the inverse decays have already ceased so that the baryon number is produced by their free decays.

More specifically, let us call X, either the gauge boson or Higgs boson, which produces the baryon asymmetry through decays. Let α be its coupling to fermions. For X a gauge boson, α will be the GUT fine structure constant, while for X a Higgs boson, $(4\pi\alpha)^{1/2}$ will be the Yukawa coupling to fermions. The decay rate for X will be

$$\Gamma_D \simeq \alpha M_X \ . \tag{54}$$

However decays can only begin occurring when the age of the Universe is longer than the X lifetime Γ_D^{-1}, i.e., when $\Gamma_D > H$

$$\alpha M_X \gtrsim N(T)^{1/2} T^2 / M_P \ , \tag{55}$$

or at a temperature

$$T^2 \lesssim \alpha M_X M_P N(T)^{-1/2}. \tag{56}$$

Scatterings on the other hand proceed at a rate $\Gamma_S \sim \alpha^2 T^3 / M_X^2$ and hence are not effective at lower temperatures. To be in equilibrium, decays must have been effective as T fell below M_X in order to track the equilibrium density of X's (and $\bar{X}$'s). Therefore, the out-of-equilibrium condition is that at $T = M_X, \Gamma_D < H$ or

$$M_X \gtrsim \alpha M_P (N(M_X))^{-1/2} \sim 10^{18} \alpha \, \text{GeV} \ . \tag{57}$$

In this case, we would expect a maximal net baryon asymmetry to be produced.

To see the role of C and CP violation, consider the two channels for the decay of an X gauge boson: $X \rightarrow (1)\bar{u}\bar{u}$, $(2)e^- d$. Suppose that the branching ratio into the first channel with baryon number $B = -2/3$ is r and that of the second channel with baryon number $B = +1/3$ is $1 - r$. Suppose in addition that the branching ratio for $\bar{X}$ into $(\bar{1})$ uu with baryon number $B = +2/3$ is $\bar{r}$ and into $(\bar{2})$ $e^+\bar{d}$ with baryon number $B = -1/3$ is $1 - \bar{r}$. Though the total decay rates of X and $\bar{X}$ (normalized to unity) are equal as required by CPT invariance, the differences in the individual branching ratios signify a violation of C and CP conservation.

The (partial) decay rate for X is computed from an invariant transition rate

$$W = \frac{s}{2^n} |\mathcal{M}|^2 (2\pi)^4 \delta^4 \left(\sum P \right) , \qquad (58)$$

where the first term is the common symmetry factor and the decay rate is

$$\Gamma = \frac{1}{2M_X} \int W d\Pi_1 d\Pi_2 , \qquad (59)$$

with

$$d\Pi = \frac{g d^3 p}{(2\pi)^3 2E} \qquad (60)$$

for g degrees of freedom. Denote the parity (P) of the states (1) and (2) by ↑ or ↓, then we have the following transformation properties:

Under CPT : $\Gamma(X \to 1 \uparrow) = \Gamma(\bar{1} \downarrow \to \bar{X}) ,$
Under CP : $\Gamma(X \to 1 \uparrow) = \Gamma(\bar{X} \to \bar{1} \downarrow) ,$ (61)
Under C : $\Gamma(X \to 1 \uparrow) = \Gamma(\bar{X} \to \bar{1} \uparrow) .$

We can now denote

$$r = \Gamma(X \to 1 \uparrow) + \Gamma(X \to 1 \downarrow) , \qquad (62)$$
$$\bar{r} = \Gamma(\bar{X} \to \bar{1} \uparrow) + \Gamma(\bar{X} \to \bar{1} \downarrow) . \qquad (63)$$

The total baryon number produced by an X, $\bar{X}$ decay is then

$$\Delta B = -\frac{2}{3} r + \frac{1}{3}(1 - r) + \frac{2}{3} \bar{r} - \frac{1}{3}(1 - \bar{r})$$
$$= \bar{r} - r = \Gamma(\bar{X} \to \bar{1} \uparrow) + \Gamma(\bar{X} \to \bar{1} \downarrow) - \Gamma(X \to 1 \uparrow) - \Gamma(X \to 1 \downarrow) .$$
$$(64)$$

One sees clearly therefore, that from eqs. (61) if *either* C *or* CP are good symmetries, $\Delta B = 0$.

In the out-of-equilibrium decay scenario [18], the total baryon asymmetry produced is proportional to $\Delta B = (\bar{r} - r)$. If decays occur out-of-equilibrium, then at the time of decay, $n_X \approx n_\gamma$ at $T < M_X$. We then have

$$\frac{n_B}{s} = \frac{(\Delta B) n_X}{s} \sim \frac{(\Delta B) n_X}{N(T) n_\gamma} \sim 10^{-2} (\Delta B) . \qquad (65)$$

The schematic view presented above can be extended to a complete calculation given a specific model [19, 20], see also [21] for reviews. It makes sense to first consider the simplest GUT, namely SU(5) (for a complete discussion of

GUTs see [22]). In SU(5), the standard model fermions are placed in a $\bar{5}$ and $\mathbf{10}$ representation of SU(5)

$$
\begin{pmatrix} d_1^c \\ d_2^c \\ d_3^c \\ e \\ \nu \end{pmatrix}_L = \bar{5} \;, \qquad
\begin{pmatrix} 0 & u_3^c & -u_2^c & -u_1 & -d_1 \\ & 0 & u_1^c & -u_2 & -d_2 \\ & & 0 & -u_3 & -d_3 \\ & & & 0 & -e^c \\ & & & & 0 \end{pmatrix}_L = \mathbf{10} \;, \tag{66}
$$

where the subscripts are SU(3)-color indices. The standard model gauge sector is augmented by the color triplet X and Y gauge bosons which form a doublet under $SU(2)_L$ and have electric charges $\pm 4/3$ and $\pm 1/3$ respectively. The full set of 24 gauge bosons are in the adjoint representation. In minimal SU(5), an adjoint of Higgs scalars, Σ, is required for the breakdown of SU(5) to the standard model $SU(3)_c \times SU(2)_L \times U(1)_Y$. The additional Higgs scalars needed to break the standard model down to $SU(3)_c \times U(1)_{em}$ requires a five-plet of scalars, H, which contains the standard model Higgs doublet in addition to a colored (charged $\pm 1/3$) triplet.

The SU(5) gauge couplings to fermions can be written as [23]

$$
\frac{1}{\sqrt{2}} g_5 X_{i\mu} \left(\bar{d}_{iR} \gamma^\mu e_R^+ + \epsilon_{ijk} \bar{u}_{kL}^c \gamma^u u_{jL} + \bar{d}_{iL} \gamma^\mu e_L^+ \right) \tag{67}
$$

$$
\frac{1}{\sqrt{2}} g_5 Y_{i\mu} \left(-\bar{d}_{iR} \gamma^\mu \nu^c_R + \epsilon_{ijk} \bar{u}_{jL}^c \gamma^u d_{kL} - \bar{u}_{iL} \gamma^\mu e_L^+ \right) \;, \tag{68}
$$

where g_5 is the SU(5) gauge coupling constant. These couplings lead to the decays shown in Fig. 5. Similar diagrams can be drawn for the decay of the Y gauge boson.

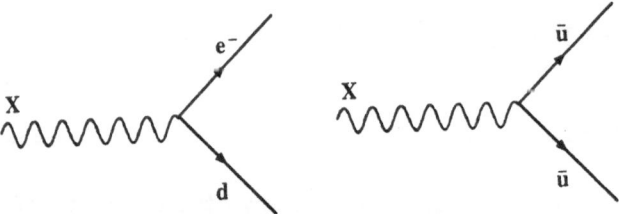

Fig. 5. Decay diagrams for the X gauge boson.

The Higgs five-plet, **H** couples to fermions via the

$$
\mathbf{H}\,\bar{5}\,\mathbf{10} \qquad \mathbf{H}\,\mathbf{10}\,\mathbf{10} \tag{69}
$$

couplings shown in Fig. 6 (shown are the couplings of the triplet relevant for baryogenesis).

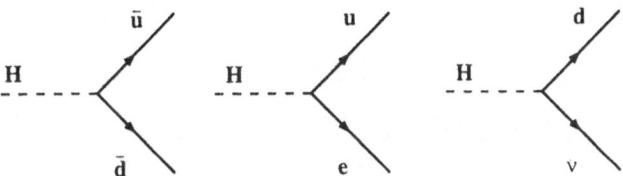

Fig. 6. Higgs couplings to fermions.

Typically, it is expected that the Higgs masses, in particular, those of the adjoint, Σ, are of order of the GUT scale, $M_X \sim 10^{15} - 10^{16}$ GeV. The five-plet is somewhat problematic however, as the the doublet in **H**, must remain light as it corresponds to the standard model electroweak Higgs doublet. The triplet can not be light because as a consequence of the diagrams in Fig. 6, it will mediate proton decay. However, because the couplings to fermions in Fig. 6 are Yukawa couplings rather than gauge couplings, the calculated rate for proton decay mediated by the triplets will be much smaller, allowing for a smaller triplet mass

$$\frac{\Gamma(p - \text{decay via X})}{\Gamma(p - \text{decay via H})} \sim \left(\frac{M_H}{M_X}\right)^4 \left(\frac{M_W}{m_q}\right)^4 , \tag{70}$$

implying that the Higgs triplet mass M_H need only be greater than about 10^{10} GeV.

From equation (65) it is clear that a complete calculation of n_B/s will require a calculation of the CP violation in the decays (summed over parities) which we can parameterize by

$$\epsilon = \bar{r} - r = \frac{\Gamma(\bar{X} \to \bar{1}) - \Gamma(X \to 1)}{\Gamma(\bar{X} \to \bar{1}) + \Gamma(X \to 1)} \sim \frac{\text{Im}\Gamma}{\text{Re}\Gamma} . \tag{71}$$

At the tree level, as one can see $\Gamma(X \to 1) \propto g_5^\dagger g_5$ is real and there is no C or CP violation. At the one loop level, one finds that the interference between the tree diagram and the loop diagram shown in Fig. 7 gives [24]

$$\epsilon \propto \text{Im} g_{X_1}^\dagger g_{Y_1} g_{X_2} g_{Y_2}^\dagger . \tag{72}$$

However in SU(5), $g_{X_1} = g_{Y_1} = g_{X_2} = g_{Y_2} = g_5$ so that

$$\epsilon \propto \text{Im}(g_5^\dagger g_5)(g_5 g_5^\dagger) = 0 . \tag{73}$$

Similarly, the exchange of the Higgs triplet at one loop also gives a vanishing contribution to ϵ.

At least two Higgs five-plets are therefore required to generate sufficient C and CP violation. (It is possible within minimal SU(5) to generate a non-vanishing ϵ at 3 loops, however its magnitude would be too small for the purpose of generating a baryon asymmetry.) With two five-plets, H and H', the interference of diagrams of the type in Fig. 8, will yield a non-vanishing ϵ

$$\epsilon \propto \text{Im}(a'^\dagger ab'b^\dagger) \neq 0 , \tag{74}$$

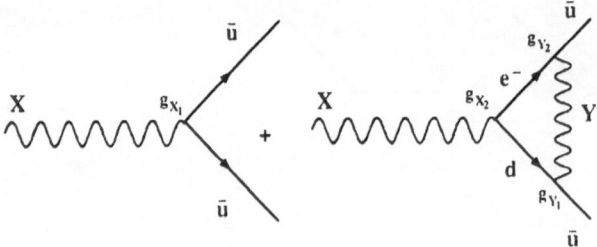

Fig. 7. One loop contribution to the C and CP violation in SU(5).

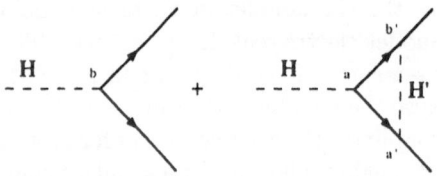

Fig. 8. One loop contribution to the C and CP violation with two Higgs five-plets.

if the couplings $a \neq a'$ and $b \neq b'$.

Given the grand unified theory, SU(5) in this case, the final task in computing the baryon asymmetry is to take into account the thermal history of the Universe and the departure from thermal equilibrium [25, 18]. A full complete numerical calculation was undertaken in [19] and these results will be briefly summarized here.

To trace the evolution of the baryon asymmetry contained in quarks, a full set of coupled differential Boltzmann equations must be computed for all relevant particle species. In general, particle number densities must satisfy

$$
\begin{aligned}
n + 3Hn = \int d\Pi_a d\Pi_b \cdots d\Pi_i d\Pi_j \cdots \\
[f_a f_b \cdots (1 \pm f_i)(1 \pm f_j) \cdots W(p_a p_b \cdots \to p_i p_j \cdots) \\
- f_i f_j \cdots (1 \pm f_a)(1 \pm f_b) \cdots W(p_i p_j \cdots \to p_a p_b \cdots)] \; , \quad (75)
\end{aligned}
$$

where

$$
n = 2 \int E f d\Pi \; , \qquad f = \frac{F(t)}{e^{E/T} \pm 1} \qquad (76)
$$

is the number density of particles and the energy distribution. In thermal equilibrium $F = 1$ and is allowed to take other values. Since we are interested in an asymmetry it is more convenient to keep track of the quantities

$$
n_{i+} = n_i + n_{\bar{i}} \; , \qquad n_{i-} = n_i - n_{\bar{i}} \; . \qquad (77)
$$

For small asymmetries, $F_+ \simeq 2$ and F_- is small. In total, it is necessary to keep track of the following 12 quantities: $U_+, D_+, L_+, \nu_+, X_+, Y_+; U_-, D_-, L_-, \nu_-,$

X_-, Y_- where these scaled functions are defined by $U(t) = n_u/(g_u A)$ with $A = [3\zeta(3)/4\pi^2]T^3$. It is also convenient to change time variables from $\frac{d}{dt}$ to

$$\frac{d}{dz} = \frac{5.8 \times 10^{17}\,\mathrm{GeV}}{M_X^2}\left(\frac{160}{N(T)}\right)^{1/2} z\frac{d}{dt}, \text{ with } z = M_X/T.$$

The full set of coupled equations can be found in [19]. For our purposes here, it will be useful to write down only a sample equation for U_-

$$\frac{1}{zK}\frac{dU_-}{dz} = -\gamma_D\left[2X_- + Y_-/2\right] - \gamma_{ID}\left[2U_+U_- + (U_+D_- + D_+U_-)/2\right.$$

$$\left. (U_+L_- + L_+U_-)/4\right] + \epsilon\left[\gamma_D X_+ - \gamma_{ID}D_+L_+/2\right] + \text{scatterings}, \tag{78}$$

where

$$\gamma_D = \frac{1}{\alpha M_X}\frac{4}{3}\frac{\int d\Pi_X d\Pi_{u_1} d\Pi_{u_2} f_X W(X \to u_1 u_2)}{g_X \zeta(3)T^3/\pi^2}, \tag{79}$$

$$\gamma_{ID} = \frac{1}{\alpha M_X}\frac{4}{3}\frac{\int d\Pi_X d\Pi_{u_1} d\Pi_{u_2} f_{u_1} f_{u_2} W(u_1 u_2 \to X)}{g_X \zeta(3)T^3/\pi^2}, \tag{80}$$

$$K = \frac{2.9 \times 10^{17}\alpha\,\mathrm{GeV}}{M_X}\left(\frac{160}{N(T)}\right)^{1/2}, \tag{81}$$

where the GUT fine-structure constant is $\alpha = g_5^2/4\pi$.

When equilibrium is maintained and all interaction rates are large compared with the expansion rate, solutions to the + equations (not shown) give $U_+, D_+, L_+, \nu_+ = 2$ and $X_+ = Y_+ = 2\gamma_{ID}/\gamma_D$. In this case, as one can see from the sample-equation in (78), all CP violation effects disappear (the coefficient of ϵ vanishes). As the ϵ term was the only one that could generate an asymmetry, the asymmetry is driven to 0 in equilibrium.

To get a feeling for the results of such a numerical integration, let us first consider the case with $\epsilon = 0$. When $B - L = 0$ initially, there is a damping of any initial baryon asymmetry as is shown in Fig. 9. The parameter z increases as a function of time ($z \sim \sqrt{t}$). In equilibrium, the asymmetries are damped until the baryon number violating interactions freeze-out. In accord with our earlier remarks, a large value of M_X (corresponding to a small value of K) results in an early departure from equilibrium and a larger final baryon asymmetry. If $B - L \neq 0$ initially, since the minimal SU(5) considered here conserves $B - L$, the asymmetry can not be erased, only reshuffled.

To generate an asymmetry, we must have $\epsilon \neq 0$. The time evolution for the generation of a baryon asymmetry is shown in Fig. 10. As one can see, for large values of M_X, i.e. values which satisfy the lower limit given in (57), the maximal value for the baryon asymmetry, $n_B/s \sim 10^{-2}\epsilon$ is achieved. This confirms numerically the original out-of equilibrium decay scenario [18]. For smaller values of M_X, an asymmetry is still produced, which however is smaller due to partial equilibrium maintained by inverse decays (γ_{ID}). The growth of the asymmetry as a function of time is now damped, and it reaches its final value when inverse decays freeze out.

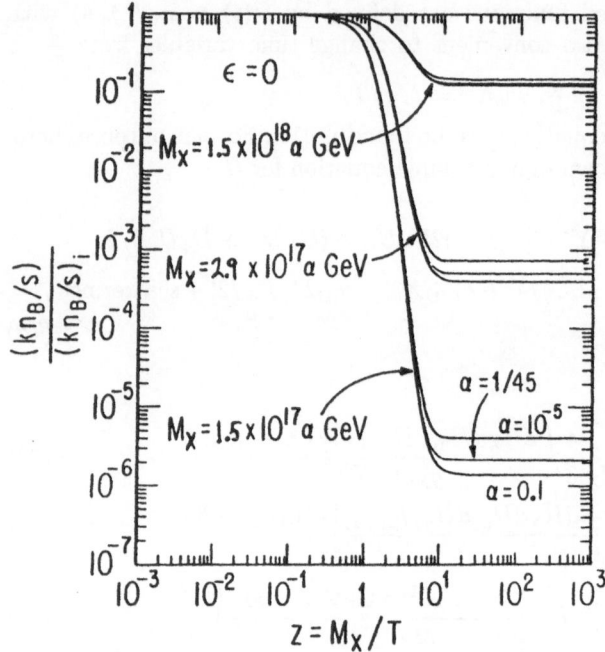

Fig. 9. The damping of an initial baryon asymmetry with $B - L = 0$ and $\epsilon = 0$.

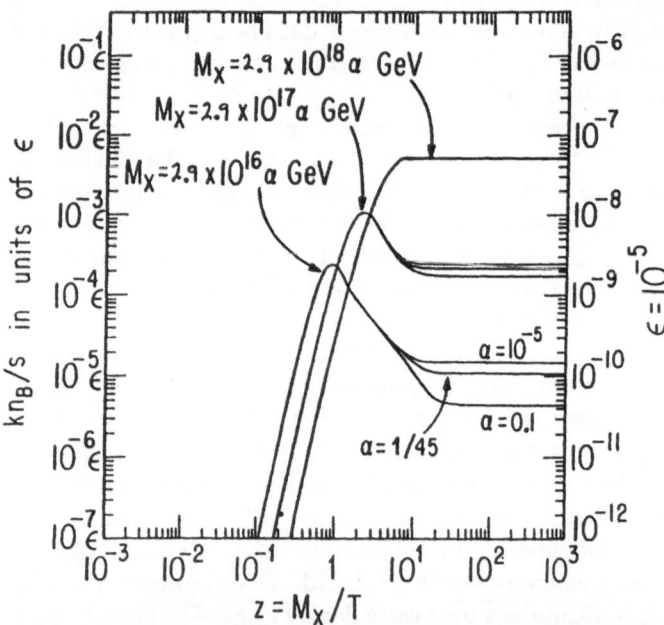

Fig. 10. The time evolution of the baryon asymmetry with $B = L = 0$ initially.

Finally it is important to note that the results for the final baryon asymmetry, as shown in Fig. 10, as a function of the mass of the X gauge boson, is in fact largely independent of the initial baryon asymmetry. This is evidenced in Fig. 11, which shows the time evolution of the asymmetry, given a large initial asymmetry. Even for large M_X, the asymmetry is slightly damped, and for smaller values of M_X, the asymmetry is damped to a level which is again determined by the freeze-out of inverse decays. This means that this mechanism of baryogenesis is truly independent of initial conditions, in particular it gives the same value for η whether or not $\eta = 0$ or 1 initially.

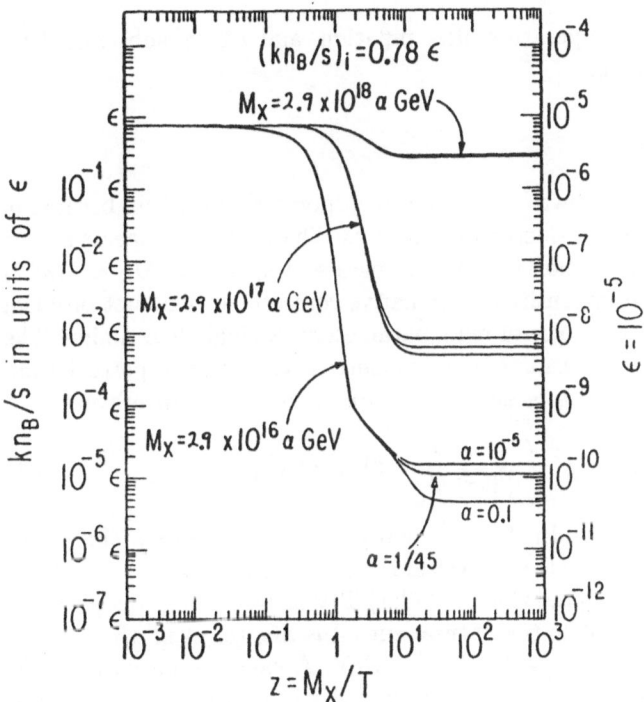

Fig. 11. The time evolution of the baryon asymmetry with a large initial baryon asymmetry.

The out-of-equilibrium decay scenario discussed above did not include the effects of an inflationary epoch. In the context of inflation, one must in addition ensure baryogenesis after inflation as any asymmetry produced before inflation would be inflated away along with magnetic monopoles and any other unwanted relic. Reheating after inflation, may require a Higgs sector with a relatively light $\mathcal{O}(10^{10} - 10^{11})$ GeV Higgs boson. The light Higgs is necessary since the inflaton, η, is typically very light ($m_\eta \sim \mu^2/M_P \sim \mathcal{O}(10^{11})$ GeV, determined from the magnitude of density perturbations on large scales as measured by COBE [5], cf. (51)) and the baryon number violating Higgs would have to be produced during inflaton decay. Note that a "light" Higgs is acceptable as discussed above due to the reduced couplings to fermions, cf. (70). The out-of-equilibrium decay scenario would now be realized by Higgs boson decay rather than gauge boson decay and a different sequence of events. First the inflaton would be required to decay to Higgs bosons (triplets?) and subsequently the triplets would decay rapidly by the processes shown in Fig. 6. These decays would be well out of equilibrium as at reheating $T \ll m_H$ and $n_H \sim n_\gamma$ [26]. In this case, the baryon asymmetry is given simply by

$$\frac{n_B}{s} \sim \epsilon \frac{n_H}{T_R{}^3} \sim \epsilon \frac{n_\eta}{T_R{}^3} \sim \epsilon \frac{T_R}{m_\eta} \sim \epsilon \left(\frac{m_\eta}{M_P}\right)^{1/2} \sim \epsilon \frac{\mu}{M_P} \sim 10^{-4}\epsilon\,, \qquad (82)$$

where T_R is the reheat temperature after inflation, and I have substituted for $n_\eta = \rho_\eta/m_\eta \sim \Gamma^2 M_P{}^2/m_\eta$.

2.2 Supersymmetry

Supersymmetry, as is well known by now, was incorporated into GUTs because of its ability to resolve the gauge hierarchy problems. There are two aspects to this problem: 1) there is a separation in physical mass scales, $M_W \ll M_X < M_P$; 2) this separation is extremely sensitive to radiative corrections. The first problem has to do with a tree-level choice of mass parameters. A single fine-tuning. The second problem requires fine-tuning at many successive orders in perturbation theory. Radiative corrections to scalar masses are quadratically divergent

$$\delta m_0^2 \sim g^2 \int \frac{d^4 k}{(2\pi^4)} \frac{1}{k^2} \sim \mathcal{O}(\alpha/\pi)\Lambda^2\,, \qquad (83)$$

where Λ is some cut-off scale. In the low energy electroweak theory, the smallness of M_W requires the mass of the physical Higgs boson to be $m_H < \mathcal{O}(1)\text{TeV}$. Requiring $\delta m_H^2 < \mathcal{O}(m_H^2)$ implies that $\Lambda < \mathcal{O}(1)$ TeV as well. The trouble comes when we move to a GUT where the natural cut-off is M_X (or even M_P) rather than $\mathcal{O}(M_W)$ and we expect $\delta m_H^2 > \mathcal{O}(10^{15})$ GeV. A cancellation may be imposed by hand, but this must be done to each order in perturbation theory. A solution to this difficulty would be to cancel the radiative corrections by including fermion loops which have the opposite sign. Then provided $|m_B^2 - m_F^2| < \mathcal{O}(1)$ TeV, the stability of the mass scales would be guaranteed. Such a cancellation occurs automatically in a supersymmetric theory (in the limit of exact supersymmetry, these radiative corrections are absent entirely). In addition, although gauge couplings still get renormalized, the Yukawa couplings of the theory, which are parameters of a superpotential do not get renormalized[27].

Standard unification (i.e. non-supersymmetric) has come across additional difficulties of late. Extrapolation of the gauge coupling constants of the standard model using the renormalization group equation with standard model inputs, does not result in the three couplings meeting at a single point. However, when the superpartners of the standard model fields are also incorporated, and the renormalization group equations are again run to high energy scales, then the gauge couplings do in fact meet at a point (within errors) at a scale of order 10^{16} GeV [28].

The field content of the supersymmetric standard model, consists of the following chiral supermultiplets: $Q, u^c, d^c, L, e^c, H_1, H_2$. The only addition is the extra Higgs doublet. The Yukawa interactions are generated by the superpotential

$$F_Y = h_u H_1 Q u^c + h_d H_2 Q d^c + h_l H_2 L e^c + \epsilon H_1 H_2 , \qquad (84)$$

leading to the Lagrangian interactions

$$\mathcal{L} \ni \left(\partial^2 F_Y / \partial \phi^i \partial \phi^j \right) \Psi^i \Psi^j . \qquad (85)$$

where Ψ^i is the fermion component of the superfield ϕ^i. (85) contains the normal fermion mass terms of the standard model.

The scalar potential in a globally supersymmetric theory can be written as

$$V(\phi^i, \phi_i^*) = \sum_i |F_i|^2 + \frac{1}{2} \sum_a g_a^2 |D^a|^2 , \qquad (86)$$

where

$$F_i = \partial F / \partial \phi^i \qquad (87)$$

for superpotential F and

$$|D^a|^2 = \left(\sum_{i,j} \phi_i^* T^{ai}{}_j \phi^j \right)^2 \qquad (88)$$

for generators $T^{ai}{}_j$ of a gauge group with gauge coupling g_a. In addition, in broken supersymmetry there will be soft supersymmetry breaking scalar masses as well as gaugino masses.

In a supersymmetric grand unified SU(5) theory, the superpotential F_Y can be expressed in terms of SU(5) multiplets

$$F_Y = h_d \mathbf{H_2} \, \bar{\mathbf{5}} \, \mathbf{10} + h_u \mathbf{H_1} \, \mathbf{10} \, \mathbf{10} \qquad (89)$$

where $\mathbf{10}$, $\bar{\mathbf{5}}$, $\mathbf{H_1}$ and $\mathbf{H_2}$ are chiral supermultiplets for the 10, and $\bar{5}$ plets of SU(5) matter fields and the Higgs 5 and $\bar{5}$ multiplets respectively.

In supersymmetric SU(5), there are now new dimension 5 operators which violate baryon number and lead to proton decay as shown in Fig. 12. The first of these diagrams leads to effective dimension 5 Lagrangian terms such as

$$\mathcal{L}_{\text{eff}}^{(5)} = \frac{h_u h_d}{M_H} (\tilde{q}\tilde{q}ql) \qquad (90)$$

and the resulting dimension 6 operator for proton decay [29]

$$\mathcal{L}_{\text{eff}} = \frac{h_u h_d}{M_H} \left(\frac{g_2^2}{M_{\tilde{W}}} \text{ or } \frac{g_1^2}{M_{\tilde{B}}} \right) (qqql) . \qquad (91)$$

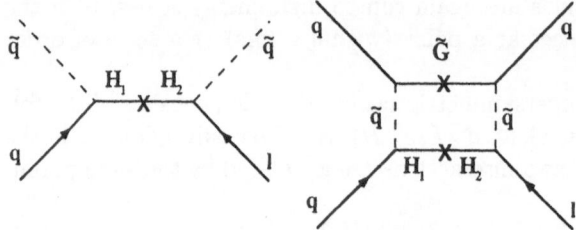

Fig. 12. Dimension 5 and induced dimension 6 graphs violating baryon number.

As a result of these diagrams the proton decay rate scales as $\Gamma \sim h^4 g^4/M_H^2 M_{\tilde{G}}^2$ where M_H is the triplet mass, and $M_{\tilde{G}}$ is a typical gaugino mass of order $\lesssim 1$ TeV. This rate however is much too large if $M_H \sim 10^{10}$ GeV.

It is however possible to have a lighter ($\mathcal{O}(10^{10} - 10^{11})$ GeV) Higgs triplet needed for baryogenesis in the out-of-equilibrium decay scenario with inflation. One needs two pairs of Higgs five-plets ($\mathbf{H_1}, \mathbf{H_2}$ and $\mathbf{H_1'}, \mathbf{H_2'}$ which is anyway necessary to have sufficient C and CP violation in the decays. By coupling one pair ($\mathbf{H_2}$ and $\mathbf{H_1'}$) only to the third generation of fermions via [30]

$$a\mathbf{H_1}\, 10\,10 + b\mathbf{H_1'}\, 10_3\, 10_3 + c\mathbf{H_2}\, 10_3\, \bar{5}_3 + d\mathbf{H_2'}\, 10\, \bar{5} \qquad (92)$$

proton decay can not be induced by the dimension five operators. Triplet decay will however generate a baryon asymmetry proportional to $\epsilon \sim \mathrm{Im}(dc^\dagger ba^\dagger)$.

2.3 The Affleck-Dine Mechanism

Another mechanism for generating the cosmological baryon asymmetry is the decay of scalar condensates as first proposed by Affleck and Dine [31]. This mechanism is truly a product of supersymmetry. It is straightforward though tedious to show that there are many directions in field space such that the scalar potential given in (86) vanishes identically when SUSY is unbroken. That is, with a particular assignment of scalar vacuum expectation values, $V = 0$ in both the $F-$ and $D-$ terms. An example of such a direction is

$$u_3^c = a\,, \qquad s_2^c = a\,, \qquad -u_1 = v\,, \qquad \mu^- = v\,, \qquad b_1^c = e^{i\phi}\sqrt{v^2 + a^2}\,, \quad (93)$$

where a, v are arbitrary complex vacuum expectation values. SUSY breaking lifts this degeneracy so that

$$V \simeq \tilde{m}^2 \phi^2\,, \qquad (94)$$

where $\tilde{m}$ is the SUSY breaking scale and ϕ is the direction in field space corresponding to the flat direction. For large initial values of ϕ, $\phi_0 \sim M_{\mathrm{GUT}}$, a large baryon asymmetry can be generated [31, 32]. This requires the presence of baryon number violating operators such as $O = qqql$ such that $\langle O \rangle \neq 0$. The decay of these condensates through such an operator can lead to a net baryon asymmetry.

In a supersymmetric GUT, as we have seen above, there are precisely these types of operators. In Fig. 13, a 4-scalar diagram involving the fields of the flat direction (93) is shown. Again, $\tilde{G}$ is a (light) gaugino. The two supersymmetry breaking insertions are of order $\tilde{m}$, so that the diagram produces an effective quartic coupling of order $\tilde{m}^2/(\phi_0^2 + M_X^2)$.

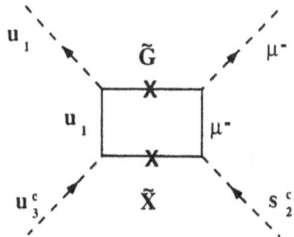

Fig. 13. Baryon number violating diagram involving flat direction fields.

The baryon asymmetry produced, is computed by tracking the evolution of the sfermion condensate, which is determined by

$$\ddot{\phi} + 3H\dot{\phi} = -\tilde{m}^2\phi \ . \tag{95}$$

To see how this works, it is instructive to consider a toy model with potential [32]

$$V(\phi, \phi^*) = \tilde{m}^2\phi\phi^* + \frac{1}{2}i\lambda[\phi^4 - \phi^{*4}] \ . \tag{96}$$

The equation of motion becomes

$$\ddot{\phi}_1 + 3H\dot{\phi}_1 = -\tilde{m}^2\phi_1 + 3\lambda\phi_1^2\phi_2 - \lambda\phi_2^3 \ , \tag{97}$$

$$\ddot{\phi}_2 + 3H\dot{\phi}_2 = -\tilde{m}^2\phi_2 - 3\lambda\phi_2^2\phi_1 + \lambda\phi_1^3 \ , \tag{98}$$

with $\phi = (\phi_1 + i\phi_2)/\sqrt{2}$. Initially, when the expansion rate of the Universe, H, is large, we can neglect $\ddot{\phi}$ and $\tilde{m}$. As one can see from (96) the flat direction lies along $\phi \simeq \phi_1 \simeq \phi_0$ with $\phi_2 \simeq 0$. In this case, $\dot{\phi}_1 \simeq 0$ and $\dot{\phi}_2 \simeq \frac{\lambda}{3H}\phi_0^3$. Since the baryon density can be written as $n_B = j_0 = \frac{1}{2}(\phi_1\dot{\phi}_2 - \phi_2\dot{\phi}_1) \simeq \frac{\lambda}{6H}\phi_0^4$, by generating some motion in the imaginary ϕ direction, we have generated a net baryon density.

When H has fallen to order $\tilde{m}$ (when $t^{-1} \sim \tilde{m}$), ϕ_1 begins to oscillate about the origin with $\phi_1 \simeq \phi_0 \sin(\tilde{m}t)/\tilde{m}t$. At this point the baryon number generated is conserved and the baryon density, n_B falls as R^{-3}. Thus,

$$n_B \sim \frac{\lambda}{\tilde{m}}\phi_0^2\phi^2 \propto R^{-3} \tag{99}$$

and relative to the number density of ϕ's ($n_\phi = \rho_\phi/\tilde{m} = \tilde{m}\phi^2$)

$$\frac{n_B}{n_\phi} \simeq \frac{\lambda\phi_0^2}{\tilde{m}^2} \ . \tag{100}$$

If it is assumed that the energy density of the Universe is dominated by ϕ, then the oscillations will cease, when

$$\Gamma_\phi \simeq \frac{\tilde{m}^3}{\phi^2} \simeq H \simeq \frac{\rho_\phi^{1/2}}{M_P} \simeq \frac{\tilde{m}\phi}{M_P} \tag{101}$$

or when the amplitude of oscillations has dropped to $\phi_D \simeq (M_P \tilde{m}^2)^{1/3}$. Note that the decay rate is suppressed as fields coupled directly to ϕ gain masses $\propto \phi$. It is now straightforward to compute the baryon to entropy ratio

$$\frac{n_B}{s} = \frac{n_B}{\rho_\phi^{3/4}} \simeq \frac{\lambda \phi_0^2 \phi_D^2}{\tilde{m}^{5/2} \phi_D^{3/2}} = \frac{\lambda \phi_0^2}{\tilde{m}^2} \left(\frac{M_P}{\tilde{m}}\right)^{1/6} \tag{102}$$

and after inserting the quartic coupling

$$\frac{n_B}{s} \simeq \epsilon \frac{\phi_0^2}{(M_X^2 + \phi_0^2)} \left(\frac{M_P}{\tilde{m}}\right)^{1/6} \tag{103}$$

which could be quite large.

In the context of inflation, a couple of significant changes to the scenario take place. First, it is more likely that the energy density is dominated by the inflaton rather than the sfermion condensate. Second, the the initial value (after inflation) of the condensate ϕ can be determined by the inflaton mass m_η, $\phi_0^2 \simeq H^3 \tau \simeq m_\eta M_P$. The sequence of events leading to a baryon asymmetry is then as follows [15]: After inflation, oscillations of the inflaton begin at $R = R_\eta$ when $H \sim m_\eta$ and oscillations of the sfermions begin at $R = R_\phi$ when $H \sim \tilde{m}$. If the Universe is inflaton dominated, $H \sim m_\eta (R_\eta/R)^{3/2}$ since $H \sim \rho_\eta^{1/2}$ and $\rho_\eta \sim \eta^2 \sim R^{-3}$. Thus one can relate R_η and R_ϕ, $R_\phi \simeq (m_\eta/\tilde{m})^{2/3} R_\eta$. As discussed earlier, inflatons decay when $\Gamma_\eta = m_\eta^3/M_P^2 = H$ or when $R = R_{d\eta} \simeq (M_P/m_\eta)^{4/3} R_\eta$. The Universe then becomes dominated by the relativistic decay products of the inflaton, $\rho_{r\eta} = m_\eta^{2/3} M_P^{10/3}(R_\eta/R)^4$ and $H = m_\eta^{1/3} M_P^{2/3}(R_\eta/R)^2$. Sfermion decays still occur when $\Gamma_\phi = H$ which now corresponds to a value of the scale factor $R_{d\phi} = (m_\eta^{7/15} \phi_0^{2/5} M_P^{2/15}/\tilde{m}) R_\eta$.

Finally, the baryon asymmetry in the Affleck-Dine scenario with inflation becomes [15]

$$\frac{n_B}{s} \sim \frac{\epsilon \phi_0^4 m_\eta^{3/2}}{M_X^2 M_P^{5/2} \tilde{m}} \sim \frac{\epsilon m_\eta^{7/2}}{M_X^2 M_P^{1/2} \tilde{m}} \sim (10^{-6} - 1)\epsilon \tag{104}$$

for $\tilde{m} \sim (10^{-17} - 10^{-16})M_P$, and $M_X \sim (10^{-4} - 10^{-3})M_P$ and $m_\eta \sim (10^{-8} - 10^{-7})M_P$.

3 Lepto-Baryogenesis

The realization [33] of significant baryon number violation at high temperature within the standard model, has opened the door for many new possibilities for the generation of a net baryon asymmetry. Indeed, it may be possible to generate the asymmetry entirely with the context of the standard model [34]. Electroweak baryon number violation occurs through non-perturbative interactions mediated by "sphalerons", which violate $B + L$ and conserve $B - L$. For this reason, any GUT produced asymmetry with $B - L = 0$ may be subsequently erased by sphaleron interactions [35].

The origin of the sphaleron interactions lies in the anomalies of the electroweak current

$$J_B^\mu = N_f \left(\frac{g_2^2}{32\pi^2} W\tilde{W} - \frac{g_1^2}{32\pi^2} B\tilde{B} \right) . \tag{105}$$

This gives rise to a non-trivial vacuum structure with degenerate vacuum states with differing baryon number. At $T = 0$, the rates for such transitions is highly suppressed [36], $\propto e^{-2\pi/\alpha_W}$. However at high temperatures, the transition rate is related to the diffusion rate over a potential barrier, $\Gamma_S \sim \left(M_W^7/\alpha_W^3 T^6 \right) \times e^{-4M_W/\alpha_W T}$ in the broken phase. In the symmetric phase, the barrier becomes very small and transitions are relatively unsuppressed, $\Gamma_S \sim \left(\alpha_W^4 T \right)$.

With $B - L = 0$, it is relatively straightforward to see that the equilibrium conditions including sphaleron interactions gives zero net baryon number [37]. By assigning each particle species a chemical potential, and using gauge and Higgs interactions as conditions on these potentials (with generation indices suppressed)

$$\mu_- + \mu_0 = \mu_W , \qquad \mu_{u_R} - \mu_{u_L} = \mu_0 , \qquad \mu_{d_R} - \mu_{d_L} = -\mu_0 ,$$
$$\mu_{l_R} - \mu_{l_L} = -\mu_0 , \qquad \mu_{d_L} - \mu_{u_L} = \mu_W , \qquad \mu_{l_L} - \mu_\nu = \mu_W , \tag{106}$$

one can write down a simple set of equations for the baryon and lepton numbers and electric charge which reduce to:

$$B = 12\mu_{u_L} ,$$
$$L = 3\mu - 3\mu_0 , \tag{107}$$
$$Q = 6\mu_{u_L} - 2\mu + 14\mu_0 ,$$

where $\mu = \sum \mu_{\nu_i}$. In (107), the constraint on the weak isospin charge, $Q_3 \propto \mu_W = 0$ has been employed. Though the charges B, L, and Q have been written as chemical potentials, since for small asymmetries, an asymmetry $(n_f - n_{\bar{f}})/s \propto \mu_f/T$, we can regard these quantities as net number densities.

The sphaleron process yields the additional condition

$$9\mu_{u_L} + \mu = 0 , \tag{108}$$

which allows one to solve for L and $B - L$ in terms of μ_{u_L}, ultimately giving

$$B = \frac{28}{79} (B - L) . \tag{109}$$

Thus, in the absence of a primordial $B - L$ asymmetry, the baryon number is erased by equilibrium processes. Note that barring new interactions (in an extended model) the quantities $\frac{1}{3}B - L_e$, $\frac{1}{3}B - L_\mu$, and $\frac{1}{3}B - L_\tau$ remain conserved.

With the possible erasure of the baryon asymmetry when $B - L = 0$ in mind, since minimal SU(5) preserves $B - L$, electroweak effects require GUTs beyond SU(5) for the asymmetry generated by the out-of-equilibrium decay scenario to survive. GUTs such as SO(10) where a primordial $B - L$ asymmetry can be generated becomes a promising choice. The same holds true in the Affleck-Dine mechanism for generating a baryon asymmetry. In larger GUTs there are baryon number violating operators and associated flat directions[38]. A specific example in SO(10) was worked out in detail by Morgan [39].

An important question remaining to be answered is whether or not the baryon asymmetry can in fact be generated during the electroweak weak phase transition. This has been the focus of much attention in recent years. I refer the reader to the review of Ref. [34]. In the remainder of these lectures, I will focus on alternative means for generating a baryon asymmetry which none-the-less makes use of the sphaleron interactions.

The above argument regarding the erasure of a primordial baryon asymmetry relied on the assumption that all particle species are in equilibrium. However, because of the extreme smallness of the electron Yukawa coupling, e_R does not come into equilibrium until the late times. The e_R decoupling temperature is determined by the rate of $e_R \rightarrow e_L + H$ transitions and comparing this rate to the expansion rate

$$\Gamma_{LR} = \frac{\pi h_e^2}{192\zeta(3)} \frac{m_H^2}{T} \sim \frac{20T^2}{M_P} \simeq H \,, \tag{110}$$

which gives $T = T_* \sim \mathcal{O}(\text{few})$ TeV. Thus one may ask the question, whether or not the baryon asymmetry may be stored in a primordial e_R asymmetry [40]. Because sphalerons preserve $B - L$, any lepton number stuck in e_R is accompanied by an equal baryon number. However, at temperatures below the e_R decoupling temperature, baryon number will begin to be destroyed so long as sphalerons are in equilibrium. Sphalerons are in equilibrium from about the electroweak phase transition to $T \sim 10^{12}$ GeV [33]. As it turns out, the e_R (baryon) asymmetry is exponentially sensitive to parameters of the model.

To clearly see the role of e_R decoupling, it is helpful to look again at the equations relating chemical potentials. Above the scale T_*, the relation $\mu_{e_R} = \mu_{e_L} - \mu_0$ does not hold. Instead there is an equilibrium solution [41]

$$\mu_0 = \frac{5}{153}(\mu_{e_R} - \mu_{e_L}) \,,$$
$$B = 12\mu_{u_L} = \frac{44}{153}(\mu_{e_R} - \mu_{e_L}) \,. \tag{111}$$

One can quickly see now that below T_*, when $\mu_{e_R} = \mu_{e_L} - \mu_0$ is respected, the only solution yields $\mu_0 = B = L = 0$. In terms of conserved quantities, above,

T_* we can write the equilibrium solution for B [41]

$$B_{eq} = \frac{66}{481} \left(3L_{eR} + (\frac{1}{3}B - L_e) \right) , \tag{112}$$

where L_{eR} is the lepton asymmetry stored in e_R's and L_e is the total lepton asymmetry. Note that this is independent of the initial baryon asymmetry. Below T_*, the baryon asymmetry drops off exponentially

$$B = B_{eq} e^{-\int_{t_*}^{t_c} \frac{711}{481} \Gamma_{LR} dt} \tag{113}$$

integrated from T_* to the electroweak phase transition where sphaleron interactions quickly freeze out. With standard model parameters, the baryon asymmetry is not preserved [40, 41].

Another possibility for preserving a primordial baryon asymmetry when $B - L = 0$ comes if the asymmetry produced by scalar condensates in the Affleck-Dine mechanism is large ($n_B/s \gtrsim 10^{-2}$) [42]. After the decay of the A-D condensate, the baryon number is shared among fermion and boson superpartners. However, in equilibrium, there is a maximum chemical potential $\mu_f = \mu_B = \tilde{m}$ and for a large asymmetry, the baryon number density stored in fermions, $n_{B_f} = \frac{g_t}{6} \mu_f T^2$ is much less than the total baryon density. The bulk of the baryon asymmetry is driven into the $p = 0$ bosonic modes and a Bose-Einstein condensate is formed [43]. The critical temperature for the formation of this condensate is given by $n_B \simeq n_{B_b} + n_{B_c} = \frac{g_b}{3} \tilde{m} T_c^2$ so that

$$n_{B_c} = \frac{g_b}{3} \left(1 - \left(\frac{T}{T_c} \right)^2 \right) T_c^2 . \tag{114}$$

At $T < T_c$, most of the baryon number remains in a condensate and for large n_B, the condensate persists down to temperatures of order 100 GeV. Thus sphaleron interactions are shut off and a primordial baryon asymmetry is maintained even with $B - L = 0$. One should note however that additional sources of entropy are required to bring η down to acceptable levels.

As alluded to above, sphaleron interactions also allow for new mechanisms to produce a baryon asymmetry. The simplest of such mechanisms is based on the decay of a right handed neutrino-like state [44]. This mechanism is certainly novel in that does not require grand unification at all. By simply adding to the Lagrangian a Dirac and Majorana mass term for a new right handed neutrino state

$$\mathcal{L} \ni M\nu^c\nu^c + \lambda H L \nu^c \tag{115}$$

the out-of-equilibrium decays $\nu^c \to L + H^*$ and $\nu^c \to L^* + H$ will generate a non-zero lepton number $L \neq 0$. The out-of-equilibrium condition for these decays translates to $10^{-3}\lambda^2 M_P < M$ and M could be as low as $\mathcal{O}(10)$ TeV. (Note that once again in order to have a non-vanishing contribution to the C and CP violation in this process at 1-loop, at least 2 flavors of ν^c are required. For the generation of masses of all three neutrino flavors, 3 flavors of ν^c are required.)

Sphaleron effects can transfer this lepton asymmetry into a baryon asymmetry since now $B - L \neq 0$. A supersymmetric version of this scenario has also been described [14, 45].

The survival of the asymmetry, of course depends on whether or not electroweak sphalerons can wash away the asymmetry. The persistence of lepton number violating interactions in conjunction with electroweak sphaleron effects could wipe out [46] both the baryon and lepton asymmetry in the mechanism described above through effective operators of the form $\lambda^2 LLHH/M$. In terms of chemical potentials, this interaction adds the condition $\mu_\nu + \mu_0 = 0$. The constraint comes about by requiring that this interaction be out of equilibrium at the time when sphalerons are in equilibrium. The additional condition on the chemical potentials would force the solution $B = L = 0$.

It is straightforward to derive a constraint [46]-[49],[37] on M/λ^2. So long as the $\Delta L = 2$ operator is out-of-equilibrium while sphalerons are in equilibrium the baryon asymmetry is safe. The out-of-equilibrium condition is

$$\Gamma_{\Delta L} = \frac{\zeta(3)}{8\pi^3} \frac{\lambda^4 T^3}{M^2} < \frac{20T^2}{M_{\rm P}} \simeq H \, , \tag{116}$$

yielding

$$\frac{M}{\lambda^2} \gtrsim 0.015 \sqrt{T_{\rm BL} M_{\rm P}} \, , \tag{117}$$

where $T_{\rm BL}$ is the temperature at which the $B - L$ asymmetry was produced or the maximum temperature when sphalerons are in equilibrium (or the temperature T_* of $e_{\rm R}$ decoupling which we will momentarily ignore) whichever is lower. Originally [46], $T_{\rm BL} \sim T_{\rm c} \sim 100$ GeV was chosen giving, $M/\lambda^2 \gtrsim 5 \times 10^8$ GeV and corresponds to a limit on neutrino masses $m_\nu \sim \lambda^2 v^2/M \lesssim 50$ keV. In [37], it was pointed out that sphalerons should be in equilibrium up to 10^{12} GeV, in which case, $M/\lambda^2 \gtrsim 10^{14}$ GeV and corresponds to $m_\nu \lesssim 1$ eV. Similarly, it is possible to put constraints on other B and/or L violating operators [47, 48] which include R-parity violating operators in supersymmetric models. For example [47], the mass scale associated with a typical dimension 3 operator is constrained to be $m \lesssim 2 \times 10^{-5}$ GeV, the quartic coupling of a dimension 4 operator, $\lambda \lesssim 7 \times 10^{-7}$ or the mass scale of a higher dimensional operator such as a dimension 9, $\Delta B = 2$, operator is $M \gtrsim 10^3 - 10^{13}$ GeV. Only the latter is dependent on the choice of $T_{\rm BL}$.

In supersymmetric models however, it has been argued by [49] that due to additional anomalies which can temporarily protect the asymmetry (until the effects of supersymmetry breaking kick in), the maximum temperature should be at $\sim 10^8$ GeV rather than $\sim 10^{12}$ GeV. Interestingly, in the context of inflation, though the reheat temperature is typically 10^8 GeV, equilibration is not achieved until about 10^5 GeV [15] thus the maximum temperature should not surpass this equilibration temperature [14]. These changes in $T_{\rm max}$ would soften the limits on the mass scales of dimension $D \geq 5$ operators. For example, for the $D = 5$ ($\Delta L = 2$) operator above, $M/\lambda^2 \gtrsim 10^9 - 10^{10}$ GeV.

There are other subtleties regarding these limits. The presence of separate lepton asymmetries combined with mass effects can protect an asymmetry as an equilibrium solution [50, 51, 52]. The rates for some operators may be small enough to leave approximately conserved quantities such as $\frac{1}{3}B - L_i$ [53]. Or, it may be possible that the asymmetry can be stored in a weakly interacting field such as the right-handed electron [40].

Indeed, it has been shown [41] that because e_R only comes into equilibrium at the relatively cool temperature $T_* \sim$ few TeV, above T_* the baryon number is safe and the picture of baryon number erasure is changed. Sphaleron erasure of the baryon asymmetry can only occur between the T_* and the decoupling temperature T_f of the additional B and/or L violating rates as seen in Fig. 14. If $T_f > T_*$, the baryon asymmetry is protected and may even be generated as shown below. Thus for limits on B and L violating operators, $T_{\max}$ should be set at T_* further relaxing the constraints on new operators.

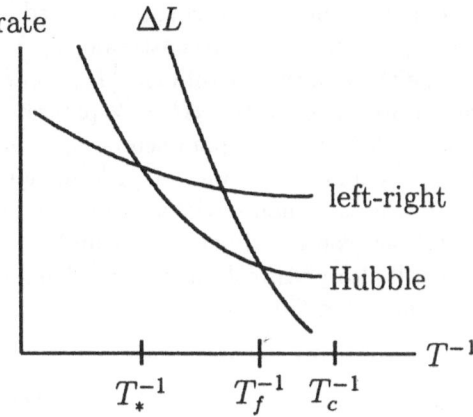

Fig. 14. Lepton-violating and left-right equilibrating rates.

How then can we generate a baryon asymmetry from a prior lepton asymmetry? In addition to the mechanism described earlier utilizing a right-handed neutrino decay, several others are now also available. In a supersymmetric extension of the standard model including a right-handed neutrino, there are numerous possibilities. Along the lines of the right-handed neutrino decay, the scalar partner [14] or a condensate [45] of $\tilde{\nu}^c$'s will easily generate a lepton asymmetry. Furthermore if the superpotential contains terms such as $\nu^{c3} + \nu^c H_1 H_2$, there will be a flat direction violating lepton number [54, 14] à la Affleck and Dine. While none of these scenarios require GUTs, those that involve the out-of equilibrium decay of either fermions, scalars or condensates must have the mass scale of the right-handed neutrino between 10^9 and about 10^{11} GeV, to avoid washing out the baryon asymmetry later (as can be seen from (117)) and to be produced after inflation respectively. In contrast the decay of the flat direction condensate (which involves other fields in addition to $\tilde{\nu}^c$) only works for $10^{11} < M < 10^{15}$ GeV.

Flavor effects may also generate a baryon asymmetry. Indeed consider the $\Delta L = 2$ operator discussed above. If all flavors are out of equilibrium, then the process for all intents and purposes can be neglected. The only baryon asymmetry that will result will be the small one due to mass effects [50, 51], unless a larger ($\gtrsim 10^{-4}$) asymmetry is produced say by the Affleck and Dine mechanism [52]. If all of the flavors are in equilibrium, then the bound (117) is not satisfied and B is driven to zero. On the other hand, if the bound is satisfied by 1 or 2 generations, then even if initially $B = L = 0$, a baryon asymmetry will be generated and will be given by

$$B = \frac{84}{247} \mu_{\frac{2}{3}B-(L_1+L_2)} \tag{118}$$

assuming that only generations 1 and 2 are out of equilibrium and satisfy the bound[40].

Once again, to see this more clearly it is helpful to write quantities in terms of chemical potentials. Below T_*, all of the quantities of interest can be expressed in terms of 5 chemical potentials: μ_{u_L}, μ_0, and μ_{ν_i}. There are two constraints: $Q = 0$ and the sphaleron constraint (108) and the three initial conditions, $\frac{1}{3}B - L_i$. If all three of these conservation laws are broken e.g. by the $\Delta L = 2$ processes discussed above and $\mu_0 = -\mu_{\nu_i}$, then we are left with two parameters, μ_{u_L} and μ_0, with two constraints: $Q = 6\mu_{u_L} - 20\mu_0 = 0$ and $9\mu_{u_L} - 3\mu_0 = 0$ yielding only the trivial solution $B = L = \mu_{u_L} = \mu_0 = 0$. Clearly a non-trivial solution will be obtained when one or two of the $\frac{1}{3}B - L_i$'s are conserved between T_c and T_*.

Finally, a pre-existing e_R asymmetry will also be transformed into a baryon asymmetry [41]. With e_R decoupled, the quantities (107) become

$$\begin{aligned} B &= 12\mu_{u_L} , \\ L &= 3\mu + \mu_{e_R} - 2\mu_H - \mu_{e_L} , \\ Q &= 6\mu_{u_L} - 2\mu - \mu_{e_R} + 13\mu_H + \mu_{e_L} , \end{aligned} \tag{119}$$

which when combined with the sphaleron condition (108), and the the $\Delta L = 2$ condition $\mu_\nu + \mu_H = 0$, one finds that above T_*

$$B_* = \frac{1}{5}\mu_{e_R} , \qquad L_* = \frac{1}{2}\mu_{e_R} , \tag{120}$$

independent of the initial value of B and L. Assuming that the $\Delta L - 2$ interactions are out of equilibrium below T_*, we now have from (109) that

$$B = \frac{28}{79}(B_* - L_*) \simeq -0.1\mu_{e_R} . \tag{121}$$

Similarly, if any other baryon and/or lepton number violating operator was in equilibrium at some point above T_* and so long as it decouples above T_*, a baryon asymmetry (or more precisely a $B - L$ asymmetry) will be produced.

In summary, I hope that it is clear that the generation of a baryon asymmetry is in principle relatively easy and that sphaleron interactions may in fact aid rather than hinder the production of an asymmetry. There are many possibilities and perhaps more than one of them are actually responsible for the final observed asymmetry.

References

1. J.C. Mather et al.: Ap.J. **420** (1994) 439

2. see, e.g., J.L.Tonry: in "Relativistic Astrophysics and Particle Cosmology", ed. by C.W. Akerlof and M. Srednicki, New York Academy of Sciences, New York (1993) p.113

3. T.P. Walker, G. Steigman, D.N. Schramm, K.A. Olive, and H.-S. Kang: Ap.J. **376** (1991) 51

4. W. Rindler: M.N.R.A.S. **116** (1956) 663;
   G.F.R. Ellis and W. Stoeger: Class. Quantum Grav. **5** (1988) 207

5. G.F. Smoot et al.: Ap.J **396** (1992) L1;
   E.L. Wright et al.: Ap.J. **396** (1992) L13

6. P.J.E. Peebles: "The Large Scale Structure of the Universe", Princeton University Press, Princeton, (1980)

7. S.K. Blau and A.H. Guth: in "300 Years of Gravitation", ed. by S.W. Hawking and W. Israel, Cambridge University Press, Cambridge (1987)

8. G. t'Hooft: Nucl. Phys. **B79** (1974) 276;
   A.M. Polyakov: JETP Lett. **20** (1974) 194

9. T.W.B. Kibble: J. Phys. **A9** (1976) 1387

10. Ya. B. Zeldovich and M.Y. Khlopov: Phys. Lett. **79B** (1979) 239;
    J.P. Preskill: Phys. Rev. Lett. **43** (1979) 1365

11. for reviews, see: A.D. Linde: "Particle Physics and Inflationary Cosmology", Harwood (1990);
    K.A. Olive: Phys.Rep. **C190** (1990) 307

12. W.H. Press: Phys. Scr. **21** (1980) 702;
    V.F. Mukhanov and G.V. Chibisov: JETP Lett. **33** (1981) 532;
    S.W. Hawking: Phys. Lett. **115B** (1982) 295;
    A.A. Starobinsky: Phys. Lett. **117B** (1982) 175;
    A.H. Guth and S.Y. Pi: Phys. Rev. Lett. **49** (1982) 1110;
    J.M. Bardeen, P.J. Steinhardt and M.S. Turner: Phys. Rev. **D28** (1983) 679

13. P.J.E. Peebles: Ap.J. Lett. **263**(1982) L1;
    L.F. Abbott and M.B. Wise: Phys. Lett. **B135** (1984) 279

14. B. Campbell, S. Davidson, and K.A. Olive: Nucl.Phys. **B399** (1993) 111

15. J. Ellis, K. Enqvist, D.V. Nanopoulos, and K.A. Olive: Phys. Lett. **B191** (1987) 343

16. G. Steigman: Ann. Rev. Astron. Astrophys. **14** (1976) 339

17. A.D. Sakharov: JETP Lett. **5** (1967) 24

18. S. Weinberg: Phys. Rev. Lett. **42** (1979) 850;
    D. Toussaint, S. B. Treiman, F. Wilczek, and A. Zee: Phys. Rev. **D19** (1979) 1036

19. J.N. Fry, K.A. Olive, and M.S. Turner: Phys. Rev. **D22** (1980) 2953; **D22** (1980) 2977; Phys. Rev. Lett. **45** (1980) 2074

20. E.W. Kolb and S. Wolfram: Phys. Lett. **B91** (1980) 217; Nucl. Phys. **B172** (1980) 224

21. for reviews, see: E.W. Kolb and M.S. Turner: Ann. Rev. Nucl.Part. Sci. **33** (1983) 645;
    A. Dolgov: Phys. Rep. in press (1993)

22. G.G. Ross: "Grand Unified Theories", Benjamin/Cummings, Menlo Park (1984)

23. A. Buras, J. Ellis, M.K. Gaillard, and D.V. Nanopoulos: Nucl. Phys. **B135** (1978) 66

24. D.V. Nanopoulos and S. Weinberg: Phys. Rev. **D20** (1979) 2484;
    S. Barr, G. Segré and H.A. Weldon: Phys. Rev. **D20** (1979) 2494;
    A. Yildiz and P.H. Cox: Phys. Rev. **D21** (1980) 906

25. M. Yoshimura: Phys. Rev. Lett. **41** (1978) 281;
    A. Ignatiev, N. Krasnikov, V. Kuzmin, and A. Tavkhelidze: Phys. Lett. **B76** (1978) 436;
    S. Dimopoulos, and L. Susskind: Phys. Rev. **D18** (1978) 4500;
    J. Ellis, M.K. Gaillard, and D.V. Nanopoulos: Phys. Lett. **B80** (1979) 360

26. A.D. Dolgov, and A.D. Linde: Phys. Lett. **B116** (1982) 329;
    D.V. Nanopoulos, K.A. Olive, and M. Srednicki: Phys. Lett. **B127** (1983) 30

27. J. Wess and B. Zumino: Phys. Lett. **B49** (1974) 52;
    J. Iliopoulos and B. Zumino: Nucl. Phys. **B76** (1974) 310;
    S. Ferrara, J. Iliopoulos, and B. Zumino: Nucl. Phys. **B77** (1974) 413;
    M.T. Grisaru, W. Siegel, and M. Rocek: Nucl. Phys. **B159** (1979) 420

28. J. Ellis, S. Kelly, and D.V. Nanopoulos: Phys. Lett. **B249** (1990) 442; Phys. Lett. **B260** (1991) 131;
    U. Amaldi, W. de Boer, and H. Furstenau: Phys. Lett. **B260** (1991) 447;
    P. Langacker and M. Luo: Phys. Rev. **D44** (1991) 817;
    G. Gambieri, G. Ridolfi, and F. Zwirner: Nucl. Phys. **B331** (1990) 331

29. J. Ellis, D.V. Nanopoulos, and S. Rudaz: Nucl. Phys. **B202** (1982) 43

30. D.V. Nanopoulos and K. Tamvakis: Phys. Lett. **B114** (1982) 235

31. I. Affleck and M. Dine: Nucl. Phys. **B249** (1985) 361

32. A.D. Linde: Phys. Lett. **B160** (1985) 243

33. V. Kuzmin, V. Rubakov, and M. Shaposhnikov: Phys. Lett. **B155** (1985) 36

34. for a review, see: A.G. Cohen, D.B. Kaplan, and A.E. Nelson: Ann. Rev. Nucl. Part. Sci. **43** (1993) 27

35. P. Arnold and L. McLerran: Phys. Rev. **D36** (1987) 581; **D37** (1988) 1020

36. G. 't Hooft: Phys. Rev. Lett. **37** (1976) 8; Phys. Rev. **D14** (1976) 3432

37. J.A. Harvey and M.S. Turner: Phys. Rev. **D42** (1990) 3344

38. B. Campbell, J. Ellis, D.V. Nanopoulos, and K. A. Olive: Mod. Phys. Lett. **A1** (1986) 389

39. D. Morgan: Nucl. Phys. **B364** (1991) 401

40. B. Campbell, S. Davidson, J. Ellis, and K.A. Olive: Phys. Lett. **B297** (1992) 118

41. J. Cline, K. Kainulainen, and K.A. Olive: Phys. Rev. Lett. **71** (1993) 2372 ; and Phys. Rev. **D49** (1994) in press

42. S. Davidson, H. Murayama, and K.A. Olive: Phys. Lett. **B** (1994) in press

43. A.D. Dolgov and D.P. Kirilova: Sov. J. Nucl. Phys. **50** (1989) 1006

44. M. Fukugita and T. Yanagida: Phys. Lett. **B174** (1986) 45

45. H. Murayama, H. Suzuki, T. Yanagida, and J. Yokoyama: Phys. Rev. Lett. **70** (1993) 1912

46. M. Fukugita and T. Yanagida: Phys. Rev. **D42** (1990) 1285

47. B. Campbell, S. Davidson, J. Ellis, and K.A. Olive: Phys. Lett. **B256** (1991) 457; Astropart. Phys. **1** (1992) 77

48. W. Fischler, G.F. Giudice, R.G. Leigh, and S. Paban: Phys. Lett. **B258** (1991) 45

49. L.E. Ibanez and F. Quevedo: Phys. Lett. **B283** (1992) 261

50. V.A. Kuzmin, V.A. Rubakov, and M.E. Shaposhnikov: Phys. Lett. **B191** (1987) 171
51. H. Dreiner and G.G. Ross: Nucl. Phys. **B407** (1993) 188
52. S. Davidson, K. Kainulainen, and K.A. Olive: University of Minnesota preprint UMN-TH-1248/94
53. A.E. Nelson and S.M. Barr: Phys. Lett. **B246** (1990) 141
54. B. Campbell, S. Davidson, and K.A. Olive: Phys. Lett. **B303** (1993) 63

Solar Neutrinos

Michel Spiro

DAPNIA, Centre d'Etudes Saclay,
F-91191 Gif-sur-Yvette Cedex, France

Abstract: Solar neutrinos have been for the past ten years a field of growing interest, both from the theoretical and experimental point of view. These lectures are intended to give a comprehensive review of the present status of the solar neutrino flux predictions confronted to the existing measurements. The implications will be discussed together with future prospects.

The most firm and solid prediction we have on the solar neutrino flux is based on energy conservation and steady state of the sun. We know that these two well admitted assumptions imply that the total power radiated by the solar surface (the luminosity L_*) should be equal to the thermonuclear power generated by the fusion of hydrogen into helium. For four protons to combine into a ^{4}He nucleus, two electrons must be involved in the initial state for electric charge conservation, and then two ν_e must be emitted in the final state. The overall reaction is then:

$$4p + 2e^- \rightarrow {}^4\text{He} + 2\nu_e + 27\text{MeV} , \tag{1}$$

where 27 MeV is the difference of the masses between the particles involved in the initial state and those involved in the final state (the energy of the neutrinos and the kinetic energy of the nuclei can be neglected in this approximate relation). It is then easy to derive the total flux of neutrinos expected to reach the earth:

$$N_\nu = \frac{2L_*}{27 \text{ MeV} \, 4\pi \, d^2} = 6.5 \; 10^{10} \text{ cm}^{-2}s^{-1} , \tag{2}$$

where d is the distance from the earth to the sun. Gallium target detectors are so far the most appropriate to measure the total number of neutrinos. This is because:

- of the very low threshold (233 keV) of the capture reaction $\nu_e + {}^{71}\text{Ga} \rightarrow {}^{71}\text{Ge} + e^-$ which makes the Gallium target detector sensitive to the bulk of the solar neutrino energy spectrum;

- of the high natural abundance of the stable ^{71}Ga isotope (40%);
- of the relatively easy identification of even a few radioactive ^{71}Ge atoms in a large quantity of Gallium (30 tons).

However, the firm prediction on the total number of solar neutrinos is not enough to compute the capture rate of solar neutrinos on a given target nucleus. To compute the energy spectrum one needs to go through solar modeling and through the exact chain of reactions which combine hydrogen into helium. There are mostly three cycles of reactions : ppI, ppII, and ppIII.

- In ppI the two neutrinos are coming from the pp$\rightarrow$ ^{2}H $+ e^{+} + \nu_e$ reaction ($2\nu_{pp}$).
- In ppII one neutrino is a ν_{pp}, the other comes from the decay (through electron capture) of ^{7}Be (ν_{7Be}).
- In ppIII one neutrino is a ν_{pp}, the other comes from the β decay of ^{8}B (ν_{8B}).

The ν_{pp} spectrum extends from 0 to 450 keV. Only the Gallium experiments are sensitive to those neutrinos. The ν_{7Be} are monoenergetic with a line at 860 keV. Both the Gallium and Chlorine ($\nu_e + {}^{37}$Cl $\rightarrow {}^{37}$Ar $+ e^-$, threshold 820 keV) are sensitive to those neutrinos.

Finally the ν_{8B} neutrino spectrum extends from 0 to 15 MeV. All the presently running experiments (Gallium, Chlorine and Kamiokande) are sensitive to them. The Kamiokande experiment is based on the detection of the recoil electron in the elastic scattering of a ν_e with an experimental threshold of about 7 MeV on the energy of the recoil electron. From the Solar Standard Model [1] the ν_{pp}, ν_{7Be} and ν_{8B} intensities are computed to be 90% , 8% and 10^{-4} of the total flux. Although the relative intensity of the ν_{8B} neutrinos is very small, they contribute significantly to the capture rate, even in the Gallium experiment, due to their high energy. Notice, however, that the ν_{7Be} and ν_{8B} fluxes are highly sensitive to the ingredients of the SSM. If, for instance, one changes the input parameters, with, as a result, a change in the central temperature T_c prediction, it has been shown that the ν_{8B} flux will vary as T_c^{18}, the ν_{7Be} as T_c^8 and the ν_{pp} flux only as $T_c^{-1.2}$.

Table 1. Standard Solar Models predictions from Turck-Chiéze et al, and Bahcall et al., for the gallium experiment.

Source	Capture rate (SNU)	
	Turck-Chiéze et al.	Bahcall et al.
pp	70.6	71.3
pep	2.795	3.07
^{7}Be	30.6	32.9
^{8}B	9.31	12.31
^{13}N	3.87	2.68
^{15}O	6.50	4.28
^{17}F	–	0.04
Total	124	127

The predictions of the SSMs are shown in Tab. 1, for the Gallium experiments, in terms of SNU (Solar Neutrino Units). One SNU corresponds to a capture rate of 10^{-36} per second per target nucleus (in this case for Ga). We see that the Bahcall et al. SSM which is generally considered as giving high SNU values, predicts fluxes only slightly higher than the Turck-Chiéze et al. SSM which is generally considered as giving low SNU values. So one might say that the predictions of the SSM for Gallium experiments are rather firm. Notice also, that although the ν_{pp} are expected to represent 90% of the total flux of solar neutrinos, their contribution to the capture rate amounts only to 71 SNU out of 127. This is due to their low energy.

Are the Predictions Right ?

ν_{8B} Flux

The Kamiokande experiment uses a water Cerenkov detector. The basic process is neutrino scattering on electrons which then give detectable Cerenkov light. They measure two quantities, the energy of the recoil electron and its direction. A clear peak can be seen in the direction of the sun and the excess in that direction is then taken as coming from solar neutrinos (Fig. 1). However, the flux of ν_{8B} they measure since January 1987 [2] is only 0.54 ± 0.08 of the Bahcall et al. SSM, so $0.5 \ 10^{-5}$ of the total flux.

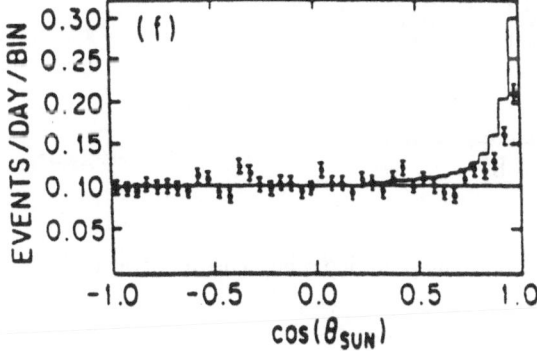

Fig. 1. Counts of the Kamiokande detector plotted against the cosines of the angle of the electron to the sun's direction ($\cos\theta_{SUN}$, 1557 days of data, electron energies greater than 9.3 MeV for 449 days, 7.5 MeV for 794 days and 7.0 MeV for 314 days).

ν_{7Be} Flux

Since 1967 Davis and co-workers have performed a pioneering experiment by extracting ^{37}Ar from a tank of 615 tons of tetrachloroethylene (C_2Cl_4). The ^{37}Ar decays by electron capture. The resulting hole in the K shell can give X rays and Auger electrons with a total energy of 2.8 keV. The counter of 0.5 cm^3 volume is designed to measure this energy. The half life of the decay is 35 days. A typical run consists of a 50 day exposure of the big tank followed by an extraction of the Argon atoms which are then introduced in the small counter. The counting lasts for 260 days. For the period 1970-1984 the data were analyzed and give 339 counts of ^{37}Ar. This gives a non-corrected ^{37}Ar counting rate of 5 per run. The data are analyzed by a maximum likelihood method assuming a flat background (as a function of time) plus a ^{37}Ar decaying component.

The result [2] is 3.6 ± 0.4 times lower (Fig. 2) than expected in the Bahcall et al. SSM. This implies, taking into account the fact that the experiment is sensitive to both the ν_{7Be} and ν_{8B} components and taking into account the Kamiokande result (reduction of a factor 2 on the ν_{8B} component) that the ν_{7Be} flux is lower by a factor > 4 than the prediction of Bahcall et al. SSM.

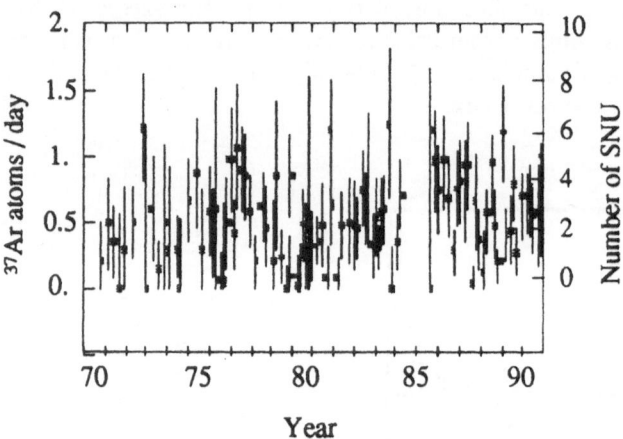

Fig. 2. ^{37}Ar production rate in the Homestake chlorine solar neutrino detector (0.4 atom/d corresponds to 2 SNU).

These deficits are the basis of the solar neutrino problem. The reductions are very hard to reconcile with any modification of the SSM since we expect that any reduction on the ν_{7Be} component should be accompanied by a stronger reduction for the ν_{8B} component [6]. Neutrino masses and mixing could reconcile these reductions with the SSM through ν_e, ν_μ, ν_τ oscillations. However, before invoking new physics in the neutrino sector, the results of the gallium experiments were eagerly awaited. The expectations are much less sensitive to

SSM and we can derive absolute lower limits for the capture rate based only on energy conservation and steady state of the sun.

Consistent Predictions for Gallium Experiments

Since we know experimentally that the flux of ν_{8B} is reduced by a factor two and that the flux of ν_{7Be} is reduced by a factor greater than 4 compared to SSM predictions, we can deduce that the number of ν_{pp} should be increased to 1.08 ± 0.02 of the SSM to conserve the total number of neutrinos insuring energy conservation. One obtains expectations which range from 80 to 105 SNU. These are not SSM predictions but rather predictions which are consistent with the basic understanding of the sun (energy conservation and steady state) and with the two experimental results coming from the Chlorine and Kamiokande experiments.

Results of the Gallium Experiments

Two experiments are now underway, SAGE in Russia which published the first results in January 1991 and GALLEX in Italy, which published their first results in June 1992. The recipes are the same: introduce 1mg of inactive stable Germanium in the 30 tons of Gallium, expose the Gallium to solar neutrinos in a low background environment, extract by a chemical method the solar neutrinos produced ^{71}Ge atoms together with the inactive Germanium, transform into a counting gas (GeH$_4$), fill a proportional counter and count the decays of ^{71}Ge (11 d. half life). The main difference is that the SAGE experiment uses a metallic liquid Gallium target while the GALLEX experiment uses an acidic aqueous Gallium Chloride solution. This induces important differences in the chemistry.

SAGE

The Soviet-American Gallium Experiment is located in the Baksan Valley in the Caucasus mountains (Russia) under about 4700 meter water equivalent. The expected rate for a 30 tons target and 132 SNU is 1.2 ^{71}Ge atoms created per day. Taking into account all the efficiencies, one expects only 3 counts per run (a run is 4 week exposure) due to ^{71}Ge K-electron capture (^{71}Ge + e_K^- → ^{71}Ga + ν + X-rays + Auger electrons). Most of the runs in 1990 have preferred values of 0 SNU. Altogether they published in 1991 [3] a preferred value of 20 SNU with upper limits of 55 SNU (68% C.L.) and 79 SNU (90% C.L.). More recently they announced the results they obtained in the last runs when they increased the total mass of Gallium from 30 tons to 60 tons. This is shown on Fig. 3. A signal seems now to emerge. In 1992 at Dallas, the quoted result was 58 ± 20 (stat.) ± 14 (sys.) SNU . It is now 70 ± 19 (stat.) ± 10 (sys.) SNU [4].

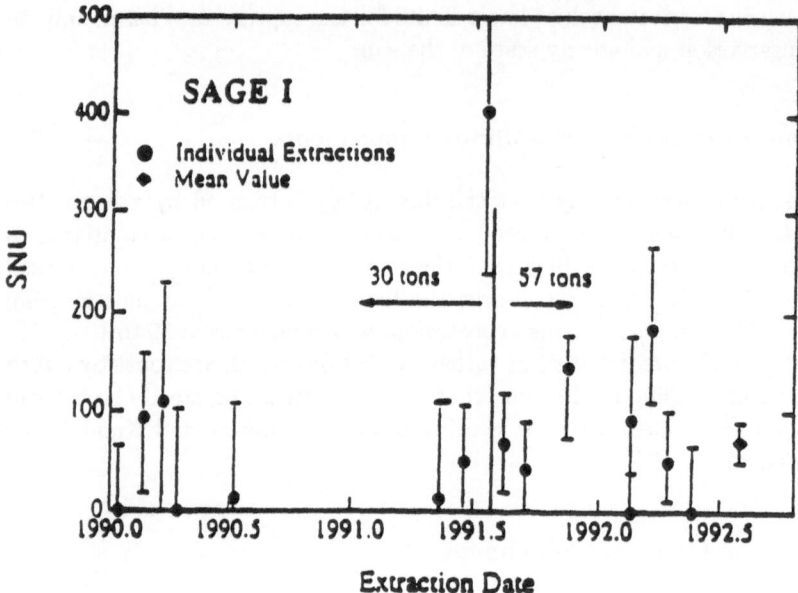

Fig. 3. Results for all runs of the SAGE experiment. The last point on the right shows the combined result.

GALLEX

This experiment is located in the Gran Sasso Underground Laboratory in Italy. The 30 tons of Gallium are in the form of a solution of $GaCl_3$ acidified in HCl. The Ge atoms form the volatile compound $GeCl_4$. At the end of 3 week exposures, these molecules are swept out by bubbling a large flow of inert gas (N_2) through the solution. The experiment is sensitive to both K-shell and L-shell electron captures in the decay of ^{71}Ge atoms. Seven counts are then expected after each run, in the K and L regions. The data used in the analysis consist of 21 runs taken from May 1991 to May 1993. They are now published [5], [7]. There is compelling evidence for a signal: the peaks in energy at 1.2 keV and 10 keV for L and K electron capture are seen, the 11.3 half life of ^{71}Ge is well identified over a flat background. Fig. 4 shows the results for all runs which have to be compared with the combined result of 83 ± 20 SNU, released in June 1992 and now updated at the level of 79 ± 13 (stat.) ± 5 (sys.) SNU [7]. Furthermore, GALLEX should be calibrated with an artificial neutrino source (2 MCi) in 1994.

Interpretations

The SSM is unable to account for the deficit of solar neutrinos as observed by the Chlorine and Kamiokande experiments. However, on the basis of these experiments it is impossible to decide whether these discrepancies come from new physics in the neutrino sector or wrong ingredients in the Solar Standard Models. The Gallium experiments are in a much better position to do so. First,

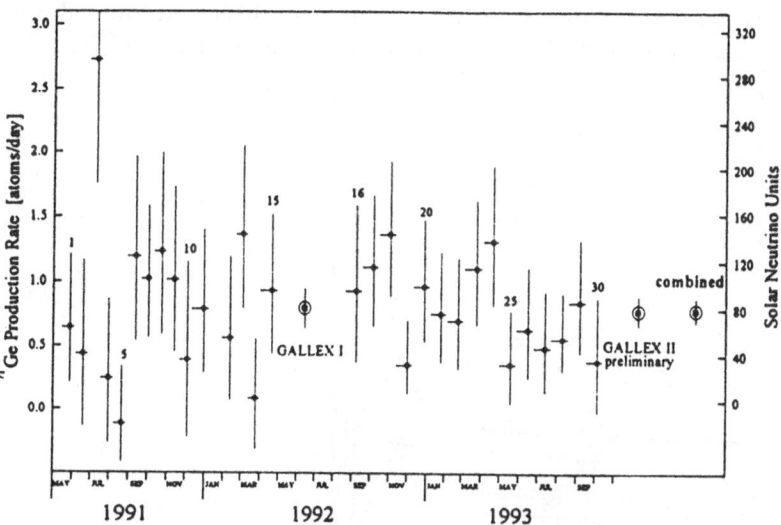

Fig. 4. Final results for the first period GALLEX I (before May 1992) and preliminary results for the second period GALLEX II (after August 1992). The combined values for GALLEX I, GALLEX II, and GALLEX I + GALLEX II are also shown.

the predictions of the SSM are more stable to changes in the ingredients (120 to 140 SNU) and second it is impossible to have predictions below 80 SNU from basic simple principles. Consistent predictions for Gallium experiments which agree with these basic principles and with the deficits of solar neutrinos observed by the Chlorine and Kamiokande experiments are in the 80-105 SNU range, in agreement with the values measured by GALLEX.

Table 2. Summary table of solar neutrino experiment results (chlorine, Kamiokande and GALLEX) with the comparison to Turck-Chiéze et al. and Bahcall et al. SSMs.

Experiment	Exp. Results	Turck-Chiéze et al.	Bahcall et al.
Chlorine (SNU) (%)	2.33 ± 0.25	6.4 ± 1.4	7.2 ± 0.9
		36 ± 4	33 ± 3
Kamiokande (%)		64 ± 8	54 ± 8
GALLEX (SNU) (%)	$87 \pm 14 \pm 7$	123 ± 7	127^{-5}_{+7}
		71 ± 13	68 ± 12

By comparing the deficits of solar neutrinos as observed by the 3 experiments (Tab. 2), the only indication which favours neutrino oscillations is the fact that the chlorine experiment has a significantly larger suppression factor than the other experiments. This would imply a more severe suppression for ^{7}Be neutrinos than for ^{8}B and pp neutrinos which cannot easily be accommodated by a modification of the SSM. A decrease of the central temperature will produce a suppression factor for ^{8}B neutrinos which is larger than for ^{7}Be or pp neutrinos [6].

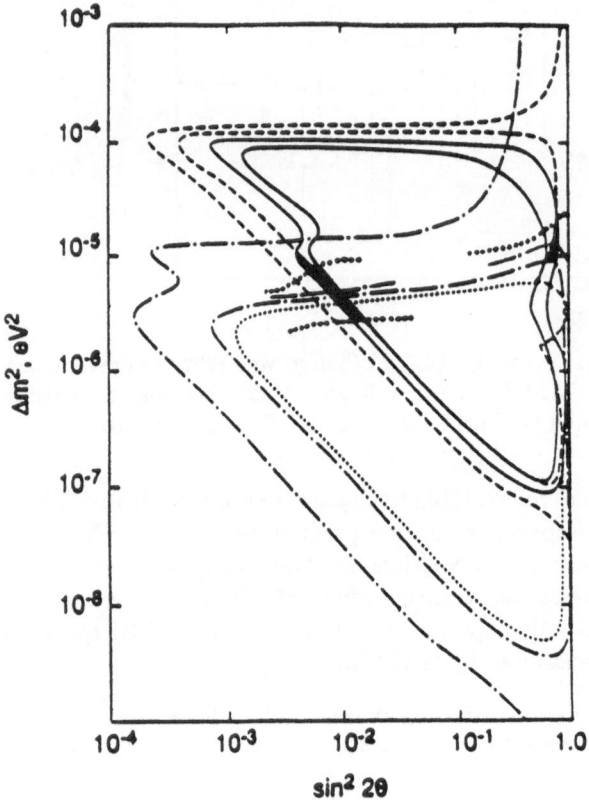

Fig. 5. MSW diagram showing the preferred solution to reconcile experimental neutrino flux measurements and Standard Solar Models.

On the contrary, oscillations (MSW effect) could reconcile the SSM with all 3 experiments. Fig. 5 shows the allowed range for neutrino masses and mixing angles. The preferred solution is for $\Delta m^2 \approx 7 \cdot 10^{-6}$ eV^2 and $\sin^2 2\theta \approx 6 \cdot 10^{-3}$. The suppression factor as a function of the neutrino energy is shown on Fig. 6 for this solution. It implies a distortion of the ^{8}B neutrino energy spectrum.

Fig. 7 shows the recoil electron energy spectrum as observed in Kamiokande [8], normalized to the SSM prediction from Bahcall and Ulrich [9]. The recoil electron energy spectrum is only a smeared reflection of the neutrino energy spectrum.

– A flat suppression (no ν oscillation) gives a χ^2 of 16.3/13.

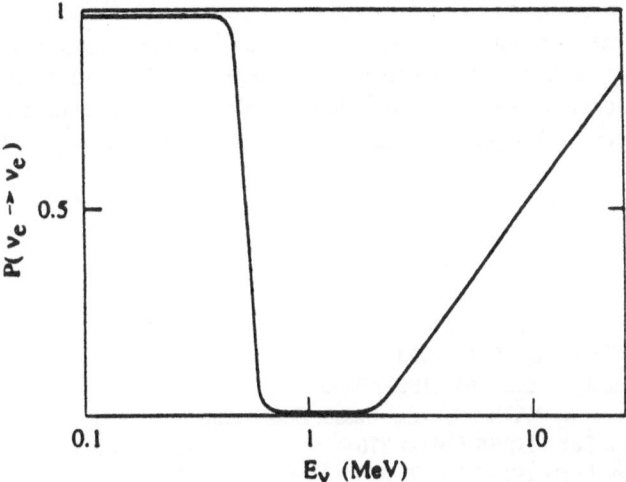

Fig. 6. Suppression factor of neutrino flux as a function of the neutrino energy for the MSW preferred solution.

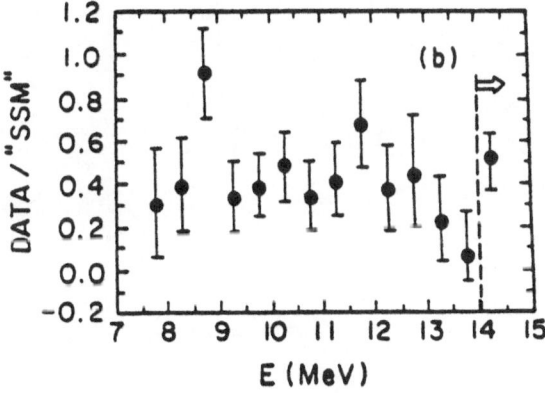

Fig. 7. Recoil electron energy spectrum as observed in Kamiokande, normalized to the SSM prediction from Bahcall and Ulrich [9].

– A suppression factor as predicted by the preferred MSW solution gives a χ^2 of 18.6/13.

It is clear that one has to wait for the Superkamiokande and SNO experiments to establish or reject the small mixing angle MSW solution.

No firm conclusion on neutrino masses can yet be drawn from the present status of solar neutrino experiments and solar modeling. This may not be the case, hopefully, in few years from now when we may expect to have more input to solar models (nuclear cross sections, helioseismology...), better understanding of running experiments (calibrations) and more experiments (SNO, SuperKamiokande...).

References

1. A. Suzuki: KEK preprint 93-96, August 1993
2. S. Turck-Chiéze et al.: Phys. Rep. **230** (1993) 59
3. A. I. Abazov et al.: Phys. Rev. Lett. **67** (1991) 3332
4. V. Gavrin: Communication at TAUP93, Gran Sasso, Sept. 1993
5. P. Anselman et al.: Phys. Lett. **B285** (1992) 376
6. P. Anselman et al.: Phys. Lett. **B285** (1992) 390
7. P. Anselman et al.: preprint DAPNIA/SPP 94-05, to appear in Phys. Lett.
8. Y. Totsuka: Proceedings of Texas/PASCOS '92, p. 344
9. J.N. Bahcall and R.K. Ulrich: Rev. Mod. Phys. **60** (1988) 297

An Introduction to the Physics of Type II Supernova Explosions

Robert Mochkovitch

Institut d'Astrophysique de Paris,
F-75014 Paris, France

Abstract: The explosion of massive stars (type II supernovae) is one of the most violent events occurring in the Universe. When a stellar core collapses to form a neutron star its density becomes larger than the density of nuclear matter and its temperature rises beyond 100 MeV. The physics of the explosion involves a variety of complex problems such as the calculation of the equation of state from 10^9 g·cm^{-3} to nuclear density, the mechanism of neutrino transport, Rayleigh-Taylor instabilities and general relativistic hydrodynamics. The purpose of these lectures is to propose a pedagogical introduction to supernova physics. After a summary of the observational data, the main theoretical results obtained during the past fifteen years are presented and discussed.

1 Introduction and Historical Overview

Bright galactic supernovae are rare events but they are very spectacular. With novae, they are among the very few astronomical events outside the Solar System which could be observed in the ancient times. One extraordinary example might have been the explosion which produced the nearby neutron star "Geminga" some 5×10^5 years ago and which may also be responsible for the local hot bubble in which the Solar System is immersed (Sol et al., 1985; Bignami et al., 1993; Gehrels and Chen, 1993). At a distance of about 100 pc, this supernova (at maximum light) was brighter than the full moon and was certainly a shock for the prehistorical witnesses who saw it from the entrance of a cave! Closer to us, five galactic supernovae have been observed in the last millenium: in 1006, 1054, 1181, 1572 and 1604. The 1006 event was the most luminous, comparable in brightness to Venus. It was a type I outburst, like the Tycho 1572 and Kepler 1604 supernovae. The famous 1054 type II explosion has been described in chinese chronicles and at its position one now sees the Crab nebula and pulsar. The 1181 supernova was dimmer, reaching only magnitude 0 at maximum light.

In modern times, the study of supernovae was initiated by Zwicky who started a systematic search for supernovae first at Mt Wilson and then at Mt Palomar. Approximately 300 supernovae were discovered in 40 years by this very successful program. Ten years earlier, Lundmark (1920) had already realized that Nova S Andromeda 1885 was really an exceptional object if the Andromeda nebula was in fact a galaxy and had to be placed outside our Milky Way some million light years away. Zwicky then invented the word super-novae to describe these "very bright novae". In 1941, Minkowski distinguished two main supernova types from their spectral characteristics. Since then, a great deal of observational data has been accumulated, but always on distant supernovae. The supernovae 1987A in the Large Magellanic Cloud, a companion of our Milky Way at a distance of $\sim$ 50 kpc only, was therefore a milestone in the story of supernova physics. It was the first supernova visible with the naked eye since Kepler 1604, and the occasion of the first detection of the neutrinos produced in core collapse, of the first observation of the X and γ-rays from the radioactivity of ^{56}Co and of many other "firsts".

On the theoretical side, Zwicky (1938) considered the possibility that super-novae arise from the collapse of a normal star to a compact object, the explosion energy being taken from the binding energy of the remnant. Hoyle and Fowler (1960) showed that nuclear burning in a degenerate stellar core should lead to an explosion. These two proposals still stand as the basic mechanisms for type II and type I supernovae, respectively.

The first detailed supernova model was constructed by Colgate and White (1966) who emphasized the importance of neutrinos to power the explosion. Freedman (1974) noticed that neutral currents play a crucial role in the physics of type II supernovae. By allowing coherent scattering of neutrinos on nuclei they increase the cross section of core material to neutrinos by a factor of about 100. As a result, neutrinos are trapped during the collapse. Bethe et al. (1979) showed that one of the consequences of trapping is that the entropy of the core remains low so that nuclei do not dissolve until nuclear density is reached at the center. Due to the stiffness of nuclear matter the inner core then bounces and a strong shock wave is generated at its boundary. The situation when the supernova is directly powered by the energy initially present in this shock corresponds to the so called "prompt mechanism". In most calculations however the shock energy is used up mainly in the photodisintegration of heavy nuclei and also in neutrino losses. The shock finally stalls at a radius of 200 – 300 km and becomes an accretion shock. The present status of the prompt mechanism is that it cannot give successful explosions except may be for very low mass iron cores and low values of the incompressibility of nuclear matter (Baron and Cooperstein, 1991).

In 1985, a possibility to give "a second chance" to the shock was discovered by Wilson (1985). In a sense, it was the modern version of the original Colgate and White (1966) idea. After the failure of the prompt mechanism, material behind the stalled shock is heated by energy deposition from the neutrinos leaving the core. In this "delayed mechanism", after a few tenths of a second, the rejuvenated shock moves on again and an explosion results. In the first calculations the explosion energy was however a bit low compared to the expected 10^{51} erg.

In the recent period, a few groups (Burrows and Fryxell, 1993; Wilson and Mayle, 1993; Herant et al., 1992, 1994) have tried to include the effects of fluid instabilities in the delayed mechanism. This was done either by adding some enhanced neutrino transport mechanism in 1-D codes or directly in 2-D calculations. It seems that at least in some cases, the explosion energy can now be increased up to the required 10^{51} erg.

At the present time, and in spite of the considerable progress made in the last fifteen years, no final consensus has been reached on the details of the explosion mechanism. The clue may come from the detection of the neutrinos from the next galactic supernova, in a hopefully close future!

This review is organized in four parts. We first summarize in Sect. 2 the observational results which fix the main constraints any theoretical model should satisfy. In Sect. 3 we discuss the physics of core collapse until the bounce at nuclear density. We expose in Sect. 4 the reasons for the failure of the prompt mechanism and we describe the role of neutrinos in the delayed mechanism. Finally Sect. 5 presents the various outputs of supernova explosions: neutrino burst, light curve, X and γ-ray emissions, gravitational radiation, nucleosynthesis.

When it was possible, we have tried to use simple (sometimes even oversimplified) models because we believe they can help to understand the results of detailed numerical calculations. Finally, we recommend some recent reviews and books about supernovae:

Supernovae: A Survey of Current Research (1982), ed. by M.J. Rees and R.J. Stoneham (Dordrecht: Reidel).

Theory of Supernovae (1988), ed. by G.E. Brown, *Phys. Rep.* **163** Nos. 1-3.

Supernova 1987A (1989) by W.D. Arnett, J.N. Bahcall, R.P. Kirshner, S.E. Woosley, *Ann. Rev. Astron. Astrophys.* **27**, 629.

Supernovae (1990), ed. by A.G. Petschek (Springer-Verlag, New York).

Supernova Mechanisms (1990) by H.A. Bethe, *Rev. Mod. Phys.* **62**, 801.

Supernovae: The Tenth Santa Cruz Summer Workshop (1991), ed. by S.E. Woosley (Springer-Verlag, New York).

Nuclear Astrophysics (1993), ed. by D.N. Schramm and S.E. Woosley, *Phys. Rep.* **227** Nos. 1-5.

Supernovae: Les Houches Summer School LIV (1994), ed. by S.A. Bludman, R. Mochkovitch, J. Zinn-Justin (Elsevier Science B.V., Amsterdam).

The interested reader is advised to consult them for details and further information.

2 Constraints from the Observations

2.1 Supernova Classification and Spectra

The classification of supernovae (hereafter SNe) is based on the characteristics of their early optical spectra in the days following the explosion (see Harkness and Wheeler, 1990 and Branch, 1990 for good reviews). The fundamental division is between SNe I which have no hydrogen lines and SNe II with hydrogen Balmer

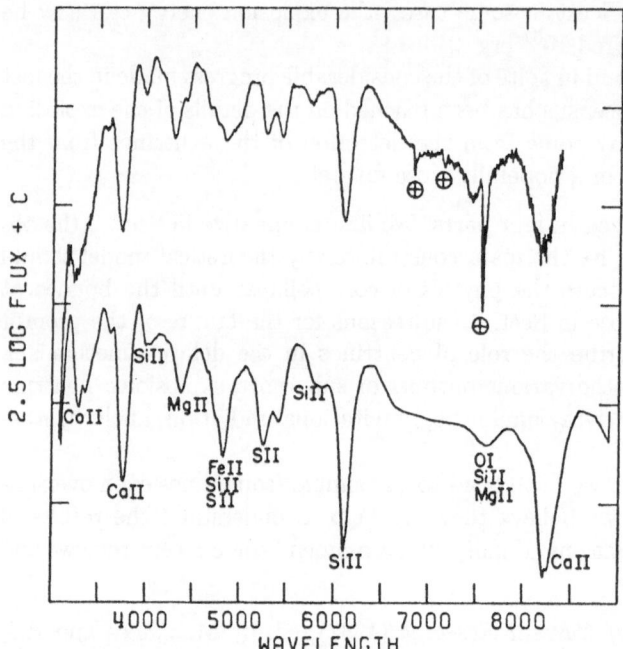

Fig. 1a. Spectrum of the SN Ia SN 1981B at maximum light. The upper curve is the observed spectrum and the lower one is a synthetic spectrum from Branch et al. (1985). The lines have a "P Cygni" profile with a blue shifted absorption and a red shifted emission which is typical of an expanding atmosphere. The narrow lines are absorption features from the Earth atmosphere.

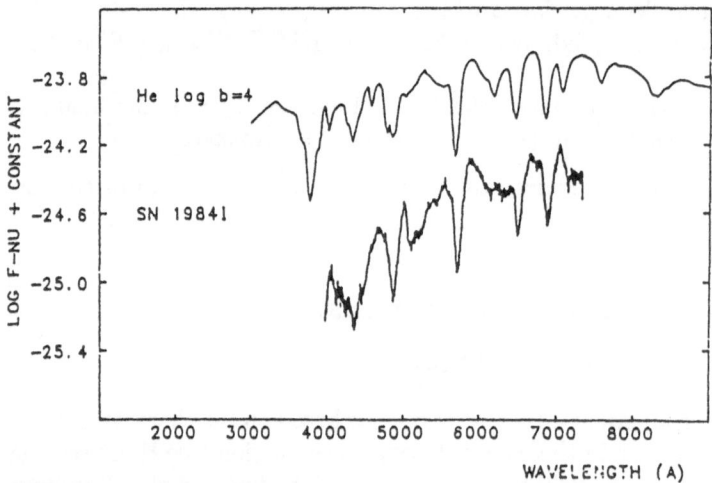

Fig. 1b. Spectrum of the SN Ib SN 1984L 20 days after maximum light compared to a synthetic spectrum computed by Harkness et al. (1987). The deepest absorption features are He lines, again with a P Cygni profile.

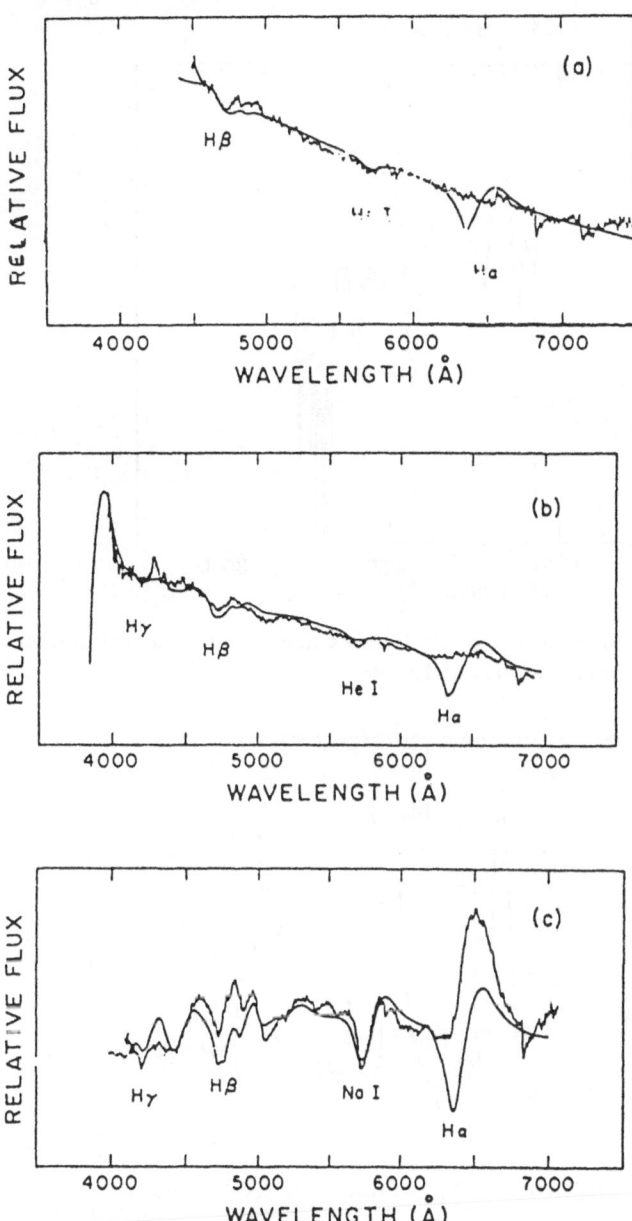

Fig. 2. Spectrum of the SN II SN 1979C at maximum light (top), 6 days after (middle) and one month after (bottom) compared to a synthetic spectrum from Branch et al. (1981). The hydrogen Balmer lines with P Cygni profiles are clearly visible one month after maximum.

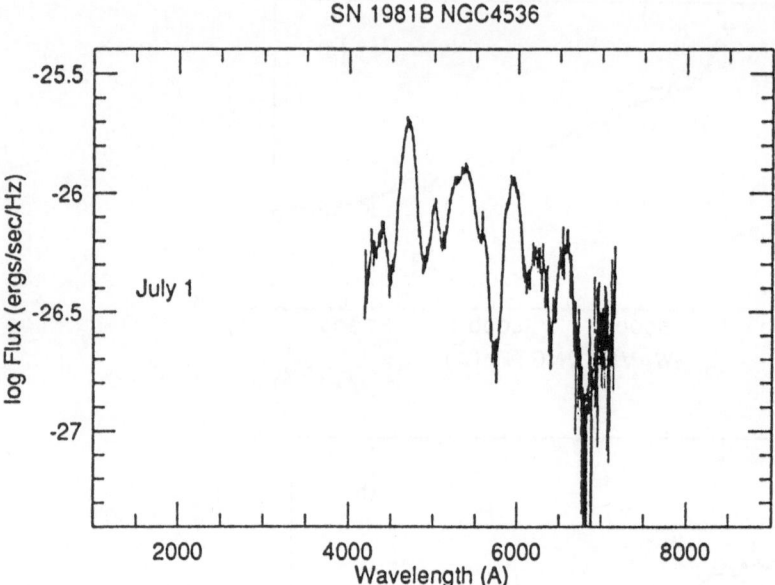

Fig. 3a. Late time spectrum of the SN Ia SN 1981B 4 months after maximum dominated by Co and Fe lines (from Harkness and Wheeler, 1990).

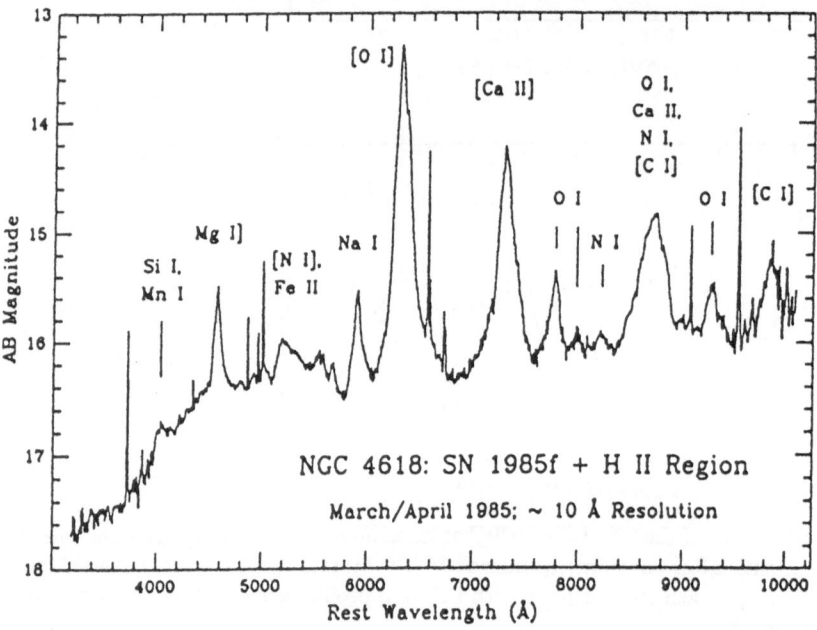

Fig. 3b. Spectrum of the SN Ib SN 1985F in the nebular phase (Filippenko and Sargent, 1986).

lines. In the case of SNe I, three more subdivisions have been introduced: SNe Ia have a strong Si II absorption line at 6150 Å with other lines of intermediate mass elements; SNe Ib show no Si II line but have a He I absorption feature at 5876 Å; in SNe Ic this He I line is much weaker while Ca II, Fe II and O I lines are now present.

At late times (a few months after the explosion), when the supernova envelope becomes transparent, lines are observed in emission. In this "nebular phase" SNe II have strong H_α, H_β and O I lines, SNe Ia are dominated by Fe and Co lines and SNe Ibc by lines of oxygen and other intermediate mass elements.

Examples of spectra for the different supernova types at different epochs are shown in Figs. 1 – 4.

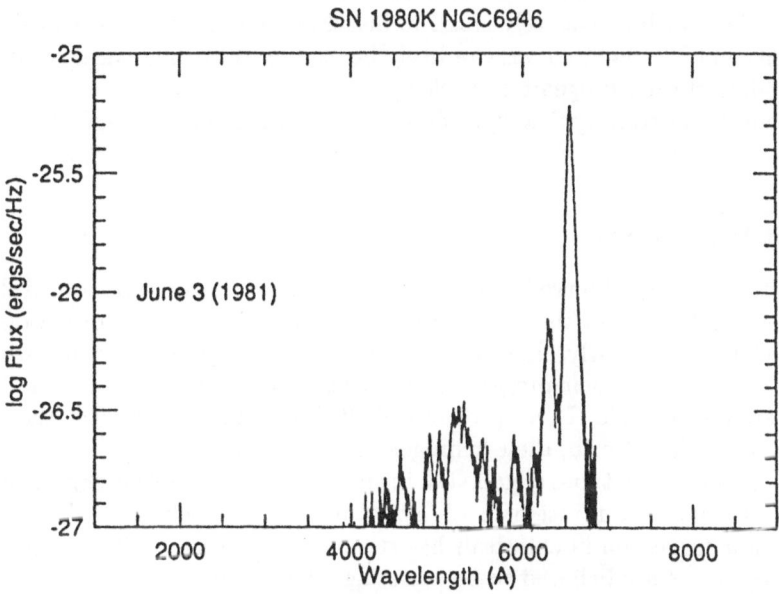

Fig. 4. Spectrum of the SN II SN 1980K in the nebular phase 6 months after maximum with proeminent H_α, H_β and O I lines (from Harkness and Wheeler, 1990).

2.2 Supernova Light Curves

Supernova light curves are powered by various sources of energy. In the case of SNe II, shock energy deposition in the extended envelope of the progenitor dominates at early times. At late times and for both SNe I and II the radioactivity of ^{56}Co is responsible for the observed exponential decline. In SNe II and possibly in SNe Ibc the radio emission of a pulsar immersed in the remnant can also contribute to the light curve.

The SN Ia and Ib light curves are rather similar in the first 50 days of evolution. Afterwards, SNe Ia decline faster than SNe Ib (0.015 and 0.01 mag·day^{-1} respectively). SNe Ib are also less luminous at maximum light by a factor of about 5, due to a smaller amount of synthetized ^{56}Ni. SNe Ia appear to form a very homogeneous class. They can be used as standard candles to determine H_o if their luminosity at maximum light is known with sufficient accuracy or q_o if they can be observed at sufficiently large cosmological redshift. SNe II have less uniform light curve shapes. SNe II-P (for plateau) have a nearly constant luminosity between day 25 and 75 after maximum before the phase of exponential decline due to radioactivity. SNe II-L (for linear) have a rapid linear decline (linear in magnitude i.e. exponential in luminosity) beginning just after maximum and lasting until day 80. After day 80 the decline becomes two times less rapid ~ 0.01 mag·day^{-1}. The supernova 1987A in the Large Magellanic Cloud had a peculiar light curve, which is now understood simply because its progenitor was a blue supergiant with a relatively small radius. More generally, it seems that the variability in SN II light curves can likely be explained by differences in the mass and radius of their progenitor envelope.

Figs. 5 and 6 illustrate typical light curves for the different supernova types and for SN 1987A.

2.3 Parent Populations

SNe II and SNe Ibc are observed in spiral and irregular galaxies only. In spiral galaxies they are located in the spiral arms where most of the star formation takes place. They are clearly associated to the young stellar population with an age $\lesssim 5 \times 10^7$ years. Their progenitors must therefore be massive stars, red or blue supergiant for SNe II and probably Wolf-Rayet stars having lost their hydrogen envelope for SNe Ib, their hydrogen and helium for SNe Ic.

SNe Ia are found in all types of galaxies. In spirals they are spread everywhere in the disk with no concentration in the spiral arms. They belong to the old population (star formation in ellipticals has stopped since several billion years) and their progenitors are believed to be accreting white dwarfs.

2.4 Supernova Rates

Several hundreds of supernovae have been observed since 50 years. This amount of data allows a statistical study of the rate of occurence of the different supernova types in galaxies (see Tammann, 1994 for a review). The results are shown in Tab. 1 where the supernova rates are expressed in SNu (supernova unit). One SNu corresponds to 1 supernova per century per 10^{10} $L_{B_\odot}$, where $L_{B_\odot}$ is the solar luminosity in the blue (4500 Å) band.

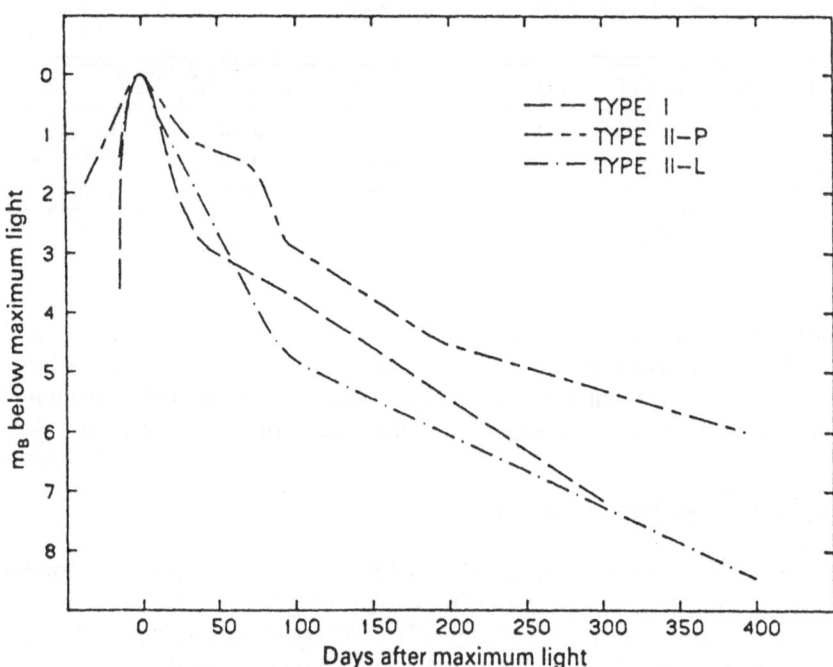

Fig. 5. Comparison of the light curves for SNe Ia, SNe II-P and SNe II-L (from Doggett and Branch, 1985).

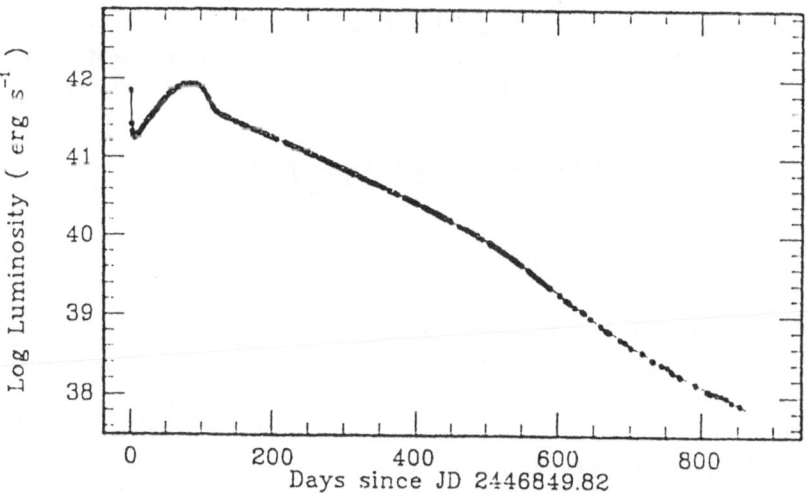

Fig. 6. Bolometric light curve of SN 1987A (from Catchpole et al., 1987).

Table 1. The absolute frequencies (in SNu) of the different supernova types along the Hubble sequence of galaxies (from Tammann, 1994).

Hubble type	E – S0	S0/a – Sa	Sab – Sb	Sbc – Sd	Sdm – Im
SN Ia	0.25	0.12	0.12	0.12	0.12
SN Ib	–	0.01	0.07	0.19	0.23
SN II	–	0.04	0.34	0.98	1.05
Total	0.25	0.17	0.53	1.29	1.40

Our Galaxy has a luminosity $L_B \approx 2.3 \times 10^{10} L_{B_\odot}$ and is generally considered to be a Sbc. This gives an expected SN rate of 3 per century with a large uncertainty, the small number of historical supernovae (5 in the last millennium) being interpreted as a consequence of dust absorption on distant objects.

2.5 Basic Energy Requirements

Supernovae provide luminous and kinetic energy, and also produce large amounts of neutrinos in the case of SNe II and (probably) SNe Ibc. The energies released in these three forms are respectively $E_{\text{ph}} \sim 10^{49}$ erg (in photons), $E_{\text{kin}} \sim 10^{51}$ erg (10^{51} erg is approximately the total energy radiated by the Sun during its entire life on the main sequence!) and $E_\nu \sim 10^{53}$ erg (in neutrinos for SNe II).

The value of E_{ph} is simply obtained from the time integration of the light curve after bolometric correction. The kinetic energy can be estimated from spectroscopy in the following way: let us assume that the supernova expands at a velocity v_{exp} after the explosion. The nebular phase begins when the ejected shell (of mass M_{ej}, radius $R \approx v_{\text{exp}}t$ and thickness ΔR) becomes transparent that is when $\rho \kappa \Delta R \approx 1$, where $\rho \approx M_{\text{ej}}/4\pi R^2 \Delta R$ is the average density and κ the opacity. Adopting for κ the Thomson opacity one gets for M_{ej} and E_{kin}

$$M_{\text{ej}} \approx 5 \left(\frac{v_{\text{exp}}}{5000 \text{ km} \cdot \text{s}^{-1}} \right)^2 \left(\frac{\tau}{1 \text{ yr}} \right)^2 M_\odot , \qquad (1)$$

$$E_{\text{kin}} = \frac{1}{2} M_{\text{ej}} v_{\text{exp}}^2 \approx 10^{51} \left(\frac{v_{\text{exp}}}{5000 \text{ km} \cdot \text{s}^{-1}} \right)^4 \left(\frac{\tau}{1 \text{ yr}} \right)^2 \text{ erg} , \qquad (2)$$

where τ is the time needed to reach the nebular phase. These estimates are naturally very crude but they nevertheless give the correct orders of magnitude for typical values of v_{exp} and τ.

The value of E_ν can be deduced from the detection of the neutrinos from SN 1987A. The Kamiokande II experiment detected 12 $\bar{\nu}_e$ of average energy $\epsilon_{\bar{\nu}_e} \sim 15$ MeV by capture on protons ($\bar{\nu}_e + \text{p} \to \text{n} + \text{e}^+$) in a tank filled by 2000 tons of water (Hirata et al., 1987). One then has

$$\frac{N_{\bar{\nu}_e}}{4\pi D^2} \times \sigma_{\bar{\nu}_e \text{p}} \times N_{\text{p}} \approx 12 , \qquad (3)$$

where $N_{\bar{\nu}_e}$ is the number of $\bar{\nu}_e$ produced by SN 1987A, $D \approx 52$ kpc is the distance to the Large Magellanic Cloud, $\sigma_{\bar{\nu}_e p} \approx 1.5 \times 10^{-41}$ cm^2 is the cross section for the capture of 15 MeV $\bar{\nu}_e$ on protons and $N_p \approx 10^{32}$ is the number of protons in 2000 tons of water (counting the protons of hydrogen only). This yields $N_{\bar{\nu}_e} \approx 2 \times 10^{57}$ and $N_\nu \approx 6N_{\bar{\nu}_e} \approx 10^{58}$ assuming an equal number of all neutrino types. Finally, the energy $E_\nu \approx N_\nu \epsilon_\nu \approx 3 \times 10^{53}$ erg is several hundred times the kinetic energy. SNe II therefore appears essentially as neutrino events from an energy point of view. We only see the very small visible part of the iceberg!

Table 2. Energies (in erg) released by the different supernova types.

	E_{ph}	E_{kin}	E_ν	E_{tot}
SN Ia	10^{49}	10^{51}	—	10^{51}
SN Ib	10^{49}	10^{51}	10^{53} (?)	10^{53} (?)
SN II	10^{49}	10^{51}	10^{53}	10^{53}

2.6 Energy Sources

In the case of SNe Ia the total explosion energy is simply the kinetic energy $E_{kin} \approx 10^{51}$ erg. In the "nuclear incineration" of a massive carbon oxygen white dwarf $\sim 40\%$ of the stellar mass undergoes complete burning to iron peak elements which releases $q_{nuc} \approx 8 \times 10^{17}$ erg·g^{-1}. In the rest of the star which is partially burned to intermediate mass elements $q_{nuc} \approx 5 \times 10^{17}$ erg·g^{-1}. For a white dwarf close to the Chandrasekhar limit, the total nuclear energy is $E_{nuc} \approx 1.8 \times 10^{51}$ erg. A first estimate of the supernova energy is then given by

$$E_{SNIa} \approx E_{nuc} - |B| , \qquad (4)$$

where $|B| \approx 10^{50}$ erg is the binding energy of the white dwarf. Since $E_{nuc} \gg |B|$, the white dwarf is completely disrupted in the explosion (Nomoto et al., 1984; Woosley, 1990). Relation (4) only gives an upper limit since the exact value of the supernova energy also depends on the amount of electron captures on burning products behind the combustion front. Electron captures decrease the pressure and the energy behind the front and can even lead to a collapse rather than an explosion if carbon has been ignited at a density $\gtrsim 10^{10}$ g·cm^{-3} (Canal et al., 1990; Nomoto and Kondo, 1991; Timmes and Woosley, 1992).

For SNe II we have $E_{SNII} \approx E_\nu \approx 10^{53}$ erg and only the formation of a compact remnant, a neutron star or a black hole can provide such a huge amount of energy. Assuming that a 1.4 $M_\odot$ neutron star of radius $R = 10$ km is formed in a SN II explosion, its binding energy will be

$$B_{NS} = -\alpha \frac{GM^2}{R} = -5\alpha \times 10^{53} \text{ erg} , \qquad (5)$$

with $\alpha \lesssim 1$. The conversion of less than 1% of this binding energy into explosion kinetic energy would be enough to power a SN II. The possible ways to realize this conversion are discussed in the next two sections.

3 Presupernova Evolution, Infall and Bounce

3.1 The Presupernova Star

In the final stages of their evolution, massive stars ($M \gtrsim 8 - 10\ M_\odot$) develop an onion-like structure where each layer is made of the burning products of its immediate (external) neighbour. Starting from the center, there is first an iron core, surrounded by silicon, oxygen, neon, carbon, and helium shells. The hydrogen envelope outside is very extended (Fig. 7).

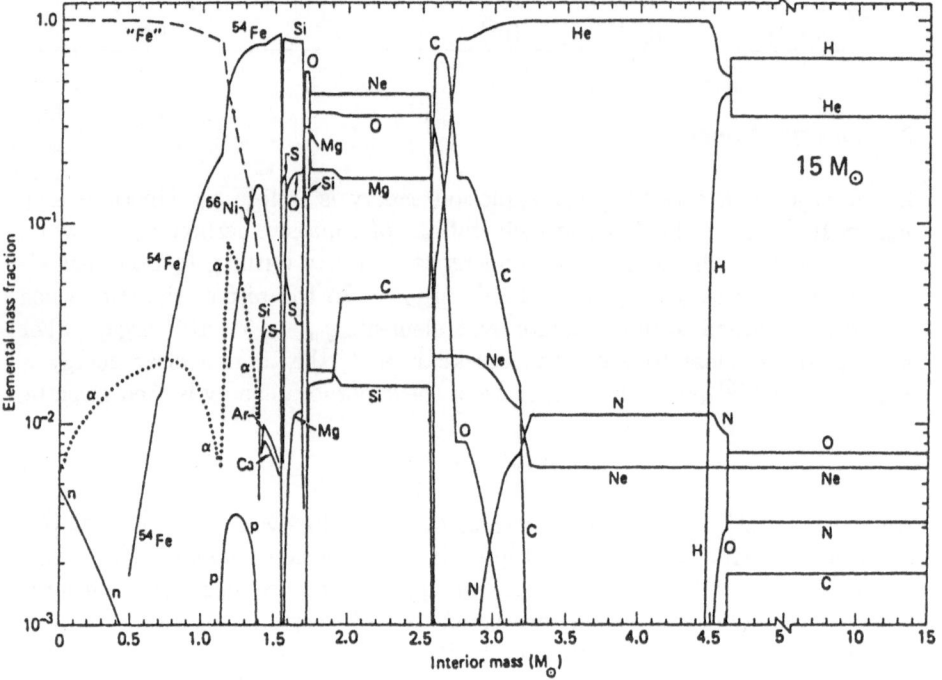

Fig. 7. Final composition of 15 $M_\odot$ star just before explosion (from Weaver et al., 1978).

If this envelope is radiative, its radius is of the order of 50 $R_\odot$ and the star is a blue supergiant (BSG). If energy transport is made by convection the radius is much larger $\sim 1000\ R_\odot$, corresponding to a red supergiant (RSG). Before the explosion of SN 1987A the common prejudice was that massive stars were ending their evolution as red supergiants. The progenitor of SN 1987A was however a blue supergiant which went through the red supergiant stage before going back to the blue. The reason for this evolution is not completely clear but it could be a consequence of the lower metallicity of the Large Magellanic Cloud (see Arnett et al., 1989, Nomoto et al., 1994 and references therein).

In a supergiant star there is an extreme contrast between the size of the iron core and that of the envelope as shown in Table 3.

Table 3. Mass and radius of the different zones — iron core, layer of intermediate mass elements, helium mantle and hydrogen envelope — in a 15 $M_\odot$ presupernova star. A reduced scale illustrates the contrast between the core and the envelope (from Weaver et al., 1978).

Composition	Fe	C, O, Ne, Si	He	H
Mass (in $M_\odot$)	1.4	1.0	2.0	10.0
Radius (in km)	10^3	$4 \cdot 10^4$	$3 \cdot 10^5$	$6 \cdot 10^8$ (RSG)/$4 \cdot 10^7$ (BSG)
Radius $\times 5 \cdot 10^{-10}$	0.5 mm	2 cm	15 cm	300 m (RSG)/ 20 m (BSG)

The iron core which will be the supernova "engine" can be viewed as a pinhead at the center of a large (BSG) to huge (RSG) balloon! The electrons are degenerate and relativistic in the core ($\epsilon_F/kT \sim \epsilon_F/m_e c^2 \sim 10$) which will then become unstable when its mass M_{Fe} reaches the Chandrasekhar limit.

3.2 The Origin of the Instability

The pressure of degenerate ultra relativistic electrons is given by

$$P_e = K(S_e)(\rho Y_e)^{4/3} , \tag{6}$$

where $S_e = \pi^2 \frac{kT}{\epsilon_F}$ is the entropy of the electron gas and Y_e the number of electron per nucleon. The Chandrasekhar mass M_{ch} which can be obtained from (6) is proportional to $(\frac{K}{G})Y_e^2$. As $M_{\mathrm{Fe}} \to M_{\mathrm{ch}}$, the central density and temperature rise which leads to electron captures and iron photodesintegration. These two processes decrease Y_e and S_e respectively and are the cause of the instability by reducing the value of M_{ch}.

The exponent 4/3 in (6) holds for ultrarelativistic (UR) particles only (of energy $\epsilon = pc$). Since 4/3 is also the critical adiabatic index γ_{crit} for the stability of newtonian stars, it could seem that the core remains stable until it reaches infinite density and zero radius! The inclusion of general relativity however adds a correction to γ_{crit} of the order of R_g/R (R_g being the gravitational radius of the object) and allows the instability to occur at a finite radius.

This can be understood from a simple one zone (uniform density) model of a star supported by the pressure of degenerate relativistic electrons. Its total energy can be written

$$E = \frac{3}{5} \frac{GM^2}{R} \left[\left(\frac{M}{M_{\mathrm{ch}}} \right)^{-2/3} - 1 \right] + A(M, Y_e)R - \frac{B(M)}{R^2} , \tag{7}$$

where M and R are the mass and radius of the star, $A(M, Y_e)$ a function of M and Y_e and $B(M) = \frac{G^2}{c^2} M^3$. The first term in (7) corresponds to an object supported by UR electrons. It goes to zero when $M = M_{ch}$. The second term is the first correction to the UR electron gas, proportional to $R^3 (\frac{m_e c^2}{\epsilon_F})^2$. The third term is the correction due to general relativity (Shapiro and Teukolsky, 1983). An equilibrium configuration is found by looking for a minimum of E. Without the relativistic term a stable solution exists up to $M = M_{ch}$ for which $R = 0$. With the relativistic term there is an energy minimum as long as the mass remains smaller than a critical value M_{crit}. In the limiting case $M = M_{crit}$, $E(R)$ has an inflexion point and M_{crit} and $R(M_{crit})$ can be obtained from the two conditions

$$\frac{dE}{dR} = \frac{d^2 E}{dR^2} = 0 . \tag{8}$$

Typically M_{crit} is a few per cent larger than the Chandrasekhar mass and $R_{crit} \approx$ 2000 km. When the iron core becomes unstable it collapses in a dynamical time $\tau_{dyn} \lesssim 0.1$ s. It is clear that the rest of the star does not react on this short time scale and remains spectator of the catastrophe going on at the center.

3.3 Initial Entropy of Core Material

Bethe et al. (1979) realized that the entropy per nucleon is low in the core. They first estimated its initial value at the moment of instability, from a presupernova model computed by Arnett (1977) giving for the central density, temperature and electron number: $\rho_c \approx 4 \times 10^9$ g·cm^{-3}, $T_c \approx 8 \times 10^9$ K ≈ 0.7 MeV and $Y_e \approx 0.45$. The total entropy is a sum of the contributions from the translational motion of nuclei, alpha particles and free neutrons (there are very few free protons), from the excited nuclear states and from the degenerate electrons (Tab. 4).

Table 4. Various contributions to the total entropy of core material (from Bethe et al., 1979).

	s per nucleon
Trans. motion of nuclei	0.30
Alpha's and neutrons	0.06
Excited states	0.09
Electrons	1.15 (per electron)

The total entropy is then

$$s = 0.30 + 0.06 + 0.09 + 0.45 \times 1.15 = 0.97 . \tag{9}$$

As will be shown in Sect. 3.5 the entropy remains low during all the collapse so that there is only a very limited amont of neutron drip and nuclei do not dissolve until they merge at nuclear density.

3.4 Nuclear Parameters

Several important quantities can be derived from the expression of the energy per nucleon in nuclei, W/A. Using a simple liquid drop model approach Bethe et al. (1979) obtained

$$\frac{W}{A}(x, A, u) = a_\mathrm{v} + a_\mathrm{s}(1 - 2x)^2 + a_\mathrm{surf}(x)A^{-1/3} + a_\mathrm{c}x^2 g(u)A^{2/3} , \qquad (10)$$

where $x = Z/A$ in nuclei (which can be different from Y_e if there are many alpha particles and free neutrons), $u = \rho/\rho_0$ (ρ_0 being the nuclear matter density) and $g(u)$ is a Coulomb form factor which accounts for the various topologies adopted by the nuclear fluid at high density (see Cooperstein and Baron, 1990 for a review). The different terms in (10) are the volume ($a_\mathrm{v} = -16$ MeV), symmetry ($a_\mathrm{s} = 32$ MeV), surface ($a_\mathrm{surf} = 290x^2(1 - x)^2$ MeV) and Coulomb ($a_\mathrm{c} = 0.75$ MeV) energies. The mass of the most bounded nucleus can be deduced from the condition

$$\left.\frac{\partial(W/A)}{\partial A}\right|_{x,u} = 0 , \qquad (11)$$

which yields $A = 194(1-x)^2/g(u)$. This most bounded nucleus will be considered as representative even if a whole collection of nuclei is naturally present in the core. Another important quantity is the difference between the neutron and proton chemical potentials $\hat{\mu}$. In the context of the liquid drop model $\hat{\mu}$ is given by

$$\hat{\mu} = \mu_\mathrm{n} - \mu_\mathrm{p} = -\left.\frac{\partial(W/A)}{\partial x}\right|_{A,u} . \qquad (12)$$

The expression (12) is an approximation which is computed for the representative nucleus. It does not take into account nuclear shell effects, which can introduce big deviations to this simple result.

3.5 Electron Captures

As the collapse proceeds the Fermi energy μ_e of the electrons increases allowing electron captures on nuclei and free protons. The effects of these captures will be (i) a reduction of Y_e in the core with important consequences on the dynamics of collapse (see Sect. 3.7) and (ii) an entropy change because the captures take place out of beta equilibrium and the daughter nucleus is produced in an excited state of energy $\Delta_\mathrm{n} \approx 3$ MeV. The entropy change is given by

$$TdS = dQ - \Sigma\mu_i dN_i = -dY_\mathrm{e}[\mu_\mathrm{e} - \hat{\mu} - \langle\epsilon_\nu\rangle] , \qquad (13)$$

where $\langle\epsilon_\nu\rangle$ is the energy carried away by the escaping neutrino (Bethe et al., 1979). In the case of captures on heavy nuclei, $\langle\epsilon_\nu\rangle \approx 0.6\Delta$ where $\Delta = \mu_\mathrm{e} - \hat{\mu} - \Delta_\mathrm{n}$ which gives an entropy increase. Captures on free protons decrease the entropy because now $\langle\epsilon_\nu\rangle \approx \frac{5}{6}\mu_\mathrm{e}$ and the difference $\frac{1}{6}\mu_\mathrm{e} - \hat{\mu}$ is negative (Van Riper and Lattimer, 1981). The overall effect of the two processes depends on their respective cross sections and on the abundance of free protons. The capture rate takes the form

$$\frac{dY_e}{dt} = -\rho Y_e \mathcal{N}_A c(\overline{\sigma_{eAZ}} Y_p + \overline{\sigma_{ep}} Y_p^f) , \qquad (14)$$

where $\mathcal{N}_A$ is the Avogadro number; $\overline{\sigma_{eAZ}}$ and $\overline{\sigma_{ep}}$ are the cross sections for electron capture on nuclei and free protons averaged on the electron and proton spectra; Y_p and Y_p^f are the fractions of protons (over the total number of nucleons) in nuclei or free, respectively. Since Y_p^f is always very small, one has $Y_p \approx Y_e$. Equation (14) then gives

$$\frac{d\text{Ln}Y_e}{dt} \approx -2\rho_{10} \left[\left(\frac{\overline{\sigma_{eAZ}}}{10^{-44}\text{cm}^{-2}} \right) Y_e + \left(\frac{\overline{\sigma_{ep}}}{10^{-44}\text{cm}^{-2}} \right) Y_p^f \right] , \qquad (15)$$

where ρ_{10} is the density in units of 10^{10} g·cm^{-3}. The time evolution of the density during the collapse can be obtained from the homologous solution (see Sect. 3.7)

$$\frac{d\text{Ln}\rho}{dt} \approx 20\rho_{10}^{1/2} . \qquad (16)$$

Combining (15) and (16) the variation of Y_e can be directly related to the density increase

$$\frac{d\text{Ln}Y_e}{d\text{Ln}\rho} \approx -0.1\,\rho_{10}^{1/2} \left[\left(\frac{\overline{\sigma_{eAZ}}}{10^{-44}\text{cm}^{-2}} \right) Y_e + \left(\frac{\overline{\sigma_{ep}}}{10^{-44}\text{cm}^{-2}} \right) Y_p^f \right] . \qquad (17)$$

The main sources of uncertainty in (17) are the cross section for capture on heavy nuclei which depends on the nature of the transition and the value of Y_p^f. Bethe et al. (1979) proposed a simplified theory assuming that captures on heavy nuclei are allowed Gamow-Teller transitions from the proton $f_{7/2}$ shell to the neutron $f_{5/2}$ shell. Fuller (1982) however pointed out that the neutron $f_{5/2}$ shell becomes full early in the collapse preventing other captures. Detailed studies (Cooperstein and Wambach, 1984) have shown that captures on protons dominate first, followed by first forbidden transitions while at high density ($\sim 10^{12}$ g·cm^{-3}) allowed transitions become important again. The fraction of free protons is controlled (exponentially) by the value of $\hat{\mu}$ which itself is related to the symmetry energy a_s, the preferred choice for a_s being generally 32 MeV.

Another process, neutrino-electron scattering, contributes to affect the final values of the entropy and Y_e (Myra et al., 1987; Myra and Bludman, 1989; Bruenn, 1989ab). Neutrinos loose energy in scattering and therefore escape more easily. More captures occur before trapping and less entropy is carried away, which leads to a Y_e smaller by a few % and a larger entropy at trapping.

The results of a calculation of the capture process during collapse are given in Tab. 5.

Table 5. Values of the temperature, Y_e, $\hat{\mu}$, μ_e, s, X_p and X_H (mass fractions in free protons and heavy nuclei) from the beginning of the collapse to the trapping density (from Cooperstein and Wambach, 1984).

Log ρ	T in MeV	Y_e	$\hat{\mu}$	μ_e	s	$10^4 X_p$	$1 - X_H$ %
10.0	0.88	0.420	3.1	8.1	1.00	2.7	3.6
10.5	1.04	0.418	3.3	11.9	0.99	5.4	4.8
11.0	1.21	0.410	4.5	17.5	0.97	5.3	4.7
11.5	1.37	0.389	7.4	25.0	0.93	1.6	4.2
12.0	1.49	0.358	12.1	36.0	0.91	0.16	5.9

3.6 Neutrino Trapping

Neutral currents allow coherent elastic scattering of neutrinos by nuclei. The cross section is essentially proportional to N^2 (where N is the neutron number) and to ϵ_ν^2 (Lamb and Pethick, 1976), which gives for the mean free path

$$\lambda_\nu \approx 10^8 \rho_{12}^{-1} \left[\left(\frac{N^2}{6A} \right) X_H + X_f \right]^{-1} \epsilon_\nu^{-2} \text{ cm} , \qquad (18)$$

where ρ_{12} is the density in units of 10^{12} g·cm^{-3} and ϵ_ν is in MeV; X_H and X_f are respectively the mass fractions in heavy nuclei and free nucleons. At $\rho_{12} = 1$, we have $A \approx 100$ and using the data of Tab. 5 we get

$$\lambda_\nu \approx 1.5 \left(\frac{10 \text{ MeV}}{\epsilon_\nu} \right)^2 \text{ km} . \qquad (19)$$

With $\langle \epsilon_\nu \rangle \approx 20$ MeV, the mean free path is 0.3 – 0.4 km and the neutrino diffusion time scale is

$$\tau_{\text{diff}} \approx \frac{3R^2}{\lambda_\nu c} \approx 30 \text{ ms} , \qquad (20)$$

which is much larger than the dynamical time scale at the same density $\tau_{\text{dyn}} \approx 5$ ms (see (16)). Following trapping, leptons and baryons get into β-equilibrium and the chemical potentials satisfy the relation

$$\mu_e \approx \hat{\mu} + \mu_\nu \qquad (21)$$

where μ_ν is the Fermi energy of the degenerate neutrinos. The entropy is now conserved as well as the total lepton number $Y_l = Y_e + Y_\nu$. Table 6 gives for two densities (10^{12} and 10^{14} g·cm^{-3}) the values of μ_e, μ_ν, $\hat{\mu}$, Y_e and Y_ν.

Table 6.

	μ_e	μ_ν	$\hat{\mu}$	Y_e	Y_ν
10^{12} g·cm^{-3}	34	22	12	0.32	0.04
10^{14} g·cm^{-3}	155	125	30	0.29	0.07

3.7 Dynamics of the Collapse

During most of the collapse until nuclear density the pressure is dominated by the contribution of relativistic degenerate electrons. The gas then behaves as a polytrope of index $4/3$ (i.e. $P \approx K\rho^{4/3}$) which leads to a simple dynamical evolution because the Jeans mass of a $4/3$ polytrope is equal to the Chandrasekhar mass. This means that the core will collapse as one piece, the sound travel time being equal to the dynamical time. Its density distribution (ρ/ρ_c as a function of r/R) will remain unchanged during the evolution (homologous collapse).

However, since Y_e is reduced from $Y_e^i \approx 0.45$ to $Y_e^f \approx 0.36$ at trapping, the relevant Chandrasekhar mass is

$$M_{ch} \approx M_{Fe} \times \left(\frac{Y_e^f}{Y_e^i}\right)^2 \approx 0.8 M_\odot . \tag{22}$$

Only this inner part of the core will collapse homologously while the rest will follow in nearly free fall. Goldreich and Weber (1980) and Yahil (1983) have obtained beautiful analytical solutions of the dynamical equations. The Yahil (1983) solution is more complete because it covers both the inner and outer core while Goldreich and Weber (1980) limit themselves to the homologous collapse of the inner core. In this simpler approach the radius and velocity of a given mass shell are written

$$r = a(t)\xi ,$$
$$v = \dot{a}(t)\xi = \frac{\dot{a}}{a}r , \tag{23}$$

where $a(t) = \rho_c^{-1/3}(\frac{K}{\pi G})^{1/2}$. Introducing these relations into the continuity and momentum equations Goldreich and Weber (1980) show that the density profile $\frac{\rho}{\rho_c}(\xi) = \phi(\xi)$ is invariant during the collapse and that the time evolution of a is given by

$$\left(\frac{\dot{a}}{a}\right)^2 = \frac{8\lambda}{3}\left(\frac{K^3}{\pi G}\right)^{1/2} a^{-3} = \frac{8\lambda}{3}\pi G \rho_c \approx 40\rho_{10} , \tag{24}$$

with $\lambda = 0.0065$. It can be seen that (16) above is a direct consequence of (24) since $\frac{dLn\rho}{dt} = -3\frac{\dot{a}}{a}$.

Outside the homologous core the velocity is approximately a constant fraction $\alpha = 0.6 - 0.7$ of the free fall velocity

$$v = \alpha v_{ff} \propto r^{-1/2} . \tag{25}$$

The velocity profile for $\rho_c \approx 2 \times 10^{13}$ g·cm^{-3} (about 1 ms before bounce) has been represented in Fig. 8.

The sound velocity

$$v_s^2 = \left(\frac{\partial P}{\partial \rho}\right)_S = \frac{4}{3}K\rho^{1/3} = \frac{4}{3}K\rho_c^{1/3}\phi^{1/3} , \tag{26}$$

is also shown in Fig. 8. The sonic radius where the infall velocity is equal to the sound velocity is close to the boundary between the inner and outer core. A very

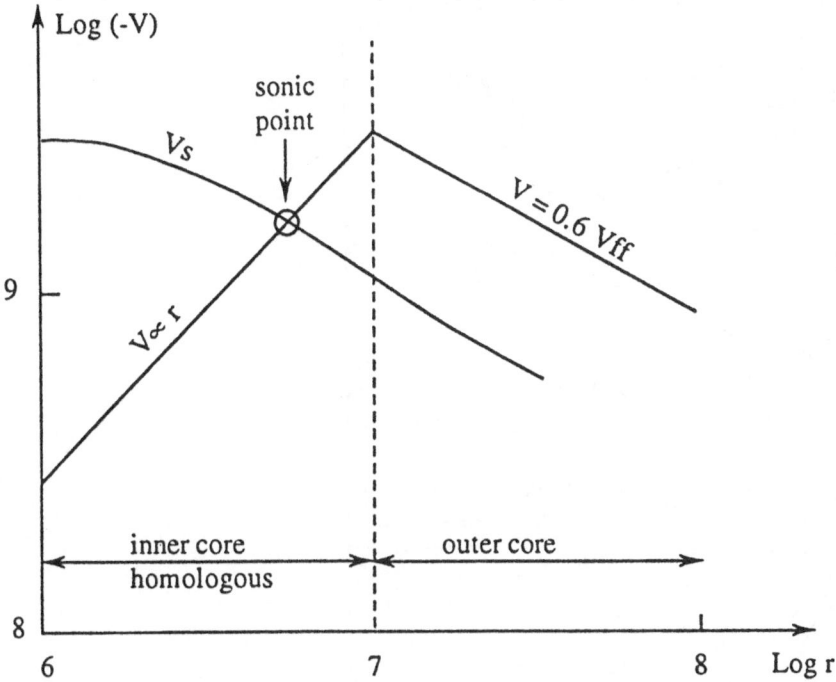

Fig. 8. Velocity profile in the collapsing core ~ 1 ms before bounce. In the inner core the velocity is proportional to r, while in the outer core $v \propto r^{-1/2}$. At the sonic radius the sound velocity is equal to the infall velocity.

important result of these studies is that the mass of the inner homologous core M_{hc} slightly exceeeds the Chandrasekhar mass

$$M_{hc} = (1+f)M_{ch} = (1+f)\left(\frac{Y_e^f}{Y_e^i}\right)^2 M_{Fe}, \qquad (27)$$

with $f = 0.05 - 0.1$. This value of f will be of prime importance in the determination of the initial energy carried by the shock after bounce (see Sect. 3.10).

In conclusion to the last three sections Fig. 9 illustrates the results of a detailed collapse calculation by Van Riper and Lattimer (1981). At different times the density, velocity, Y_e and Y_ν profiles are given as a function of the mass coordinate. One sees the homologous core of mass $M_{hc} \approx 0.7\ M_\odot$ progressively formed during collapse due to the decrease of Y_e and at late times the presence of a fraction of trapped neutrinos.

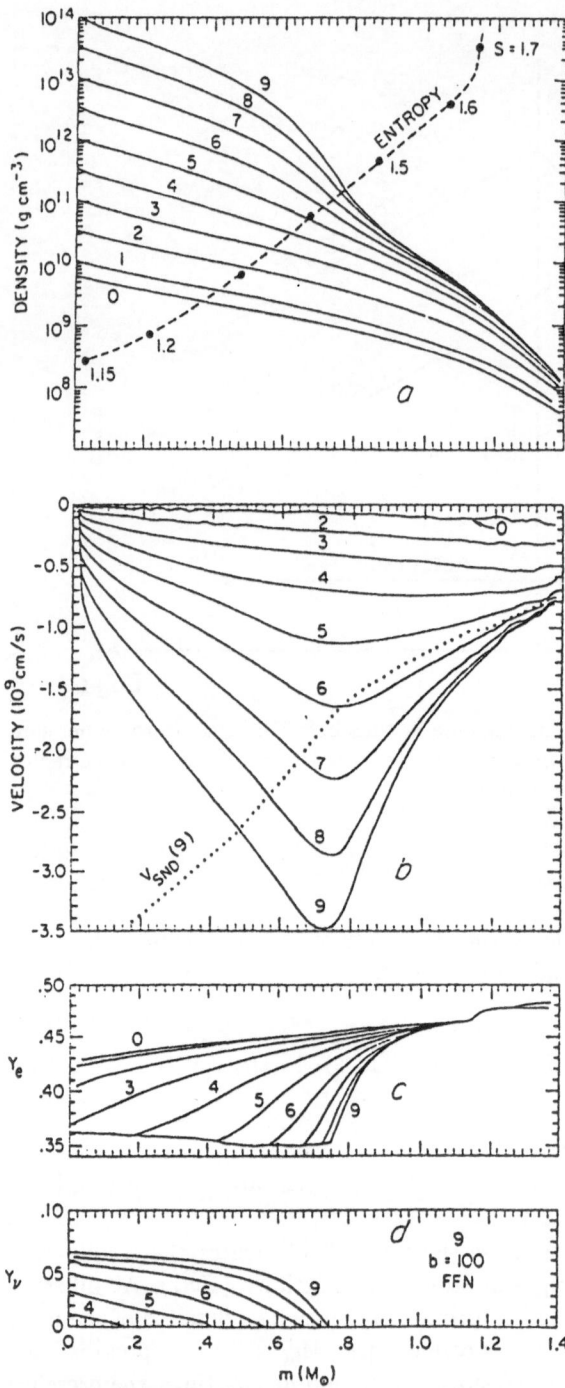

Fig. 9. Density, velocity, Y_e and Y_ν profiles in the core for several times during the collapse, 0 being the initial model (from Van Riper and Lattimer, 1981).

3.8 The Equation of State

The detailed computation of the equation of state (EOS) from $\sim 10^9$ to several times 10^{14} g·cm^{-3} is naturally a very difficult task but the basic features of the EOS are nevertheless quite simple. Due to the low entropy maintained all along the collapse, nucleons stay into nuclei until nuclei themselves merge at nuclear density. This means that until that moment the pressure is dominated by the contribution of degenerate electrons and the adiabatic index γ is close to $4/3$. At $\rho \gtrsim 10^{14}$ g·cm^{-3}, the stiffness of nuclear matter increases γ to ~ 2 which causes the bounce of the inner core.

Below nuclear density the EOS can be computed with various methods (see Vautherin, 1994 for a review). The most detailed calculations solve the mean field equations coupling the nuclear potential and the nucleon density microscopically, using Thomas-Fermi or Hartree-Fock techniques, the latter giving the most detailed results (Marcos et al., 1982; Bonche and Vautherin, 1981, 1982; Hillebrandt et al., 1984). The neutron and proton distributions are then directly obtained in a Wigner-Seitz cell as shown in Fig. 10. In the simpler macroscopic approach nuclei and free nucleons form two phases in equilibrium. The free energy of the nuclei is derived from a finite temperature generalization of the compressible liquid drop model (Lamb et al., 1978, 1981; Lattimer et al., 1985). Nuclei make a "dense phase" in equilibrium with a "dilute phase" made of free nucleons and alpha particles, the chemical potentials of the neutrons and protons being continuous between the two phases.

An analytical EOS has even been proposed by Cooperstein (1985) where the thermal part of the free energy of the nuclei (which is related to the density of excited states) takes the form

$$F_{\text{th}} = -\frac{a}{A}\frac{m^*}{m}T^2 , \tag{28}$$

with

$$\frac{a}{A} - \frac{\pi^2}{4E_{\text{f}}} = \left(\frac{\rho}{0.16 \text{ fm}^{-3}}\right)^{-2/3}(14.9)^{-1} \text{ MeV} , \tag{29}$$

and where $\frac{m^*}{m}$ is a ratio correcting for the effective nucleon mass in nuclei. Cooperstein (1985) adopts for $\frac{m^*}{m}$ an interpolation formula between the value for iron group nuclei and that of a saturated uniform medium.

The EOS is then obtained from a minimization of the free energy. At a given entropy and Y_e the equilibrium equations can be solved to find the mass fractions of the different species and the temperature as a function of density. Some results are shown in Fig. 11 and 12. Fig. 11 is a density temperature diagram where lines of constant entropy and constant X_H have been represented for a given $Y_e = 0.35$. In the shaded area there is a "swiss cheese" of nuclear matter with bubbles and to the right of it lies uniform nuclear matter. It can be seen that the line $s = 1$, which can represent the track followed by the core, remains in the region of nuclei, with very few free nucleons until nuclear density. Fig. 12 give the mass fractions in neutrons, protons, alpha particles and heavy nuclei as a

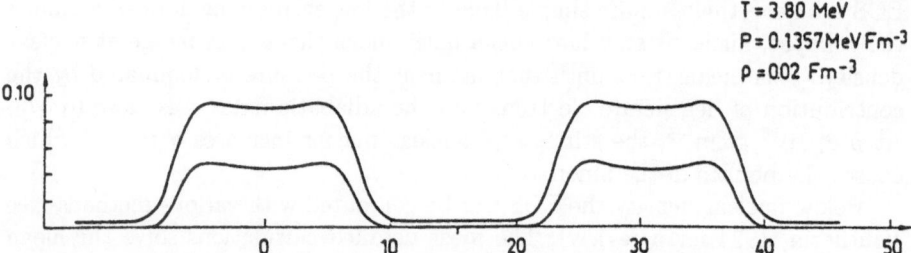

Fig. 10. Neutron and proton density profiles along the line joining the centers of neighbouring cells (from Vautherin, 1994).

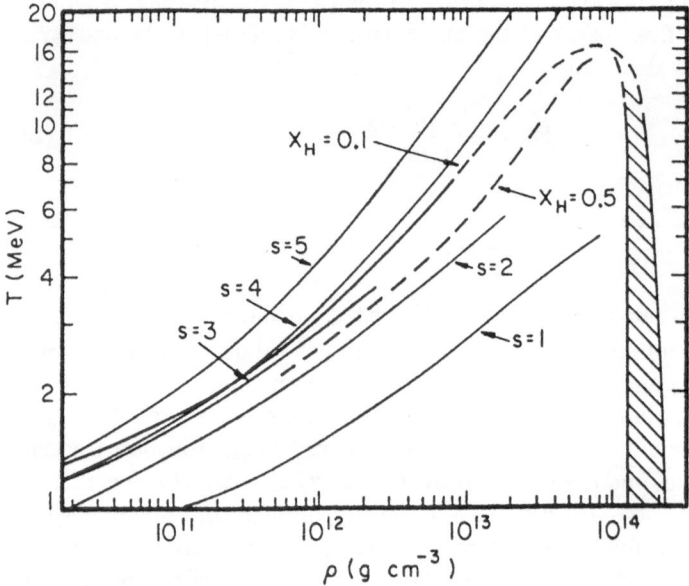

Fig. 11. Lines of constant entropy and constant mass fraction in heavy nuclei (from Lamb et al., 1978).

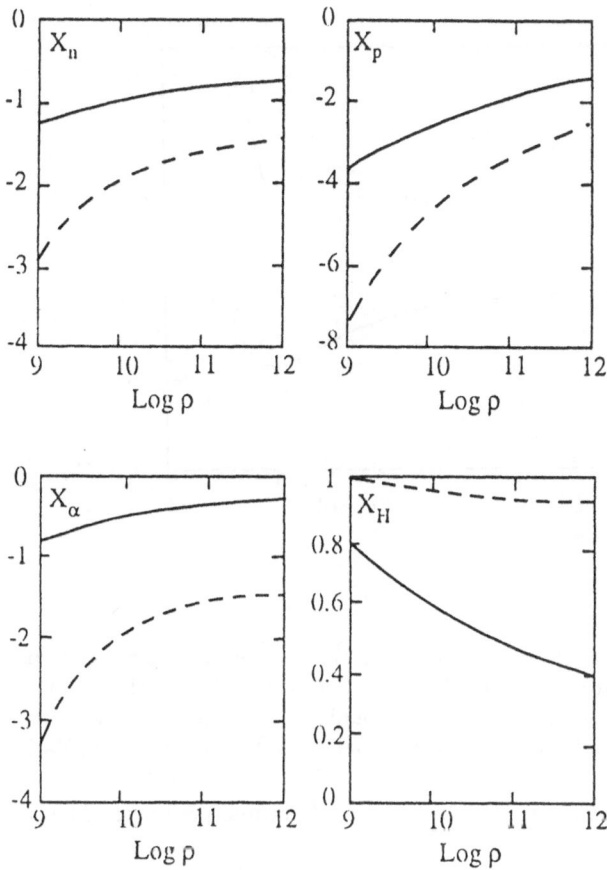

Fig. 12. Mass fractions in neutrons, protons, alpha particles and heavy nuclei for $s = 1$ (solid line) and $s = 2.5$ (dashed line) (from Cooperstein and Baron, 1990).

function of density for two values of the entropy and for $Y_e = 0.4$. At the larger entropy there are much more free nucleons and alpha particles and less heavy nuclei.

The adiabatic index $\gamma = (\frac{\partial \mathrm{Log} P}{\partial \mathrm{Log} \rho})_S$ can be computed from the EOS,

$$
\begin{aligned}
\gamma &= 1 + \left(\frac{\partial \mathrm{Log}(P/\rho)}{\partial \mathrm{Log}\rho} \right)_S \\
&= 1 + \frac{P_e}{P_e + P_N} \times \left(\frac{\partial \mathrm{Log}(P_e/\rho)}{\partial \mathrm{Log}\rho} \right)_S + \frac{P_N}{P_e + P_N} \times \left(\frac{\partial \mathrm{Log}(P_N/\rho)}{\partial \mathrm{Log}\rho} \right)_S
\end{aligned}
\tag{30}
$$

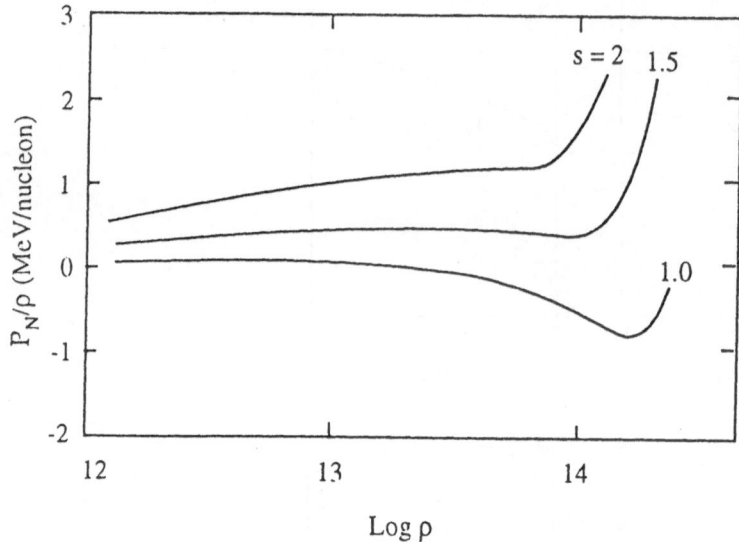

Fig. 13. P_N/ρ in MeV per nucleon. P_N represents the pressure of the baryonic part of the EOS i.e. the contributions of both the nuclei and free nucleons. The results are shown for three values of the entropy: $s = 1.0$, 1.5 and 2. For comparison, the electron pressure is given by $P_e/\rho \approx 3.5\rho_{12}^{1/3}$ MeV/nucleon (adapted from Cooperstein, 1985).

where P_e and P_N are respectively the electron and baryonic pressures. The first derivative, relative the electron gas, is equal to 1/3.

The contribution P_N from the nuclear part of the EOS has been represented in Fig. 13 for different values of the entropy per nucleon. For $s = 1$, and below 2×10^{14} g·cm^{-3}, P_N is negative since it is dominated by the Wigner-Seitz energy of nuclei due to the very small number of free nucleons. The results for the adiabatic index γ (30) are shown in Fig. 14. For $s = 1$, γ remains smaller than 4/3 until nuclear density.

Beyond nuclear density the problem is more complicated and controversial. At maximum compression when the inner core bounces the central density reaches $\sim 5\rho_o$. Informations on the properties of matter in this density range can be extracted from three different experimental areas: (i) analysis of the Pb Giant Monopole Resonance breathing mode; (ii) relativistic heavy ion collisions and (iii) neutron star masses. The results of Blaizot et al. (1976) and Blaizot (1980) on the monopole resonance have yielded for the incompressibility of symmetric nuclear matter

$$K_o^{sym} = 9 \left(\frac{dP}{d\rho} \right)_{\rho_o} = 210^{+}_{-}30 \text{ MeV} , \qquad (31)$$

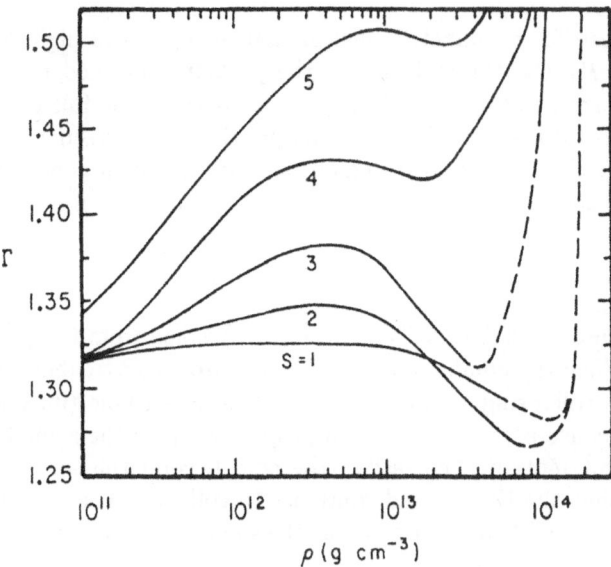

Fig. 14. Adiabatic index as a function of density for different values of the entropy. For $s = 1$, $\gamma < 4/3$ until nuclear density (Cooperstein, 1985).

which is the commonly accepted value. However Brown (1988) using the Landau forward-scattering sum rule (Friman and Dhar, 1979) obtained a much lower incompressibility $K_o \sim 120$ MeV, arguing that the monopole resonance in Pb does not give the value of K_o appropriate to nuclear matter. Heavy ion collisions are indicative of a stiff EOS (Stöcker and Greiner, 1986) even if it has been shown that the data can also be reconciled with a soft EOS when the momentum dependence of the potential is taken into account (Gale et al., 1987; Aichelin et al., 1987). Finally, any EOS should lead to a maximum neutron star mass of at least 1.4 $M_\odot$, a constraint which does not allow a too small value of the incompressibility (Swesty et al., 1994).

In any case, and for computational convenience a very simple parametrized EOS (Baron et al., 1985ab) has been used by many supernova modellers. The pressure is given by the relation

$$P = \frac{K_o(Y_e)\rho_o(Y_e)}{9\gamma_o}\left[\left(\frac{\rho}{\rho_o(Y_e)}\right)^{\gamma_o} - 1\right] + P_{th} , \qquad (32)$$

where

$$\rho_o(Y_e) = 0.16[1 - 3(0.5 - Y_e)^2] \text{ fm}^{-3} \qquad (33)$$

is the saturation density for asymmetric matter and

$$K_o(Y_e) = K_o^{sym}[1 - 2(1 - 2Y_e)^2] \tag{34}$$

is the incompressibility for asymmetric nuclear matter; γ_o is the high density adiabatic index and P_{th} the thermal part of the pressure obtained from (28) with $m^*/m = 0.7$, corresponding to uniform nuclear matter. The full adiabatic index $\Gamma = d\mathrm{Log}P/d\mathrm{Log}\rho$ resulting from this simple EOS is between 1.5 and 2 for γ between 2 and 3. Due to this stiffening of the EOS, the inner core stops its collapse at nuclear density and bounces.

3.9 The Bounce

The bounce is illustrated in the velocity plot (Fig. 15) taken from Cooperstein and Baron (1990). When the adiabatic index increases from 4/3 to about 2, the inner core which was collapsing homologously bounces as a whole (its velocity becomes positive) because pressure waves can propagate up to the sonic radius and impose a coherent motion. The rest of the core beyond the sonic radius is not "informed" about the bounce and continues to collapse. A large velocity discontinuity – a shock – is therefore formed at the sonic radius and propagates outward.

The shock takes its energy from the binding energy of the inner core which rapidly goes into hydrostatic equilibrium. This kind of "gravitational bomb" can be illustrated with a simple experiment using two hard rubber balls one being about three times more massive than the other. When the two balls are dropped (the lighter one being laid on top of the massive one) there is a shock when they reach the floor between the bouncing massive ball and the still falling lighter ball. The massive ball communicates most of its kinetic energy to the lighter ball which is strongly projected upward!

3.10 The Initial Energy of the Shock

An analytic expression of the shock energy has been obtained by Yahil and Lattimer (1982). It is assumed that the shock carries most of the binding energy of the hydrostatic object which is formed after bounce. Its mass M_{static} corresponds to the fraction of the core which was collapsing homologously

$$M_{static} \approx M_{hc} \approx (1 + f)M_{ch} , \tag{35}$$

where M_{ch} is the Chandrasekhar mass at trapping (27). Since $M_{static} > M_{ch}$, the hydrostatic core must have a central region supported by nuclear pressure while electron pressure remains dominant outside. The binding energy $B(M_{static})$ of the hydrostatic core can be obtained from an expansion of B around $M = M_{ch}$

$$B(M) = B(M_{ch}) + \frac{dB}{dM}\bigg|_{M_{ch}} (M - M_{ch}) , \tag{36}$$

where $B(M_{ch}) = 0$. Then

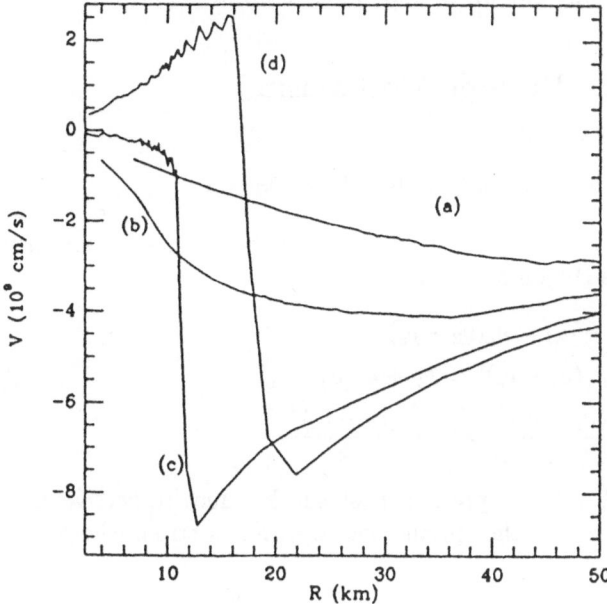

Fig. 15. Velocity as a function of radius at different times: a) last good homology; b) central density $= \rho_o$; c) maximun scrunch; d) the shock is formed (from Cooperstein and Baron, 1990).

$$B(M_{\text{static}}) - \frac{dB}{dM}\bigg|_{M_{\text{ch}}} f M_{\text{ch}} , \qquad (37)$$

with

$$M_{\text{ch}} \approx 0.7 \left(\frac{Y_l}{0.35} \right)^2 (1 + 0.1 s^2) . \qquad (38)$$

The derivative $\frac{dB}{dM}\big|_{M_{\text{ch}}}$ can be computed analytically (Yahil and Lattimer, 1982) or numerically (Lattimer et al., 1985). Yahil and Lattimer (1982) give

$$\frac{dB}{dM}\bigg|_{M_{\text{ch}}} \approx -70 \left(\frac{\rho}{\rho_o} \right)^{1/3} \left(\frac{Y_l}{0.35} \right)^{4/3} \text{foe} \cdot M_{\odot}^{-1} , \qquad (39)$$

where one "foe" is 10^{51} erg. This finally yields for the initial shock energy

$$E_s^o = -B(M_{\text{static}}) \approx 5 \left(\frac{f}{0.1} \right) \left(\frac{\rho}{\rho_o} \right)^{1/3} \left(\frac{Y_l}{0.35} \right)^{10/3} (1 + 0.1 s^2) \text{ foe} . \qquad (40)$$

This is more than what is needed to account for the kinetic energy of a supernova. If this shock energy can be carried out of the core and deposited in the stellar envelope an explosion will result. This scenario is the so called prompt mechanism for SNe II. However we shall see below that the shock undergoes severe energy losses in its way out which may finally stop its propagation.

4 The Failure of the Prompt Mechanism — The Delayed Mechanism

4.1 The Physical Conditions behind the Shock Wave

The physical parameters in the shocked material can be deduced from their pre-shock values using the Hugoniot relations

$$
\begin{aligned}
&\rho_1(v_1 - v_s) = \rho_2(v_2 - v_s) \,, \\
&P_1 + \rho_1(v_1 - v_s)^2 = P_2 + \rho_2(v_2 - v_s)^2 \,, \\
&h_1 + \frac{(v_1 - v_s)^2}{2} = h_2 + \frac{(v_2 - v_s)^2}{2} \,,
\end{aligned}
\tag{41}
$$

where the indices 1 and 2 refer to pre and post shock values respectively; v_s is the shock velocity and $h_{1,2}$ is the specific enthalpy. For a strong shock, the Hugoniot relations give

$$
\begin{aligned}
&v_s = v_1 + \frac{R}{R-1}(v_2 - v_1) \,, \\
&e_2 - e_1 = \frac{1}{2}(v_2 - v_1)^2 \,,
\end{aligned}
\tag{42}
$$

where $R = \rho_2/\rho_1$ and $e_{1,2}$ is the specific internal energy. Typical values for v_s and v_1 are 5 and -5×10^4 km·s^{-1} so that for $R = 7$ (corresponding to a gas dominated by relativistic particles) $v_2 \approx 3.5 \times 10^4$ km·s^{-1} and $\Delta e = e_2 - e_1 \approx$ 40 MeV/nucleon. This value of Δe has to be compared to the energy for the complete photodesintegration of nuclei, $e_{\text{diss}} \approx 8$ MeV/nucleon. For example, at a density of 10^{12} g·cm^{-3}, Δe will be distributed in the following way: 8 MeV for dissociation, $0.3\Delta e \approx 12$ MeV in relativistic particles (e$^+$e$^-$, $\nu\bar{\nu}$, γ) and the remaining 20 MeV in thermal energy of the nucleons. The resulting temperature is $T_2 \approx \frac{20}{1.5} \approx 13$ MeV and the entropy per nucleon is now $s = 5 - 10$ (see Brown, 1982).

4.2 The Photodesintegration of Nuclei

The break up of nuclei represents a dramatic energy sink for the shock. The dissociation energy of 8 MeV/nucleon or 1.5×10^{52} erg·$M_\odot^{-1}$ means that the complete photodesintegration of only a few tenths of solar mass will exhaust the shock energy. However, below 3 MeV, dissociation is not complete producing more and more alpha particles as the temperature decreases. At 1 MeV the material is broken mostly into alpha particles which costs only 0.9 MeV/nucleon (see Tab. 7).

Table 7. Density, temperature and mass fraction in alpha particles for ^{56}Fe along the $s = 7$ adiabat (from Bethe, 1982).

$Log\rho$	T (MeV)	X_α
11	2.9	0.17
10.6	2.2	0.29
10.2	1.8	0.41
9.8	1.5	0.51
9.4	1.3	0.61
9.0	1.1	0.70

The photodesintegration is essentially complete for $Log\rho^{post} \gtrsim 11$ (where ρ^{post} is the post shock density) which corresponds to a dissociated mass $M_{diss} \gtrsim 0.4 \, M_\odot$ in typical models. More than 6 foe are therefore lost in the photodesintegration of nuclei which then appears as the main cause of failure for the prompt mechanism.

4.3 Neutrino Losses at Shock Breakout

In the shock heated material a thermal distribution of neutrinos is rapidly established. Pairs of neutrinos and antineutrinos of the three flavors are produced by e^+e^- annihilation, fast enough to reach equilibrium with the reverse process in 0.1 ms, which is shorter than the dynamical time. In addition, electron captures on free protons resulting from the photodesintegration of nuclei produce additional electron neutrinos. All these neutrinos remain trapped until the shock reaches the "neutrinosphere" from which they can escape freely (shock breakout). The energy loss at shock breakout reaches a few foe (Bethe et al., 1980; Burrows and Mazurek, 1983) and represents an additional cause of failure for the prompt mechanism.

4.4 Criterion of Success for the Prompt Mechanism

Burrows and Lattimer (1983) have proposed a criterion of success for the prompt mechanism which simply requires that the energy remaining in the shock after dissociation and neutrino losses must still be large enough to power a SN II explosion i.e.

$$E_s^o - M_{diss} \times e_{diss} - E_\nu \approx 1 \text{ foe} . \qquad (43)$$

The initial shock energy is given by (40) above and it is assumed that all the mass from the outer edge of the homologous core (where the shock is formed) to the limit of the iron core is dissociated which gives using (35) for M_{hc}

$$M_{diss} \approx M_{Fe} - M_{hc} \approx M_{Fe} - (1+f) \times 0.7 \left(\frac{Y_l}{0.35}\right)^2 (1 + 0.1 s^2) . \qquad (44)$$

The minimum value of Y_l for which the prompt mechanism can lead to a successful explosion is shown in Fig. 16 for two masses of the iron core (1.3 and 1.5 $M_\odot$).

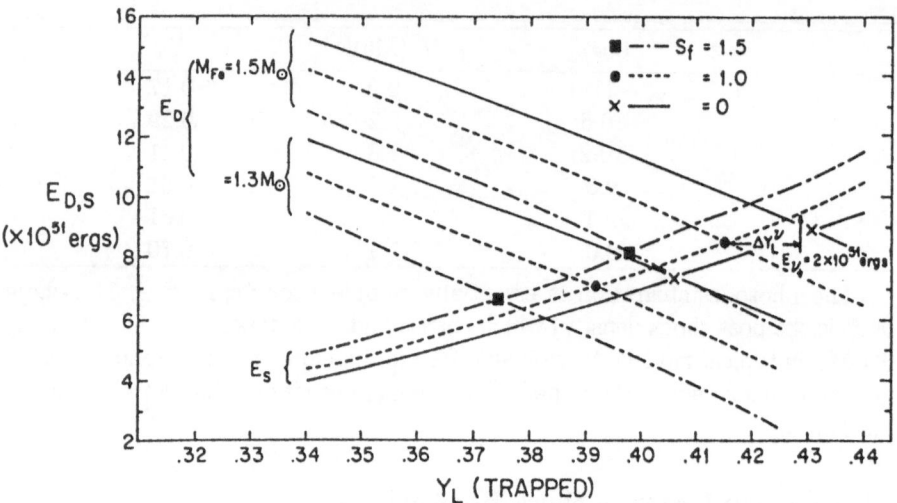

Fig. 16. Initial energy of the shock and dissociation energies for $M_{Fe} = 1.3\ M_\odot$ and 1.5 $M_\odot$ and three values of the entropy $s = 0$, 1 and 1.5. The minimum Y_l for a successful explosion are marked by filled crosses, circles and squares respectively. The neutrino losses have been neglected, except for $s = 1$, $M_{Fe} = 1.5\ M_\odot$ where arrows indicate the shifts due to an assumed $E_\nu = 2$ foe (from Burrows and Lattimer, 1983).

The initial shock energy and the mass of the homologous core are both very sentitive to Y_l. More energy and a smaller amount of material to dissociate are obtained for large Y_l. It appears that only very low mass iron cores ($M_{Fe} \lesssim 1.2$ $M_\odot$) can explode by the prompt mechanism if $Y_l \approx 0.35$. Stars of $10 - 15\ M_\odot$ may indeed produce iron cores of low mass due to the effect of Coulomb interactions in a dense degenerate plasma which reduce the Chandrasekhar mass (Nomoto and Hashimoto, 1988). But SK −69 202, which was the progenitor of SN 1987A, was a 25 $M_\odot$ star with an iron core of ∼ 1.5 $M_\odot$. If such stars indeed explode by the prompt mechanism it would mean that the amount of electron captures has been grossly overestimated. When Fuller (1982) realized that Gamow-Teller transitions from the proton $f_{7/2}$ shell to the neutron $f_{5/2}$ shell were forbidden because the neutron shell becomes full early in the collapse it was hoped that the lepton fraction at trapping might be closer to 0.4 than to 0.35, allowing the explosion of stars more massive than 15 $M_\odot$. However the most detailed studies of the capture process which now include the "poison" effect of νe scattering (Myra et al., 1987; Myra and Bludman, 1989; Bruenn, 1989ab) have brought Y_l back to 0.36 which restricts the prompt mechanism to the less massive iron cores.

4.5 General Relativity

General relativity was first included in supernova calculations by Van Riper (1979). The role of general relativity is complicated because it has two opposite effects on the dynamics of core collapse as discussed by Cooperstein and Baron (1990). General relativity is "stronger" than newtonian gravity leading to a more bounded hydrostatic core and therefore to a larger initial shock energy for the same initial conditions. On the negative side, general relativity increases the critical adiabatic index for stability. With $\gamma_{\text{crit}} = \frac{4}{3} + \mathcal{O}(\frac{GM}{Rc^2})$ the homologous core is smaller and more energy is lost in the dissociation of nuclei. Cooperstein and Baron (1990) have shown that the net result is favorable when a soft EOS $(K_o(Y_e = \frac{1}{3}) \approx 140 \text{ MeV})$ is adopted at nuclear densities because it amplifies the effect of general relativity at maximum scrunch.

4.6 The Accretion Shock

In most calculations, except those which use a soft EOS and adopt a very low mass iron core $(M_{\text{Fe}} \lesssim 1.2 \ M_\odot)$, the shock stalls after $10 - 20$ ms at a radius $R_s = 200 - 400$ km and becomes an accretion shock (Baron and Cooperstein, 1991). The Hugoniot relations with $v_s = 0$ yield $v_2 = v_1/R$ and

$$\Delta e = \frac{1}{2}(v_2 - v_1)^2 = \frac{1}{2}(1 - \frac{1}{R})^2 v_1^2 , \tag{45}$$

where $R = \rho_2/\rho_1$ and $v_1 = \alpha v_{\text{ff}}$. With $R = 7$ and $\alpha = 0.7$, $\Delta e \approx 2$ MeV at $R_s = 300$ km. This value of Δe shows that the shock stalls in the region of partial dissociation. If the density in the presupernova star can be approximated by

$$\rho_o(r_o) \approx \frac{H}{r_o^3} , \tag{46}$$

The accretion rate through the shock is simply given by

$$\dot{M} = \frac{8\pi}{3} H t^{-1} \approx 2.5 \left(\frac{0.1}{t}\right) \ M_\odot \cdot \text{s}^{-1} , \tag{47}$$

with $H \approx 0.03 \ M_\odot$ which is representative of the region just outside the iron core (Woosley and Weaver, 1988; Bethe, 1993).

4.7 Neutrino Processes

Absorption and emission by free nucleons

The material behind the shock is partially dissociated and electron neutrinos and antineutrinos outflowing from the core can heat it via the reactions

$$\begin{aligned} n + \nu_e &\rightleftharpoons p + e^- \\ p + \overline{\nu}_e &\rightleftharpoons n + e^+ \end{aligned} \tag{48}$$

The heating rate depends on the opacity for neutrino absorption $\kappa_\nu = \kappa_o \langle \epsilon_\nu \rangle^2 = \kappa'_o T_\nu^2 \approx 4 \times 10^{-19} \, T_\nu^2$ cm$^2\cdot$g^{-1} where T_ν is the temperature of the neutrinosphere (in MeV). Let $\mathcal{L} = L_{\nu_e} + L_{\bar{\nu}_e}$ be the luminosity in $\nu_e \bar{\nu}_e$ pairs. The heating rate at a radius R is

$$H(R) = \kappa_\nu \frac{\mathcal{L}}{4\pi R^2} = \kappa'_o \left(\frac{R_\nu}{2R}\right)^2 \times \frac{7}{8} a c T_\nu^6 , \tag{49}$$

where we have used $\mathcal{L} = 4\pi R_\nu^2 \times \frac{7}{8} \sigma T_\nu^4$, R_ν being the radius of the neutrinosphere. The cooling rate resulting from the neutrino emission processes can be readily obtained from the Kirchhoff's law

$$C(R) = \kappa'_o \times \frac{7}{8} a c T^6 , \tag{50}$$

where T is the local temperature at radius R. The net effect of (48) is

$$H(R) - C(R) = \frac{7}{8} a c \kappa'_o \left[\left(\frac{R_\nu}{2R}\right)^2 T_\nu^6 - T^6 \right] \tag{51}$$

and can be heating or cooling depending on the temperature distribution (see Sect. 4.8 below).

Neutrino-antineutrino annihilation

Neutrinos can also inject energy into the shocked material by annihilation

$$\nu\bar{\nu} \rightleftharpoons e^+ e^- . \tag{52}$$

All three neutrino flavors can participate to the process (52). The rate of energy release by $\nu\bar{\nu}$ annihilation is given by

$$L_{\nu\bar{\nu}} = \frac{1}{27} \frac{\sigma_o}{(m_e c^2)} D \frac{L_\nu L_{\bar{\nu}}}{c R_\nu} \frac{(\omega_\nu + \omega_{\bar{\nu}})}{8} , \tag{53}$$

where $\sigma_o = 1.7 \times 10^{-44}$ cm^2 and ω_ν is an average neutrino energy. The factor D is equal to $1 \pm 4 \sin^2 \theta_W + 8 \sin^4 \theta_W$, θ_W being the Weinberg angle and the signs $+$ and $-$ corresponding to ν_e and $\nu_\mu \nu_\tau$ respectively (Goodman et al., 1987; Cooperstein et al., 1987). For typical values of the parameters $L_{\nu\bar{\nu}}$ can reach $\sim 10^{51}$ erg$\cdot$s^{-1}, releasing a few tenths of foe. However it must be noticed that neutrino annihilation will be efficient only if the reverse process in (52) is negligible. This requires a low density medium so that a steep density gradient must be present outside the neutrinosphere.

νe scattering

Neutrinos interacting with $e^+ e^-$ pairs can deposit some energy by elastic scattering. This process can add another 0.15 foe to the total energy transferred from neutrinos to matter.

4.8 The Gain Radius

The main neutrino process is capture on free nucleons. Cooling dominates until a "gain radius" R_{gain}, and there is net heating beyond. To determine the gain radius the temperature distribution from R_ν to R_s has to be known. This region is in quasi hydrostatic equilibrium since the gravity and pressure gradient terms are much larger than the acceleration (Bethe, 1993). Then

$$\frac{1}{\rho}\frac{dP}{dr} \approx -\frac{GM}{r^2}, \tag{54}$$

where the pressure $P = P_g + P_r$, P_g being the perfect gas contribution and P_r a relativistic part including the radiation and the e^+e^- pairs. If the chemical potential of the electrons can be neglected, $P_r \propto T^4$. Moreover, the ratio P_g/P_r does not vary a lot between R_ν and R_s so that approximately $P \propto T^4 \propto \rho^{4/3}$. This yields $T \approx \frac{K}{R} + cst$, with $K \approx 2.5$ if T is in MeV and R in units of 10^7 cm. The constant can be obtained from the post shock conditions at $R_s = 300$ km where $\Delta e \approx 2$ MeV/nucleon (see Sect. 4.6). Assuming that the material is broken mostly into alpha particles, the internal energy is distributed in the following way: 0.9 MeV/nucleon for dissociation, 0.5 MeV/nucleon in relativistic particles and 0.6 MeV/nucleon in thermal energy of the alpha's. The resulting temperature is $T(R_s) \approx \frac{0.6 \times 4}{1.5} = 1.6$ MeV and between R_ν and R_s

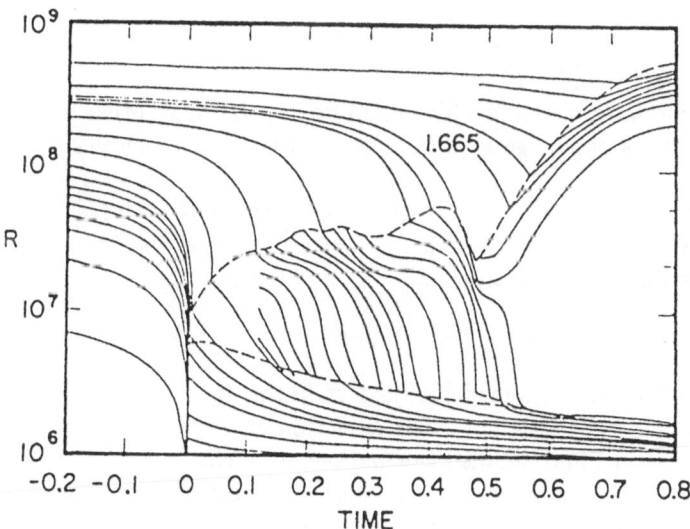

Fig. 17. Shell trajectories in the delayed mechanism. After 0.5 s the stalled shock is pushed forward by the entropy bubble generated by neutrino heating. The upper dashed line represents the shock while the lower one is the neutrinosphere (from Wilson, 1985).

$$T_{\mathrm{MeV}} \approx \frac{2.5}{R_7} + 0.75 . \tag{55}$$

The gain radius can now be obtained from (51) and (55). With $R_\nu = 30$ km and $T_\nu = 5$ MeV, one finds $R_{\mathrm{gain}} = 165$ km. Heating between R_{gain} and R_s will occur simultaneously to a thinning of the infalling material. The density at R_s

$$\rho_s = \frac{\dot{M}}{4\pi R_s^2 v_s} \approx 10^9 \left(\frac{100 \text{ km}}{R_s}\right)^{3/2} \left(\frac{0.1s}{t}\right) \text{ g} \cdot \text{cm}^{-3} , \tag{56}$$

decreases as $1/t$. This leads to the formation of a "radiation bubble" of relatively low density and very high entropy ($s \gtrsim 100$). The high pressure in the bubble can push the shock forward (Fig. 17) and eventually cause the explosion (Wilson, 1985).

4.9 Criterion of Success for the Delayed Mechanism

Burrows and Goshy (1993) have introduced a simple criterion for the re-start of the stalled shock. For a given neutrino luminosity and mass inflow they compute the shock radius R_s from the stationary hydrodynamic equations

$$\begin{aligned}
4\pi r^2 \rho v &= \dot{M} , \\
v\frac{dv}{dr} &= -\frac{1}{\rho}\frac{dP}{dr} - \frac{GM}{r^2} , \\
v\frac{d\epsilon}{dr} &- \frac{v}{\rho^2}P\frac{d\rho}{dr} = H - C ,
\end{aligned} \tag{57}$$

where ϵ is the specific internal energy. The main feature of the solution is that for a given M the shock radius goes to infinity for a critical luminosity $L_{\nu_e}^{\mathrm{max}}$ (Fig. 18).

For $L > L_{\nu_e}^{\mathrm{max}}$ no stationary solution with an accretion shock can exist and Burrows and Goshy (1993) define the critical $L_{\nu_e}^{\mathrm{max}}(\dot{M})$ line as the "explosion line". The explosion line can be approximated by a power law

$$L_{\nu_e}^{\mathrm{max}} \approx 5 \times 10^{52} \dot{M}^{0.435} \text{ erg} \cdot \text{s}^{-1} , \tag{58}$$

with $\dot{M}$ in units of $M_\odot \cdot \text{s}^{-1}$. If the model trajectory in the $\dot{M}$, L_{ν_e} plane crosses this line, an explosion results (Fig. 19). The explosion energy is naturally not given by this simple model and detailed calculations have shown that it is not easy to reach the required 1 foe. Several recent works have therefore emphasized the possible role of fluid instabilities to increase the efficiency of energy deposition and obtain stronger explosions.

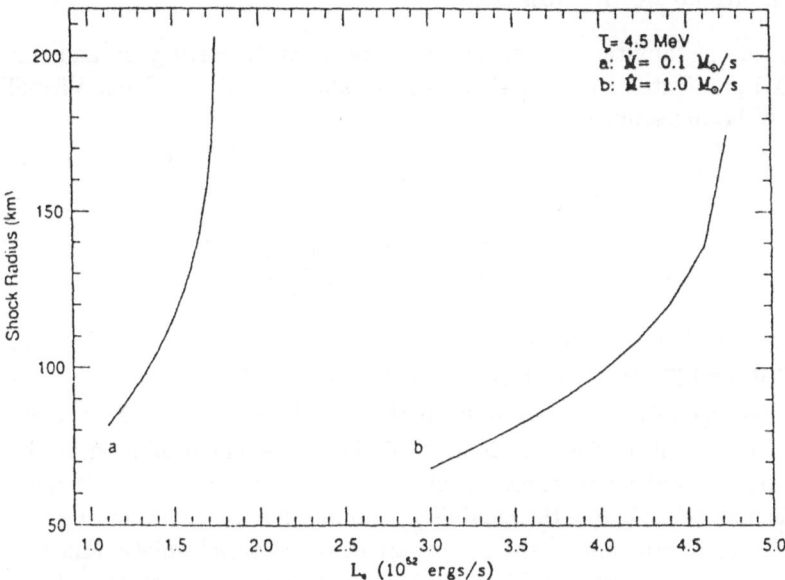

Fig. 18. Shock radius as a function of the neutrino luminosity for two values of the accretion rate. The temperature of the neutrinosphere is fixed at $T_\nu = 4.5$ MeV. There is no stationary solution for $L_{\nu_e} > 1.75 \times 10^{52}$ erg·s^{-1} ($\dot{M} = 0.1 \ M_\odot \cdot$ s^{-1}) and $L_{\nu_e} > 4.9 \times 10^{52}$ erg·s^{-1} ($\dot{M} = 1 \ M_\odot \cdot$ s^{-1}) (from Burrows and Goshy, 1993).

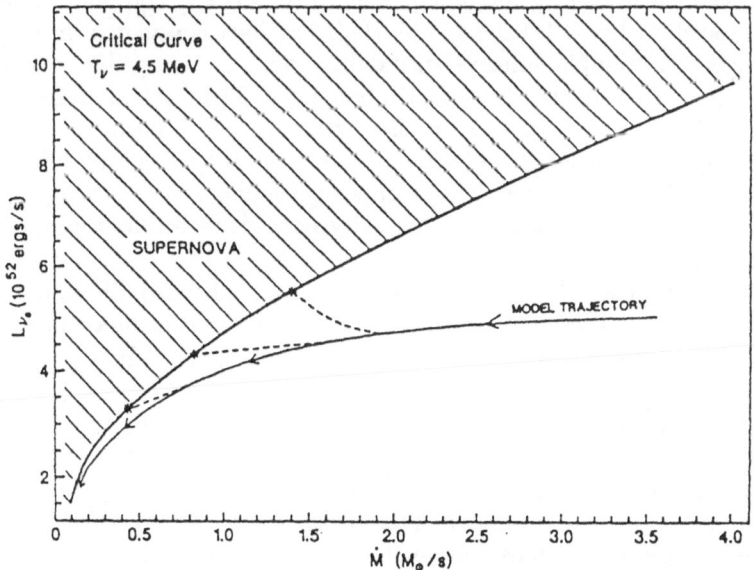

Fig. 19. The explosion line in the $\dot{M}$, L_{ν_e} plane (from Burrows and Goshy, 1993).

4.10 Hydrodynamic Instabilities

Hydrodynamic instabilities can be the result of unstable entropy or/and composition (Y_l) gradients. The condition for instability is that the Brunt-Väissälä frequency N becomes imaginary i.e.

$$
\begin{aligned}
N^2 &= -|g| \left(\frac{d\mathrm{Ln}\rho}{dr} - \frac{1}{\gamma} \frac{d\mathrm{Ln}P}{dr} \right) \\
&= \frac{|g|}{\gamma P} \frac{\partial P}{\partial S}\bigg|_{\rho,Y_l} \left[\frac{dS}{dr} - \frac{\partial S}{\partial Y_l}\bigg|_{\rho,P} \frac{dY_l}{dr} \right] < 0 ,
\end{aligned}
\tag{59}
$$

where g is the local gravity and γ is the adiabatic index, all other symbols having their usual meanings. Since the signs of the two partial derivatives $\frac{\partial P}{\partial S}\big|_{\rho,Y_l}$ and $\frac{\partial S}{\partial Y_l}\big|_{\rho,P}$ are respectively $+$ and $-$ it appears that negative entropy or Y_l gradients can drive an instability. An examination of the numerical results show that several possible unstable situations can be found at different times after bounce.

There is first an early mantle instability ~ 30 ms after bounce resulting from a negative entropy gradient in the outer part of the shocked mantle (Burrows and Lattimer, 1988; Burrows and Fryxell, 1993). The entropy decrease is due to photodesintegration and neutrino losses. In the same region there is a positive Y_l gradient (see Fig. 20) but not large enough to restore a stable stratification. The turn-over time scale at a radius $r \approx 100$ km (where $g \approx 10^{12}$ cm·s^{-2} and $\frac{ds}{dr} \approx -0.1$ km^{-1}) is $5 - 10$ ms. The convective motions will bring heat and neutrinos to the neutrinosphere and produce a spike in the neutrino luminosity which may boost the explosion.

If this convective instability fails and does not drive an explosion (for example because it is too short-lived) Wilson and Mayle (1993) have considered the effect of a "salt-finger" or "double diffusive" instability which is expected to develop on a longer time scale at the core mantle boundary. The fluid motions here are oscillating due to an unstable Y_l gradient stabilized by a positive entropy gradient (see Fig. 20). Since heat diffusion is more rapid than chemical diffusion the amplitude of the oscillations will be growing leading to the formation of "salt-fingers" as in the situation where hot salty water is placed on top of cold fresh water (Fig. 21).

The "salt-finger" instability can be considered as an enhanced diffusion process which can increase the neutrino luminosity. When it was included in their SN calculations by Wilson and Mayle (1993) they obtained successful explosions releasing ~ 1 foe .

Later in the evolution ($t > 0.1$ s after bounce) another instability can take place outside the neutrinosphere in the entropy bubble generated by neutrino heating. The entropy in the bubble reaches $s > 100$ and convection carries heat (and pressure) to the shock. Herant et al. (1992) have performed 2-D calculations of this convective phase and have obtained a weak explosion releasing ~ 0.35 foe, about one third of the required value.

Finally, and very recently Herant et al. (1994) have reconsidered in details the hydrodynamics of core collapse, in a series of beautiful realistic 2-D SPH

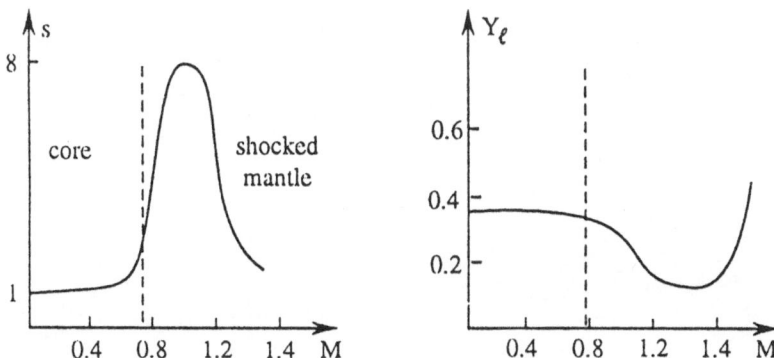

Fig. 20. Entropy (per nucleon) and Y_l profiles in the core $\sim$ 30 ms after bounce.

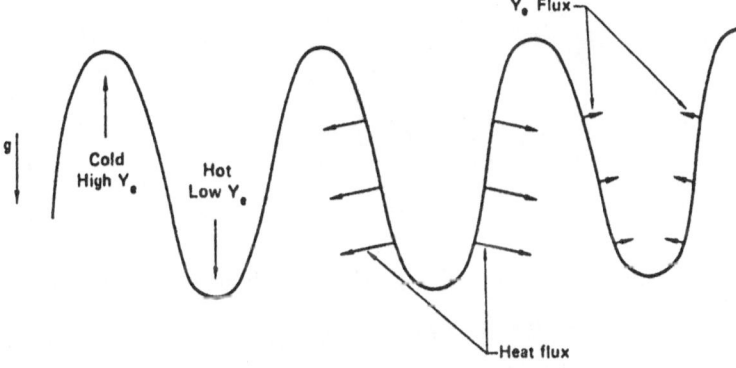

Fig. 21. Formation of "salt-fingers" at the boundary between regions with low Y_l, high S and high Y_l, low S (from Wilson and Mayle, 1988).

calculations. They find early core instabilities driven both by composition and entropy gradients. These instabilities bring heat to the shock and cold material to the neutrinosphere which increases the efficiency of neutrino energy deposition. Successful explosions are obtained, releasing 1.2 to 1.3 foe (Fig. 22)

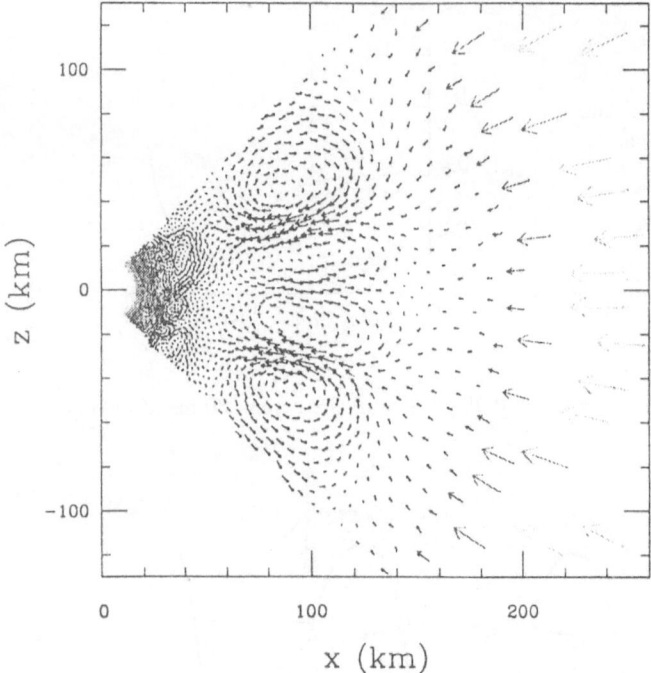

Fig. 22. Convective motions in the core of a 25 $M_\odot$ star, 25 ms after bounce (from Herant et al., 1994).

5 The Supernova Outputs

5.1 The Neutrino Emission

Even if some questions remain in the physics of supernova explosions it is sure that a neutrino burst of several 10^{53} erg is produced in core collapse as was confirmed by the Kamioka and IMB detections of SN 1987A. It is therefore of major interest to understand the properties of this neutrino burst to be able to interpret the signal from the next galactic SN II. The basic features of the neutrino burst have been described by Burrows et al. (1992) and their results are summarized below.

The neutrino emission can be divided in three distinct periods: (i) the infall and flash (at shock breakout); (ii) the period of accretion until the explosion; (iii) the neutronization and cooling of the proto-neutron star.

Infall and flash

During core collapse electron captures produce electron neutrinos which can escape freely until trapping. These neutrinos are non thermal and have an average energy $\langle \epsilon_{\nu_e} \rangle \approx 15$ MeV. The energy lost during infall is given by

$$E_{\nu_e}^{\text{infall}} \approx \frac{M_{Fe}}{m} \Delta Y_e \, \epsilon_{\nu_e} \approx 2 \text{ foe} , \tag{60}$$

where m is the nucleon mass and $\Delta Y_e \approx 0.05 - 0.1$ is the decrease of the lepton fraction before trapping. When most of the electron captures take place, the dynamical timescale is $\tau^{\text{infall}} \approx 5$ ms and the resulting luminosity is

$$L_{\nu_e}^{\text{infall}} \approx 4 \times 10^{53} \text{ erg} \cdot \text{s}^{-1} . \tag{61}$$

After the formation of the shock material is heated and nuclei are dissociated. This leads to rapid captures on free protons and to the build up of a thermal distribution of neutrinos at the post shock temperature (ν_e but also ν_μ and ν_τ). When the shock reaches the neutrinosphere all these neutrinos escape on a time scale $\tau_{\text{breakout}} \approx 5$ ms. About 1 foe in ν_e is lost from the captures on free protons while ν_μ and ν_τ carry ~ 0.5 foe, giving for the breakout luminosities

$$L_{\nu_e}^{\text{breakout}} \approx 2 \times 10^{53} \text{ erg} \cdot \text{s}^{-1} ,$$
$$L_{\nu_{\mu\tau}}^{\text{breakout}} \approx 10^{53} \text{ erg} \cdot \text{s}^{-1} . \tag{62}$$

The electron neutrino luminosities from infall and at breakout merge to reach a peak value of $\sim 6 \times 10^{53}$ erg·s^{-1}. After breakout all neutrino spectra rapidly become thermal with respective temperatures $T_{\nu_e} = 3 - 5$ MeV, $T_{\overline{\nu}_e} = 4 - 5$ MeV and $T_{\nu_{\mu\tau}} = 6 - 8$ MeV. The temperatures are different because the neutrinospheres are located at different radii depending on the process which controls the neutrino mean free path: captures on free neutrons (resp. protons) for ν_e (resp. $\overline{\nu}_e$), scattering for $\nu_{\mu\tau}$.

Accretion and explosion

After $10 - 30$ ms the shock stalls and the infalling material flows throughout to be accreted by the hydrostatic core. A rough estimate of the accretion luminosity is then

$$L^{\text{acc}} \approx \frac{GM}{R} \dot{M} , \tag{63}$$

where $\dot{M} \approx \frac{0.25}{t} M_\odot \cdot \text{s}^{-1}$ (see (47)). With $M = 1.4 \, M_\odot$ and $R = 30$ km (63) gives

$$L^{\text{acc}} \approx \frac{3 \times 10^{52}}{t} \text{ erg} \cdot \text{s}^{-1} , \tag{64}$$

which is (approximately) equally shared between all neutrino species. During the accretion phase the neutrinosphere moves downwards into the proto-neutron star due to the thinning of the infalling material (see Sect. 4.8). Its temperature (and the energy of the neutrinos) therefore slowly increases. At the moment of explosion the mantle is ejected and there is a sudden increase of both the luminosity and the neutrino energy. One could therefore use the neutrino "light curve" and spectrum of a galactic SN to obtain the duration of the accretion phase which would be of great help to decide between the prompt and delayed mechanisms. The dynamics of the proto-neutron star (such as oscillations with periods in the range 10 – 100 ms or the various instabilities discussed in Sect. 4.10 may also affect the neutrino luminosity but the amplitude of the possible signatures is very uncertain.

Proto-neutron star neutronization and cooling

Accretion stops after explosion but at that time the amount of energy lost

$$E_\nu \approx 3 \times 10^{52} \int_{t_{sf}}^{t_{ex}} \frac{dt}{t} = 3 - 5 \times 10^{52} \text{ erg} , \qquad (65)$$

is only 10% – 20% of the neutron star binding energy (t_{sf} and t_{ex} are respectively the times of shock formation and of explosion). The bulk of the energy will be released in a deleptonization and cooling phase lasting a few seconds. Neutrinos interact with matter through scattering (on neutrons, protons or nuclei) or absorption (by neutrons or protons) processes. Scattering is possible for all neutrino types while absorption is restricted to electron neutrinos. A representative cross section for all these processes is given by

$$\sigma \approx \frac{\sigma_o}{F} \left(\frac{\epsilon_\nu}{m_e c^2} \right)^2 , \qquad (66)$$

where $\sigma_o \approx 10^{-44}$ cm^2 is a typical weak interation cross section and F is a factor accounting for Pauli blocking and Fermi liquid corrections when the neutrons become degenerate. The neutrino mean free path resulting from these cross sections is 10 – 100 cm for electron neutrinos and about ten times larger for μ or τ neutrinos which only have scattering interactions. The duration of the cooling phase is controlled by the neutrino diffusion time scale

$$\tau_{\text{diff}} \approx \frac{3R^2}{\lambda_\nu c} \approx 10 \left(\frac{R}{10 \text{ km}} \right)^2 \left(\frac{10 \text{ cm}}{\lambda_\nu} \right) \text{ s} . \qquad (67)$$

After 30 s to 1 mn, F increases (to 5 – 10) due to neutron degeneracy and τ_{diff} decreases. The proto-neutron star becomes transparent and the neutrino luminosity plummets.

All these results are summarized in Fig. 23 and 24 below which show the luminosity and average energy of the different neutrino types during the first second and first fifty seconds of evolution.

Some information on neutrino masses can also be obtained from supernovae because for massive neutrinos the velocity depends on energy and the arrival time from a supernova at a distance D is delayed by

$$\Delta t \approx \left(\frac{D}{10 \text{ kpc}} \right) \left(\frac{m_{\nu_e}}{10 \text{ eV}} \right)^2 \left(\frac{10 \text{ MeV}}{\epsilon_\nu} \right)^2 \text{ s} . \qquad (68)$$

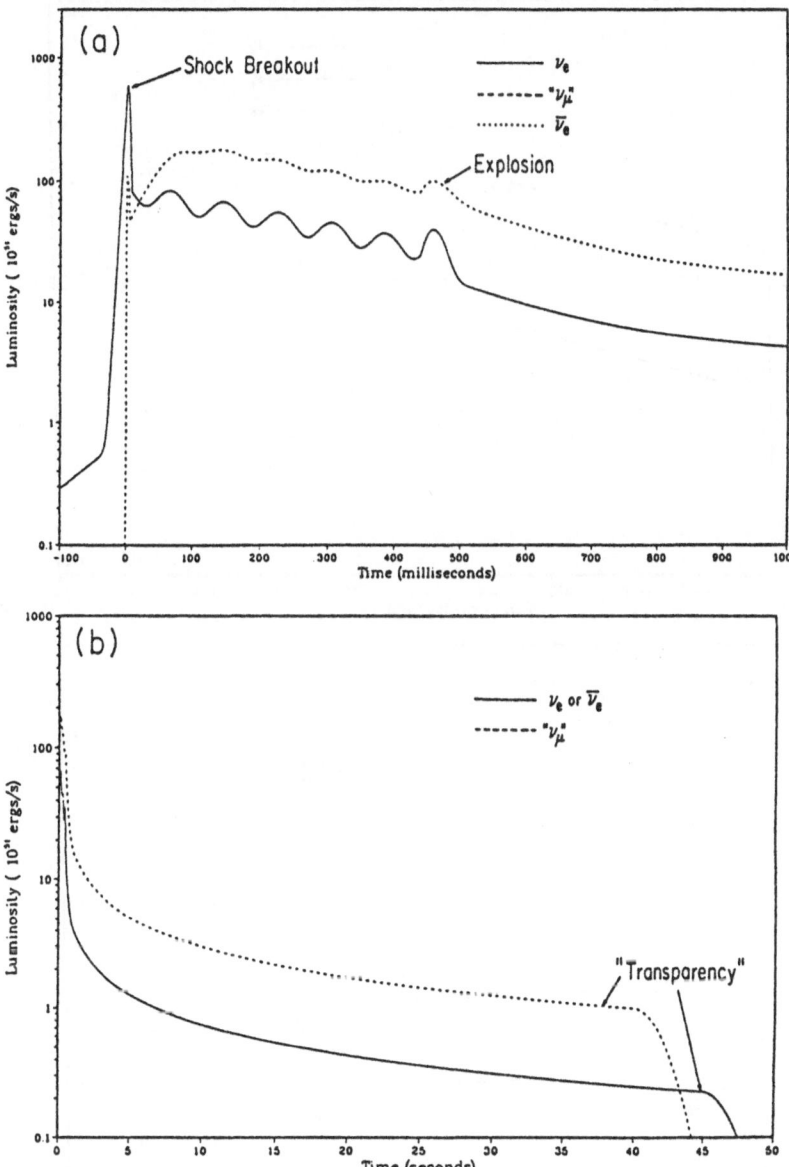

Fig. 23. Neutrino luminosity during (a) the first second after shock breakout and (b) the first fifty seconds. The line "ν_μ" represent both the $\nu_\mu \overline{\nu}_\mu$ and the $\nu_\tau \overline{\nu}_\tau$. Oscillations are possible during the accretion period from shock breakout to explosion. The cooling lasts about 45 s until the luminosity plummets at transparency (from Burrows et al., 1992).

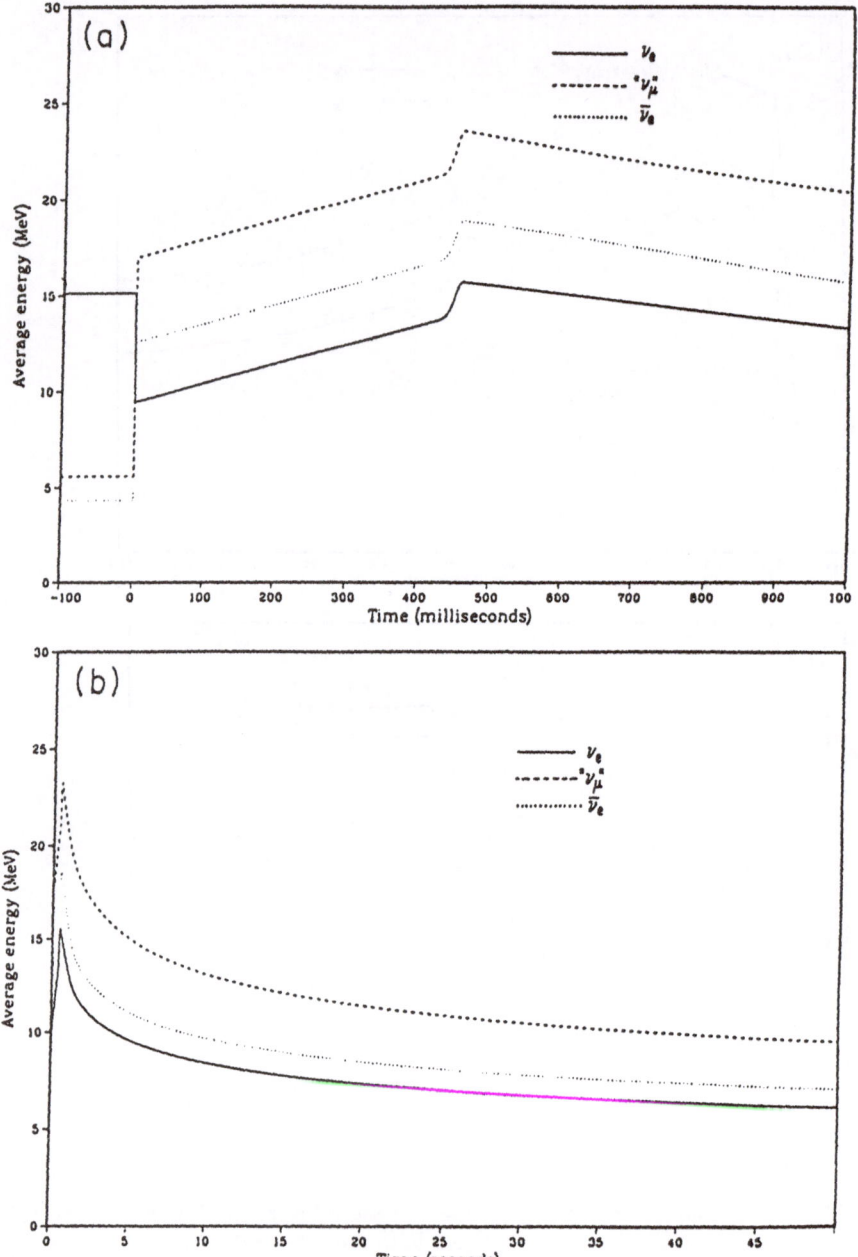

Fig. 24. Average energy of the different neutrino types (from Burrows et al., 1992).

The fact that the signal from SN 1987A (at a distance of 52 kpc) was not spread on more than 10 seconds gave an upper limit on m_{ν_e} of the order of 20 eV, the best limit being now 9.3 eV at the 95% confidence level from laboratory experiments (Robertson et al., 1991). Neutrinos of 10 eV would also increase the rise time of the signal from about 50 to 300 ms, an effect which could be measurable on a galactic supernova. Finally, if oscillations of the proto-neutron star produce a modulation it will be smeared out if $m_{\nu_e} > 10$ eV.

Mass determination for the $\nu_{\mu\tau}$ will be more problematic since they are only detected via neutral current reactions which are much less frequent. In addition, there is no signature of the neutrino type in neutral current reactions so that the extraction of the $\nu_{\mu\tau}$ signal implies the substraction of the ν_e contribution. It will have to be modelled from the $\bar{\nu}_e p$ reaction counts, which may turn to be a difficult task.

If neutrino are massive vacuum and/or matter oscillations are possible. Concerning matter oscillations it is important to notice that for neutrinos of ~ 10 eV they only occur below 100 g·cm^{-3}. Therefore there is no neutrino mixing in the core and the explosion mechanism is not affected. Moreover, the electron antineutrino, which is the most detected species, is not converted during its travel in the supernova envelope ($\bar{\nu}_e$ can only exchange with positrons which are not normally present in the envelope). Vacuum oscillations can be sensitive down to $\Delta m^2 \gtrsim 10^{-19}$ eV^2 for a supernova at 10 kpc, the signature of the $\bar{\nu}_\mu \to \bar{\nu}_e$ conversion being a harder $\bar{\nu}_e$ spectrum.

Both for supernova theory and neutrino physics the next galactic supernova at a typical distance of 10 kpc (i.e. five times closer than SN 1987A) will be an exceptional opportunity. In Kamiokande II 350 $\bar{\nu}_e p$ and 10 νe interactions can be expected and a factor of 15 more in Super Kamiokande which may be operating within the next decade. Such large numbers will allow a detailed neutrino spectroscopy and the construction of the supernova neutrino "light curve"! The estimated galactic SN rate of 3 per century may let us hope to witness such an extraordinary event....

5.2 The Light Curve

Several energy sources can contribute to the light curve depending on the SN type. If the progenitor star has a large radius shock energy deposition in the envelope powers the early light curve with thermal and kinetic energies initially in equipartition i.e.

$$E_{th}^o \approx E_{kin}^o \approx \frac{E_{SN}}{2} \ . \tag{69}$$

At later times (in SNe II) or from the beginning (in SNe I) ^{56}Ni and ^{56}Co radioactivity

$$^{56}\text{Ni} \to{}^{56}\text{Co} \to{}^{56}\text{Fe} \tag{70}$$

dominates. The total energy available from these two reactions is $q_{Ni} \approx 7 \times 10^{49} M_{Ni}$ erg and $q_{Co} \approx 1.5 \times 10^{50} M_{Ni}$ erg respectively, where M_{Ni} is the mass of synthetized ^{56}Ni in solar mass. Finally, if a pulsar is formed in the explosion its luminosity

$$L_{\text{pulsar}} \approx 5 \times 10^{38} \left(\frac{33 \text{ ms}}{P}\right)^4 \text{ erg} \cdot \text{s}^{-1}, \qquad (71)$$

can be important after about 500 days (in (71) P is the pulsar period, 33 ms being the period of the Crab pulsar).

The observed light curve results from a convolution between the different energy sources and the process of diffusive transport. An order of magnitude of the diffusion time scale is given by

$$\tau_{\text{diff}} \approx \frac{3R_{\text{env}}^2}{\lambda c} \approx 9\frac{\kappa_{\text{Th}}M_{\text{env}}}{4\pi c R_{\text{env}}}, \qquad (72)$$

where $\lambda = (\kappa_{\text{Th}}\rho)^{-1}$ is the mean free path for Thomson scattering, M_{env} and R_{env} being the mass and radius of the expanding envelope (a uniform density has been assumed to obtain (72)). If the envelope expands at a constant velocity v_{exp} one has

$$R_{\text{env}} \approx R_{\text{o}} + v_{\text{exp}}t, \qquad (73)$$

where R_{o} is the radius of the progenitor star. The diffusion time scale essentially behaves as $1/t$ and after about 150 days it becomes smaller than the radioactive period of ^{56}Co, $\tau_{\text{Co}} = 77$ days (corresponding to a radioactive e-folding time of $77/\text{Ln}2 \approx 111$ days). One then directly sees (without delay) the energy input from ^{56}Co radioactivity,

$$L_{\text{Co}} = 1.5 \times 10^{43} M_{\text{Ni}} \exp(-t/111 \text{ days}) \text{ erg} \cdot \text{s}^{-1}, \qquad (74)$$

which explains the observed exponential decline.

The plateau phase of SNe II-P is another consequence of (72). During the early evolution of the expanding envelope the evolution remains quasi-adiabatic. Since the envelope is radiation dominated its temperature decreases as $1/R_{\text{env}}$ so that the thermal energy $E_{\text{th}} \approx aT^4 R_{\text{env}}^3$ also behaves as $1/R_{\text{env}}$. Then,

$$E_{\text{th}} \approx E_{\text{th}_{\text{o}}}\frac{R_{\text{o}}}{R_{\text{env}}} \approx \frac{E_{\text{SN}}}{2}\frac{R_{\text{o}}}{R_{\text{env}}}, \qquad (75)$$

which leads to the following rough estimate of the plateau luminosity

$$L_{\text{P}} \approx \frac{E_{\text{th}}}{\tau_{\text{diff}}} \approx \frac{2\pi c}{9\kappa_{\text{Th}}}\frac{E_{\text{SN}}}{M_{\text{env}}}R_{\text{o}}. \qquad (76)$$

Equation (76) shows that L_{P} is directly proportional to the progenitor radius R_{o}. It is now straightforward to understand why SN 1987A which had a blue supergiant progenitor of radius 50 $R_{\odot}$ was underluminous compared to a standard SN II.resulting from the explosion of a red supergiant twenty times larger.

The non adiabatic evolution of the radiation dominated envelope can be obtained from the energy equation

$$\frac{dE}{dt} + 4\pi p R_{\text{env}}^2 \frac{dR_{\text{env}}}{dt} = -L + L_*, \qquad (77)$$

where $E = \frac{4\pi}{3} R_{env}^3 a T^4$ is the thermal energy and $p = \frac{1}{3} a T^4$ is the pressure. We assume that $L \approx \frac{E_{th}}{\tau_{diff}}$ and $L_* = L_{*_o} \exp(-t/\tau_*)$ is the radioactive energy input. We then have

$$\frac{R_o}{R_{env}} \frac{dL}{dt} = (-L + L_*) \frac{1}{\tau_{diff_o}} , \qquad (78)$$

where τ_{diff_o} is the initial diffusion time scale when $R_{env} = R_o$. Equation (77) and (78) correspond to a "one zone" model of the envelope where the density and temperature are uniform. Arnett (1980, 1982) has found an analytic solution to these equations which can be extended to more realistic density and temperature distributions

$$L = L_P \exp[-u(x)] + L_{*_o} \Omega(x, y, w) , \qquad (79)$$

where $u(x) = wx + x^2$, $\Omega(x, y, w) = \exp[-u(x)] \int_0^x (w + 2z) \exp[-2zy + u(z)] dz$, $x = t/\tau_m$, $y = \tau_m/2\tau_*$, $\tau_m^2 = 2\tau_{diff_o} R_o/v_{exp}$ and $w = \tau_m/\tau_{diff_o}$. The function $\Lambda = L/L_{*_o}$ has been represented in Fig. 25 for different values of the initial radius. It can be seen that the light curve changes with increasing initial radius from a typical SN I to a SN II shape. The intermediate case with $R_o = 10^{13}$ cm bears some resemblance with SN 1987A (the radius of SK -69 202 was 4×10^{12} cm).

These light curves are the result of the slow diffusion of energy through the envelope. In addition, there is a transient UV flash which occurs when the shock breaks out at the stellar surface (see Fig. 6) and which cannot be obtained from Arnett's theory. About 10^{47} erg are released in the UV flash and its duration does not exceed 30 s (for an analytical description of shock breakout see Nadyozhin, 1994)

5.3 X and γ Emission

In the first months following the explosion the envelope remains optically thick to γ-rays from ^{56}Ni and ^{56}Co radioactivity (produced by positron annihilation and decay of nuclear excited states). They undergo successive Compton scattering until their energy is reduced to ~ 100 keV after which they can be absorbed. During this early stage the energy from radioactivity ultimately powers the visible light curve as discussed above. However as the envelope expands, its transparency increases and partially comptonized photons are first emitted in the X-ray range (1 – 100 keV) followed shortly after by γ photons which directly escape from the remnant. In SN 1987A these X and γ-ray emissions have been detected for the first time. Of particular interest are the ^{56}Fe 847 and 1238 keV lines (from the decay of ^{56}Co) which directly show the print of a "freshly synthetized" element. A very naïve estimate of the time evolution of the 847 keV line intensity is simply given by

$$I_{847} \propto \exp(-t/111 \text{ days}) \times \exp[-(R_{env}/\lambda_{847})] , \qquad (80)$$

λ_{847} being the Compton mean free path for 847 keV photons. The optical thickness of the envelope is given by

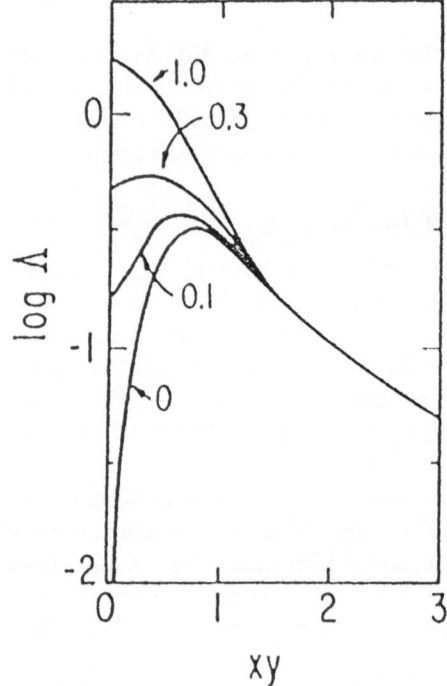

Fig. 25. Analytical SN light curves. The luminosity is given as a function of time ($xy = t/2\tau_*$) for four values of the initial radius in units of 10^{14} cm (from Arnett, 1982).

$$\frac{R_{\mathrm{env}}}{\lambda_{847}} \approx \frac{3\kappa_{\mathrm{KN}}M_{\mathrm{env}}}{4\pi R_{\mathrm{env}}^2} \approx \frac{3\kappa_{\mathrm{KN}}M_{\mathrm{env}}}{4\pi v_{\mathrm{exp}}^2 t^2} \;, \tag{81}$$

κ_{KN} being the Klein-Nishina opacity. The intensity is maximum for

$$t_{\mathrm{max}} \approx \left(\frac{3\kappa_{\mathrm{KN}}M_{\mathrm{env}}t_{\mathrm{Co}}}{2\pi v_{\mathrm{exp}}^2}\right)^{1/3} \approx 1 \;\; \mathrm{yr}\;, \tag{82}$$

where $t_{\mathrm{Co}} = 111$ days. The detailed evolution of the X and γ-ray emission must naturally be obtained by Monte-Carlo techniques (see Sutherland, 1990 and Cassé and Lehoucq, 1994 for good reviews). Compared to theoretical predictions, the observational results from SN 1987A have been a surprise because the detection of both the X-rays and the γ-ray lines occurred much earlier than expected. It was soon realized that Rayleigh-Taylor instabilities can develop in the expanding envelope, mixing radioactive nuclei much further out than the place where they are produced which can then explain the early detections (see Müller, 1994 and references therein).

5.4 Gravitational Radiation

Gravitational waves will be radiated during the collapse and neutron star formation only if the core has some non zero quadrupolar moment, i.e. if the collapse is non spherical. If the core has a large angular momentum the effect of centrifugal forces will increase as its radius shrinks and a deviation from spherical symmetry can be expected. Gravitational waves are emitted in two stages: (i) during the collapse and (ii) as a result of the oscillations of the newly formed neutron star. The amount of energy released in gravitational radiation has been obtained either from detailed 2-D codes (Müller, 1982) or from simpler semi-analytical models of homogeneous rotating spheroids (Saenz and Shapiro, 1981). For a core of angular momentum $J = 3 \times 10^{48}$ erg·s (which would produce a millisecond pulsar) Müller (1982) obtains $\frac{E_{GW}}{Mc^2} \approx 10^{-7}$ at a typical frequency $f = 1$ kHz for (i) and 10^{-6} at 10 kHz for (ii). For the same angular momentum Saenz and Shapiro (1981) find a larger efficiency of 5×10^{-5}. The amplitude of the strain on Earth for a supernova at a distance D is given by

$$h = 3 \times 10^{-20} \left(\frac{E_{GW}}{Mc^2} \right)^{1/2} \left(\frac{1\,\mathrm{kHz}}{f} \right) \left(\frac{10\,\mathrm{Mpc}}{D} \right) , \qquad (83)$$

which has to be compared with the expected sensitivity of the VIRGO/LIGO interferometers $h_{\mathrm{lim}} \approx 10^{-22}$. Assuming a wave frequency of 1 kHz a supernova in the Virgo cluster will be detected if $\frac{E_{GW}}{Mc^2} > 10^{-5}$ while for a galactic supernova an efficiency of only 10^{-11} will be enough. Since there is no evidence that the core of massive stars is rapidly rotating the detection of gravitational radiation from a supernova in the Virgo appears very uncertain. Conversely a core collapse in our Galaxy would be clearly within the reach of VIRGO/LIGO but the rate of such events is only ~ 3 per century!

5.5 Nucleosynthesis

Explosive nucleosynthesis occurs in SNe when the shock wave travels across the C, Ne, O and Si shells of the presupernova star. The nature of the burning products depends on the temperature reached behind the shock. For $T_9 \gtrsim 5$ ($T_9 = T/10^9$ K) there is essentially complete Si burning to statistical equilibrium giving ^{56}Ni since the neutron excess of the burned material is small. For $5 \gtrsim T_9 \gtrsim 4$ Si burning is uncomplete, while for $4 \gtrsim T_9 \gtrsim 3.3$ there is oxygen burning and for $3.3 \gtrsim T_9 \gtrsim 2.5$ carbon/neon burning. The temperature at a given radius can be simply estimated assuming that thermal energy is uniformly distributed in the shocked material

$$E_{\mathrm{th}} \approx E_{\mathrm{SN}} \approx \frac{4\pi}{3} R^3 a T^4 , \qquad (84)$$

or

$$T_9 \approx 10 \left(\frac{E_{SN}}{10^{51}\text{erg}} \right)^{1/4} R_8^{-3/4} \text{ K} , \tag{85}$$

where R_8 is the radius in units of 10^8 cm. The list of the main burning products in the different zones is given in Tab. 8 and the detailed results of an explosive nucleosynthesis calculation are shown in Fig. 26.

Table 8. Burning products in explosive nucleosynthesis (from Thielemann et al., 1994).

R_8	3.5	5.0	6.4	11.5
T_9	4	3	2.5	1.6
burning site	complete Si	incomplete Si	O	C/Ne
burning products	^{56}Ni	S, Ar, Ca, Fe	S,Si	O, Mg, Si
$M(R)$ (in $M_\odot$)	1.7	1.75	1.8	2.0

Apart from shock propagation which provides the bulk of the burning products SNe II can also make neutron rich isotopes by "r-process" nucleosynthesis. The r-process nuclei are formed by rapid neutron captures on iron-peak elements which implies high neutron fluxes and therefore a large neutron excess

$$\eta = \frac{n_n - n_p}{n_n + n_p} = 1 - 2Y_e \gtrsim 0.1 , \tag{86}$$

i.e. $Y_e \lesssim 0.45$. Woosley and Hoffman (1992) have suggested that the hot bubble which forms in the delayed mechanism can be a very promising site for the r-process. Starting at a very high entropy the material is entirely made of free nucleons with a neutron excess $\eta = 0.1$, corresponding to the initial Y_e value of the core. As it expands and cools down alpha particles are formed first, followed by heavy nuclei built from the alpha's (α-rich freeze-out) which then undergo rapid neutron captures. Recent calculations have shown that r-process nuclei can be produced in that way with abundances which nicely match the solar values (Witti et al., 1994; Takahashi et al., 1994).

Finally, some fragile nuclei which are not formed by standard ways might be the result of a "neutrino induced nucleosynthesis". Neutrino induced nucleosynthesis which was first proposed by Domogatskii and Nadyozhin (1977, 1978) considers the possibility of neutrino-nucleus interactions during the few seconds of intense neutrino emission. There is a large number of possible reaction channels, the most important being

$$\begin{aligned} \nu + (A, Z) &\to \nu' + (A - 1, Z) + \text{n} \\ \nu + (A, Z) &\to \nu' + (A - 1, Z - 1) + \text{p} \end{aligned} \tag{87}$$

where the neutrino can be of any type. For example the synthesis of ^{19}F, an element ordinary difficult to produce, can follow the path: ^{20}Ne$(\nu, \nu'\text{p})^{19}$F or ^{20}Ne$(\nu, \nu'\text{n})^{19}$Ne$(e^+\nu_e)^{19}$F. The averaged cross section of these reactions is $\sigma \approx 8\times10^{-42}$ cm^2 while the neutrino flux in the neon shell reaches $F \approx 10^{38}$ cm$^{-2}\cdot$s^{-1}

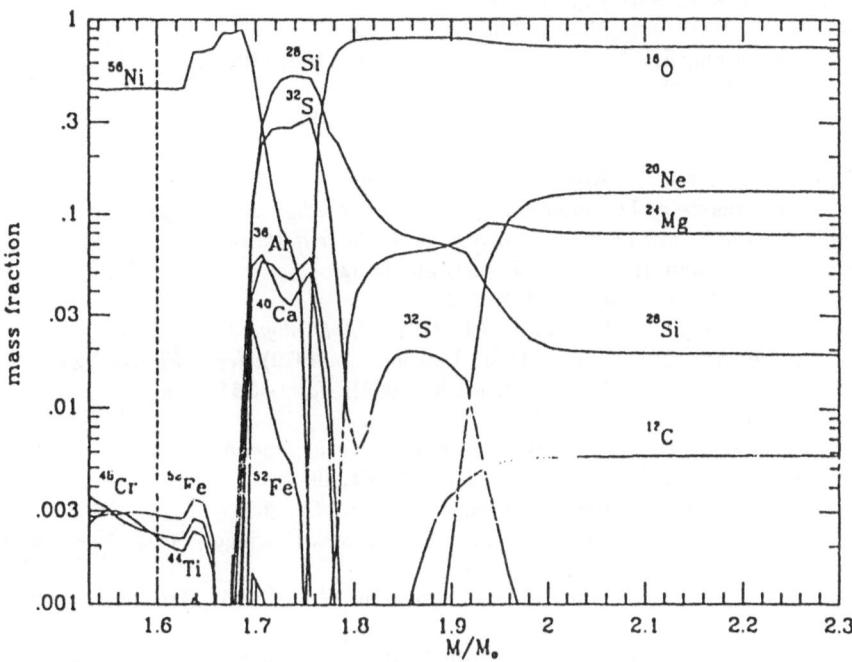

Fig. 26. Mass fractions of the different burning products as a function of the lagrangian mass in a 20 $M_\odot$ star. In SN 1987A the inner 1.6 $M_\odot$ formed the remnant and 0.07 $M_\odot$ of ^{56}Ni have been ejected (from Thielemann et al., 1991).

for a few seconds. The amount of ^{19}F synthetized from ^{20}Ne can then be expected to be $(^{19}F/^{20}Ne) \approx \sigma F \Delta t \approx 10^{-3}$ with $\Delta t = 2 - 3$ seconds. This value has to be compared to $(^{19}F/^{20}Ne)_\odot \approx 3 \times 10^{-4}$ and shows that neutrino induced nucleosynthesis can indeed be an original way to make ^{19}F and some other fragile nuclei (see Woosley and Weaver, 1994 for a review).

Acknowledgments

It is a pleasure to thank the organizers of the Schladming Winter School. Their kindness and efficiency, together with the presence of a number of interested students contributed to make of this session an excellent and successful school.

References

Aichelin, J., Rosenhauer, A., Peilert, G., Stöcker, H., Greiner, W. (1987): *Phys. Rev. Lett.* **58**, 1926.

Arnett, W.D. (1977): *Astrophys. J.* **218**, 815.

Arnett, W.D. (1980): *Astrophys. J.* **237**, 541.

Arnett, W.D. (1982): *Astrophys. J.* **253**, 785.

Arnett, W.D., Bahcall, J.N., Kirshner, R.P., Woosley, S.E. (1989): *Ann. Rev. Astron. Astrophys.* **27**, 629.

Baron, E., Cooperstein, J. (1991): in *Supernovae*, ed. by S.E. Woosley (Springer-Verlag, New York) pp. 342-351.

Baron, E., Cooperstein, J., Kahana, S. (1985a): *Phys. Rev. Lett.* **55**, 126.

Baron, E., Cooperstein, J., Kahana, S. (1985b): *Nucl. Phys.* **A440**, 744.

Bethe, H.A. (1982): in *Supernovae: A Survey of Current Research*, ed. by M.J. Rees and R.J. Stoneham (Dordrecht: Reidel) pp. 35-52.

Bethe, H.A. (1993): *Astrophys. J.* **412**, 192.

Bethe, H.A., Applegate, J.H., Brown, G.E. (1980): *Astrophys. J.* **241**, 343.

Bethe, H.A., Brown, G.E., Applegate, J., Lattimer, J. (1979): *Nucl. Phys.* **A324**, 487.

Bignami, G.F., Caraveo, P.A., Mereghetti, S. (1993): *Nature* **361**, 704.

Blaizot, J.P. (1980): *Phys. Rep.* **64**, 171.

Blaizot, J.P., Gogny, D., Grammaticos, B. (1976): *Nucl. Phys.* **A265**, 315.

Bonche, P., Vautherin, D. (1981): *Nucl. Phys.* **A372**, 496.

Bonche, P., Vautherin, D. (1982): *Astron. Astrophys.* **112**, 268.

Branch, D. (1990): in *Supernovae*, ed. by A.G. Petschek (Springer-Verlag, New York) pp. 30-58.

Branch, D., Doggett, J.B., Nomoto, K., Thielemann, F.K. (1985): *Astrophys. J.* **294**, 619.

Branch, D., Lacy, C.H., McCall, M.L., Sutherland, P.G., Uomoto, A.K., Wheeler, J.C., Wills, B.J. (1981): *Astrophys. J.* **244**, 780.

Brown, G.E. (1982): in *Supernovae: A Survey of Current Research*, ed. by M.J. Rees and R.J. Stoneham (Dordrecht: Reidel) pp. 13-33.

Brown, G.E. (1988): *Nucl. Phys.* **A488**, 689.

Bruenn, S. (1989a): *Astrophys. J.* **340**, 955.

Bruenn, S. (1989b): *Astrophys. J.* **341**, 385.

Burrows, A., Fryxell, B. (1993): *Astrophys. J. Lett.* **418**, L33.

Burrows, A., Goshy, J. (1993): *Astrophys. J. Lett.* **416**, L75.

Burrows, A., Klein, D., Gandhi, R. (1992): *Phys. Rev. D* **45**, 3361.

Burrows, A., Lattimer, J.M. (1983): *Astrophys. J.* **270**, 735.

Burrows, A., Lattimer, J.M. (1988): *Phys. Rep.* **163**, 51.

Burrows, A., Mazurek, T.J. (1983): *Nature* **301**, 315.

Canal, R., Garcia, D., Isern, J., Labay, J. (1990): *Astrophys. J. Lett.* **356**, L51.

Cassé, M., Lehoucq, R. (1994): in *Supernovae*, ed. by S.A. Bludman, R. Mochkovitch, J. Zinn-Justin (Elsevier Science B.V., Amsterdam) pp. 589-628.

Catchpole, R.M. and 19 others (1987): *Monthly Notices Roy. Astron. Soc.* **229**, 15p.

Colgate, S.A., White, R.H. (1966): *Astrophys. J.* **143**, 626.

Cooperstein, J. (1985): *Nucl. Phys.* **A438**, 722.

Cooperstein, J., Baron, E.A. (1990): in *Supernovae*, ed. by A.G. Petschek (Springer-Verlag, New York) pp. 213-266.

Cooperstein, J., van den Horn, L.J., Baron, E. (1987): *Astrophys. J. Lett.* **321**, L129.

Cooperstein, J., Wambach, J. (1984): *Nucl. Phys.* **A420**, 591.

Doggett, J.B., Branch, D. (1985): *Astron. J.* **90**, 2303.

Domogatskii, G.V., Nadyozhin, D.K. (1977): *Monthly Notices Roy. Astron. Soc.* **178**, 33p.

Domogatskii, G.V., Nadyozhin, D.K. (1978): *Sov. Astr.* **22** (3), 297.

Filippenko, A.V., Sargent, W.L.W. (1986): *Astron. J.* **91**, 691.

Freedman, D.Z. (1974): *Phys. Rev. D* **9**, 1389.

Friman, B.L., Dhar, A.K. (1979): *Phys. Lett. B* **85**, 1.

Fuller, G.M. (1982): *Astrophys. J.* **252**, 741.

Gale, C., Bertsch, G., Das Gupta, S. (1987): *Phys. Rev. C* **35**, 1666.

Gehrels, N., Chen, W. (1993): *Nature* **361**, 706.

Goldreich, P., Weber, S.V. (1980): *Astrophys. J.* **238**, 991.

Goodman, J., Dar, A., Nussimov, S. (1987): *Astrophys. J. Lett.* **314**, L7.

Harkness, R.P. and 10 others (1987): *Astrophys. J.* **317**, 355.

Harkness, R.P., Wheeler, J.C. (1990): in *Supernovae*, ed. by A.G. Petschek (Springer-Verlag, New York) pp. 1-29.

Herant, M., Benz, W., Colgate, S. (1992): *Astrophys. J.* **395**, 642.

Herant, M., Benz, W., Hix, W.R., Fryer, C.L., Colgate, S. (1994): *Astrophys. J.* in press.

Hillebrandt, W.H., Nomoto, K., Wolff, R.G. (1984): *Astron. Astrophys.* **133**, 175.

Hirata, K. and 22 others (KII collaboration) (1987): *Phys. Rev. Lett.* **58**, 1494.

Hoyle, F., Fowler, W.A. (1960): *Astrophys. J.* **132**, 565.

Lamb, D.Q., Lattimer, J.M., Pethick, C., Ravenhall, D.G. (1978): *Phys. Rev. Lett.* **41**, 1623.

Lamb, D.Q., Lattimer, J.M., Pethick, C., Ravenhall, D.G. (1981): *Nucl. Phys.* **A360**, 459.

Lamb, D.Q., Pethick, C.J. (1976): *Astrophys. J. Lett.* **209**, L77.

Lattimer, J.M., Burrows, A., Yahil, A. (1985): *Astrophys. J.* **288**, 644.

Lattimer, J.M., Pethick, C., Ravenhall, D.G., Lamb, D.Q. (1985): *Nucl. Phys.* **A432**, 646.

Lundmark, K. (1920): *Sven. Velekapsakad. Handlingar* **60**, No. 8.

Marcos, S., Barranco, M. Buchler, J.R. (1982): *Nucl. Phys.* **A381**, 507.

Minkowski, R. (1941): *Publ. Astron. Soc. Pac.* **53**, 224.

Müller, E. (1982): *Astron. Astrophys.* **114**, 53.

Müller, E. (1994): in *Supernovae*, ed. by S.A. Bludman, R. Mochkovitch, J. Zinn-Justin (Elsevier Science B.V., Amsterdam) pp. 393-488.

Myra, E.S., Bludman, S.A. (1989): *Astrophys. J.* **340**, 384.

Myra, E.S., Bludman, S.A., Hoffman, Y., Lichtenstadt, I., Sack, N., Van Riper, K.A. (1987): *Astrophys. J.* **318**, 744.

Nadyozhin, D.K. (1994): in *Supernovae*, ed. by S.A. Bludman, R. Mochkovitch, J. Zinn-Justin (Elsevier Science B.V., Amsterdam) pp. 569-587.

Nomoto, K., Hashimoto, M. (1988): *Phys. Rep.* **163**, 13.

Nomoto, K., Kondo, Y. (1991): *Astrophys. J. Lett.* **367**, L19.

Nomoto, K., Shigeyama, T., Kumagai, S., Yamaoka, H., Suzuki, T. (1994): in *Supernovae*, ed. by S.A. Bludman, R. Mochkovitch, J. Zinn-Justin (Elsevier Science B.V., Amsterdam) pp. 489-568.

Nomoto, K., Thielemann, F.K., Yokoi, K. (1984): *Astrophys. J.* **286**, 644.

Robertson, R.G.H., Bowles, T.J., Stevenson, G.J., Wark, D.L., Wilkerson, J.F., Knapp, D.A. (1991): *Phys. Rev. Lett.* **67**, 957.

Saenz, R.A., Shapiro, S.L. (1981): *Astrophys. J.* **244**, 1033.

Shapiro, S.L., Teukolsky, S.A. (1983): *Black Holes, White Dwarfs and Neutron Stars* (Wiley & Sons, New York).

Sol, H., Tarenghi, M., Vanderriest, C., Vigroux, L., Lelièvre, G. (1985): *Astron. Astrophys.* **144**, 109.

Stöcker, H., Greiner, W. (1986): *Phys. Rep.* **137**, 277.

Swesty, F.D., Lattimer, J.M., Myra, E.S. (1994): *Astrophys. J.* **425**, 195.

Takahashi, K., Witti, J., Janka, H.T. (1994): *Astron. Astrophys.* in press.

Tammann, G.A. (1994): in *Supernovae*, ed. by S.A. Bludman, R. Mochkovitch, J. Zinn-Justin (Elsevier Science B.V., Amsterdam) pp. 1-29.

Thielemann, F.K., Hashimoto, M., Nomoto, K., Yokoi, Y. (1991): in *Supernovae*, ed. by S.E. Woosley (Springer-Verlag, New York) pp. 609-618.

Thielemann, F.K., Nomoto, K., Hashimoto, M. (1994): in *Supernovae*, ed. by S.A. Bludman, R. Mochkovitch, J. Zinn-Justin (Elsevier Science B.V., Amsterdam) pp. 629-676.

Timmes, F.X., Woosley, S.E. (1992): *Astrophys. J.* **396**, 649.

Van Riper, K.A. (1979): *Astrophys. J.* **232**, 558.

Van Riper, K.A., Lattimer, J.M. (1981): *Astrophys. J.* **249**, 270.

Vautherin, D. (1994): in *Supernovae*, ed. by S.A. Bludman, R. Mochkovitch, J. Zinn-Justin (Elsevier Science B.V., Amsterdam) pp. 345-391.

Weaver, T.A., Zimmerman, Z.B., Woosley, S.E. (1978): *Astrophys. J.* **225**, 1021.

Wilson, J.R. (1985): in *Numerical Astrophysics*, ed. by J.M. Centrella, J.M. Leblanc, R.L. Bowers (Jones and Bartlett, Boston) pp. 422.

Wilson, J.R., Mayle, R.W. (1988): *Phys. Rep.* **163**, 63.

Wilson, J.R., Mayle, R.W. (1993): *Phys. Rep.* **227**, 97.

Witti, J., Janka, H.T., Takahashi, K. (1994): *Astron. Astrophys.* in press.

Woosley, S.E. (1990): in *Supernovae*, ed. by A.G. Petschek (Springer-Verlag, New York) pp. 182-212.

Woosley, S.E., Hoffman, R.D. (1992): *Astrophys. J.* **395**, 202.

Woosley, S.E., Weaver, T.A. (1988): *Phys. Rep.* **163**, 79.

Woosley, S.E., Weaver, T.A. (1994): in *Supernovae*, ed. by S.A. Bludman, R. Mochkovitch, J. Zinn-Justin (Elsevier Science B.V., Amsterdam) pp. 63-154.

Yahil, A. (1983): *Astrophys. J.* **265**, 1047.

Yahil, A., Lattimer, J.M. (1982): in *Supernovae: A Survey of Current Research*, ed. by M.J. Rees and R.J. Stoneham (Dordrecht: Reidel) pp. 53-70.

Zwicky, F. (1938): *Phys. Rev.* **54**, 242.

Exotic QED Processes and Atoms in Strong External Fields

Günter Wunner

Theoretische Physik I, Ruhr-Universität Bochum,
D–44780 Bochum, Germany

Abstract: The discovery of very intense magnetic fields in compact cosmic objects (white dwarf stars with $B \approx 10^2 - 10^5$ Tesla, neutron stars with $B \approx 10^7 - 10^9$ Tesla) has opened the possibility of studying the properties of matter under extreme conditions which can never be realized in terrestrial laboratories. Since in neutron star magnetic fields the characteristic quantum energy, the cyclotron energy, becomes of the order of the electron rest energy, it is obvious that even *quantum electrodynamical processes* are strongly influenced by these magnetic fields. In white dwarf magnetic fields the cyclotron energy is on the order of the binding energy of atoms, and therefore at these field strengths the *properties of atoms* will be drastically altered by the presence of the fields. I report on the progress that has been made in recent years in recalculating the properties of QED processes and atoms in these intense fields and the way this has helped in understanding the physical processes that occur in the vicinity of white dwarfs and neutron stars.

Highly excited atoms in strong laboratory field strengths ($B \sim$ several Tesla) are exposed to an intense-field situation, because of their small binding energies, in the same way as are low-lying states of atoms in strong cosmic magnetic fields. Moreover, from a classical point of view these atoms serve as a real and physical example of simple nonintegrable systems with classical *chaos*. Therefore in this lecture I also elaborate on the important rôle studies of the properties of Rydberg atoms in strong terrestrial magnetic fields have played in answering the exciting question as to the existence of *"quantum"* chaos.

1 Introduction

1.1 Magnetic Fields of Compact Cosmic Objects

The largest magnetic field strengths measured in nature so far have been found in white dwarf stars ($B \approx 10^2$–10^5 T) and neutron stars ($B \approx 10^7$–10^9 T). (Rapidly time-variable magnetic fields over nuclear dimensions with peak values up to $B \approx 10^{11}$ T are assumed to occur in heavy-ion collisions, cf. Rafelski and Müller [1]). White dwarf stars and neutron stars represent final stages of stellar evolution. Neutron stars are formed from normal stars in a dramatic cosmic event, when the star has consumed its nuclear energy supply and becomes unstable against its own gravitational forces. The catastrophic collapse to a neutron star normally goes along with a supernova explosion, in which the star becomes almost as bright as a whole galaxy consisting of a hundred billion suns. Typical values of relevant physical parameters are listed in Table 1 in comparison with those of our sun.

Table 1. Physical parameters of compact cosmic objects, compared with the sun

	Sun	Magnetized white dwarf	Pulsar = magnetized rotating neutron star
Mass	2×10^{30} kg $= m_\odot$	$\sim 1\, m_\odot$	$\sim (1\text{--}2)\, m_\odot$
Radius	7×10^5 km $= R_\odot$	$\sim 10^4$ km $\approx 10^{-2} R_\odot$	~ 10 km $\approx 10^{-5} R_\odot$
Mean density	$1.4\, \text{g/cm}^3$	$\sim 10^6\ \text{g/cm}^3$	$\leq 10^{15}\ \text{g/cm}^3$
Period	27 d	100 s–days?	$10^{-3}\text{--}10^3$ s
Magnetic field	$\sim 10^{-4}\text{--}10^{-3}$ T	$\sim 10^2\text{--}10^5$ T	$\sim 10^7\text{--}10^9$ T

The existence and the physical properties of these compact objects had been predicted theoretically from the knowledge of the equation of state long before they were actually observed. In particular, the expected order of magnitude of the magnetic field strength can easily be estimated by the following consideration. Because of the high conductivity of stellar matter, the magnetic flux is conserved, in good approximation, during the gravitational collapse. As a consequence, the magnetic field is compressed along with the stellar matter and is boosted, taking the radius reduction from Table 1, by a factor of 10^4 for white dwarfs and 10^{10} for neutron stars. With possible initial field strengths of about $10^{-2}\text{--}10^{-1}$ T, fields up to 10^3 T or 10^9 T, respectively, are automatically achieved. Physically, these magnetic fields are generated by electric currents in the neutron stars, which are induced — according to the generally accepted theory — during the gravitational collapse and are stable against ohmic dissipation on time scales of 10^7 years. (Note that a neutron star must possess a fraction of electrons and protons of a few per cent in order to be stable against beta decay.)

For *white dwarf stars*, the discovery of these fields was based on the observation of the circular polarization of the continuum radiation in the optical and the ultraviolet range (Kemp et al. [2]; Angel et al. [3]). The determination of absolute sizes of the magnetic field strengths up to $B \approx 10^3$ T was possible by measuring the splitting of spectral lines in the range of validity of the linear and quadratic Zeeman effect (Kemic [4], [5]; Garstang and Kemic [6]; Garstang [7]). For magnetic field strengths beyond that, it was only on the basis of the results presented in this report that a quantitative interpretation of the spectra, and thus the pinning down of the values of B, could be made.

The evidence of field strengths of the order of 10^8 T in neutron stars is not quite as immediate, and the observations suggest a distinction between pulsars that predominantly emit in the radio or the X-ray range. Radio pulsars can be identified as isolated, rapidly rotating, strongly magnetized neutron stars, whereas X-ray pulsars are rapidly rotating, strongly magnetized neutron stars, which are the compact components in close binary systems pulling matter from the companion star via gravitation onto their own surface. (For comprehensive reviews of the physics of compact objects see, e.g., Shapiro and Teukolsky [8]; Helfand and Huang [9]; Nagase [10]).

The determination of the magnetic field strength of *radio pulsars* is made indirectly by observing the braking of the rotation of the pulsars, which is caused by the emission of extremely intense electromagnetic waves associated with the rotation of the magnetic moment of the pulsar. From classical electrodynamics it is easy to derive the relation $B^2 \propto P\dot{P}$ between the magnetic field strength at the pole, and the period P and its time derivative $\dot{P}$. The values measured for $P\dot{P}$ thus lead, in a more indirect way, to values of B, which are distributed around 2×10^8 T for typical radio pulsars. At these magnetic field strengths, by unipolar induction huge potential differences of up to 10^{17}–10^{18} V occur between the pole and the equator, which correspond, on an atomic length scale of 10^{-10} m, to an electric potential drop of about 3000 V!

The electric forces at the surface thus still outweigh the enormous gravitational forces on electrons — 10^{11} times stronger than on earth — by a factor of 10^{12}, which unavoidably gives rise to field emission. As a consequence, every rapidly rotating highly magnetized neutron star must necessarily be surrounded by a highly relativistic plasma, in which the huge unipolar induction is transformed into kinetic energy. The charges and currents in the plasma change the vacuum fields, and the self-consistent solution of this magnetospheric problem represents an extremely complicated task that has not been solved in a satisfactory way until now. It is evident that, for the whole complex of questions connected with radio pulsars, the accurate knowledge of the properties of matter in superstrong magnetic fields is a prerequisite to a quantitative understanding.

I now turn to a short discussion of the class of *X-ray pulsars*. To date, approximately 35 objects of this kind, with pulse periods from 69 ms – 835 s, are known. Conclusive evidence was found that these X-ray pulsars are magnetized neutron stars in close binary systems, where, in contrast to radio pulsars, the rotational frequencies usually increase due to accreting gas from their normal companion stars.

Probably the most thoroughly investigated, and understood, system is the X-ray source *Hercules X-1*, whose story is briefly retold in Fig. 1. The figure also includes the scenario in the vicinity of one magnetic polar cap of the neutron star, which has been deduced from the total observed information.

Channelled by the extremely strong magnetic field, ionized matter splashes down with a free-fall velocity of about half the velocity of light on an area of roughly one square kilometer. The matter is stopped by a shock, the kinetic energy is thermalized, and from the observed X-ray spectra one infers temperatures on the order of 10^8 K for this emitting hot spot. From the measured total X-ray luminosity of $\approx 2 \times 10^{30}$ W (about 10^4 times larger than the total radiative

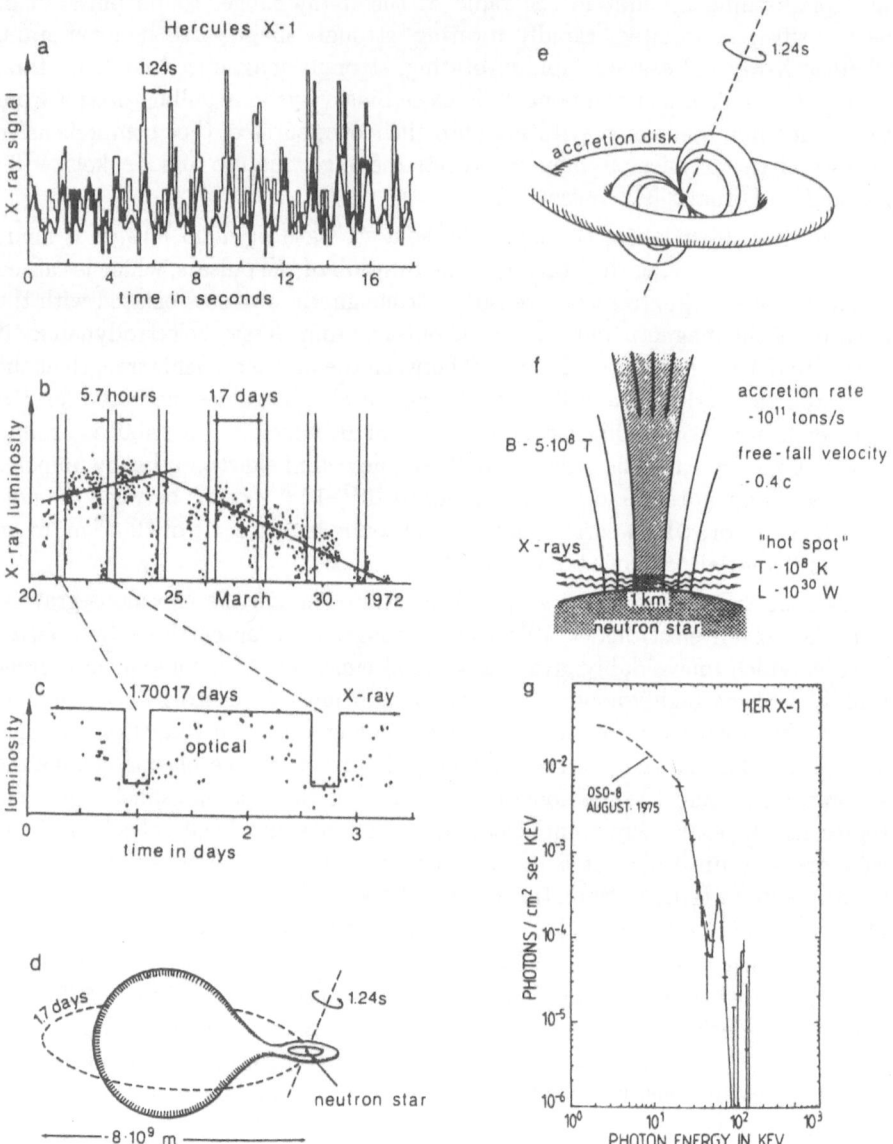

Fig. 1. The story of the pulsating X-ray source Hercules X-1: (a) observed X-ray signal with a mean pulse period of 1.24 s; (b) every 1.70017 days the X-ray source turns off for 5.7 hours; (c) correlated luminosity variation between the X-ray source and the optical companion star; (d) schematic view of the close binary system, the neutron star pulls over matter from the normal star; (e) the material is collected in an accretion disk around the neutron star; (f) a close-up of one polar cap of the neutron star, where the observed X-rays are produced; (g) the X-ray spectrum of Hercules X-1 (Trümper et al. [11]), note the two cyclotron line features at about 50 keV and 100 keV.

power of our sun) of this huge cosmic X-ray tube, it can be concluded that the enormous amount of 10^{11} tons is accreted per second.

Further information can be gained by carefully evaluating the spectral distribution of the observed X-ray quanta. Fig. 1 also shows the X-ray spectrum of Hercules X-1 in the range of 2–150 keV obtained by Trümper et al. [11] in 1976 using a balloon-borne detector. The spectrum exhibits two line features near 50 keV and 100 keV. In the more pronounced feature near 50 keV, about 1% of the total energy radiated is contained, which still corresponds to 100 times the radiation of our sun over the whole spectrum. Conclusive discussions ruled out an atomic or nuclear origin of these line structures; as the only natural explanation there remained the interpretation in terms of cyclotron transitions between quantized Landau states of electrons in the hot, strongly magnetized plasma above the polar cap. Identifying the energy of the first line feature with the cyclotron energy of an electron, $\hbar\omega_c = \hbar eB/m_e$, led to $B \approx 5 \times 10^8$ T, which represented the first direct measurement of such a huge magnetic field strength. The other feature can then be explained without difficulty as the second harmonic. (It should be mentioned that, in the meantime, cyclotron features have been observed in the spectra of several other neutron stars.)

It is obvious that a quantitative theoretical description of this second class of pulsating sources again requires an accurate knowledge of the properties of matter in superstrong magnetic fields. Of particular interest are, because of the prevailing high temperatures up to 20 keV, cross sections for ionization, recombination, and bremsstrahlung, the propagation of photons in these highly magnetized plasmas, and the formation of cyclotron line features. The results of such efforts serve to construct self-consistent models of accretion columns. From the comparison between computed and observed X-ray spectra, reliable conclusions as to the physical conditions in the emitting regions can then be drawn.

1.2 Reference Magnetic Field Strengths

We now introduce units for the magnetic field strength which allow us defining the term "strong magnetic field" more precisely. The appropriate *atomic unit* is the field strength $B_0 = 2\alpha^2 m_e^2 c^2/(e\hbar) \approx 4.70108 \times 10^5$ T, ($\alpha \approx 1/137$ is the fine structure constant), which is the field strength where the characteristic quantum length associated with an electron (or positron) in a magnetic field, the Larmor length $a_L = \sqrt{2\hbar/(eB)}$, is equal to the Bohr radius $a_0 = \hbar/(\alpha m_e c) \approx 0.529177 \times 10^{-10}$ m, or equivalently, where the cyclotron energy is equal to four times the Rydberg energy $E_\infty = \alpha^2 m_e c^2/2 \approx 13.6058$ eV. It is therefore convenient to introduce the dimensionless magnetic field parameter $\beta = B/B_0$, in terms of which the Larmor length at some magnetic field strength B is given by $a_L = a_0 \beta^{-1/2}$. The cyclotron energy can be expressed in the form $\hbar\omega_c = 4\beta E_\infty$, whence it is seen that the cut $\beta \approx 1$ marks, for low-lying states, the transition region from the Coulomb-dominated regime to the Lorentz-dominated regime. The strongest laboratory fields reach up to $\beta \approx 10^{-4}$, magnetic white dwarf stars lie between $\beta \approx 10^{-3}$–10^{-1}, and neutron stars between $\beta \approx 10$–10^3.

In quantum electrodynamics a more appropriate unit is $B_{crit} = m_e^2 c^2/(e\hbar) \approx 4.41405 \times 10^9$ T, which is the field strength where the quantity $a_L/\sqrt{2}$ equals the Compton wavelength $\lambda_C = \hbar/m_e c = \alpha a_0$, or, where the cyclotron energy of an electron (or positron) becomes equal to the electron *rest energy*. Since the electron's discrete excitation energies perpendicular to the field occur in quantized increments of $\Delta E \approx \hbar\omega_c$, B_{crit} sets the scale on which to gauge the importance of quantizing effects of strong magnetic fields on quantum electrodynamical processes involving electrons and positrons.

In the magnetic field strengths characteristic of neutron stars, the Lorentz forces are larger by 2 to 3 orders of magnitude than the Coulomb binding forces acting on an electron in the ground state of a hydrogen atom. Coulomb binding forces are dominated by the Lorentz forces. Therefore, a recalculation of atomic structure is necessary; this task represents, apart from its astrophysical applications, a new and fascinating chapter of quantum mechanics. By complete analogy with the development of field-free quantum mechanics, it is the one-electron problem which here also can serve as a useful guide to working out a new atomic physics in strong magnetic fields. I shall demonstrate that great difficulties occur even in this simplest atomic system if a total understanding of the complete level scheme, wave functions, and electromagnetic transitions in their continuous dependencies on the magnetic field is desired. In particular, it is evident that the transition regime from Coulomb dominance to Lorentz dominance causes the greatest problems. It is, however, exactly this region of magnetic field strengths which occurs in strongly magnetized white dwarf stars. Since the atmospheres of these stars mainly consist of hydrogen, the knowledge of the accurate atomic data in this regime is absolutely necessary for the understanding and modeling of the observed spectra.

The same situation of comparable Lorentz and Coulomb forces can be realized in terrestrial laboratories with magnetic field strengths of several Teslas for highly excited (*Rydberg*) states. These serve as fascinating objects of studies of "quantum" chaos.

Because of space limitations I can report only a few highlights of the investigations of the properties of matter in strong magnetic fields. A comprehensive covery of the subject, including many more results and references to original literature, can be found in the recent book by Ruder et al. [12].

2 Exotic QED Processes in Strong Magnetic Fields

2.1 Landau Representation of the Electron Field

We have just seen that the remarkable feature of the magnetic field strengths in the vicinity of neutron stars is that the cyclotron energy becomes of the order of the electron rest energy and, consequently, in calculating quantum electrodynamical processes the quantization of the electron states into discrete Landau levels has to be fully taken into account. In other words, a perturbative treatment of the effects of the magnetic field using the field-free electron propagator fails, and all relevant quantum electrodynamical cross sections have to be recalculated using the correct relativistic electron states and the corresponding propagator in a magnetic field. More specifically, the electron field is expanded in terms of the solutions of Dirac's equation for a charged particle in a uniform magnetic field, viz.

$$\Psi = \psi_{n,\tau} \, \exp\left(i \, p_z \, z \, /\hbar\right), \tag{1}$$

which are characterized by the (continuous) momentum component p_z, the (discrete) Landau quantum number $n \; (= 0, 1, 2, \cdots)$, and a spin quantum number $\tau = \pm 1$, and the energy eigenvalues are given by εE, with

$$E = m_e c^2 \, \sqrt{1 + (p_z/m_e c)^2 + 2n \, B/B_{\mathrm{crit}}} \,. \tag{2}$$

The wave functions satisfy

$$-i\hbar \frac{\partial}{\partial z}\psi = \varepsilon p_z \psi \qquad \text{and} \qquad i\hbar \frac{\partial}{\partial t}\psi = \varepsilon E \psi \tag{3}$$

with $\varepsilon = 1$ for electron and $\varepsilon = -1$ for positron eigenstates, respectively. The spinor states can be taken to be eigenstates of the conserved z-component of the magnetic moment

$$\boldsymbol{\mu} := m_e \, c \, \boldsymbol{\Sigma} \, + \, i\gamma_5\beta\left(\boldsymbol{\Sigma} \times \mathbf{P}\right), \tag{4}$$

with the kinetic momentum in the external field

$$\mathbf{P} = (\hbar/i)\nabla + e\mathbf{A} \,. \tag{5}$$

The spin quantum number $\tau = \pm 1$ is therefore defined through

$$\mu_z\psi = \tau(E_0/c)\psi \,, \tag{6}$$

where E_0 corresponds to the energy for vanishing p_z-momentum

$$E_0 = m_e c^2 \, \sqrt{1 + 2nB/B_{\mathrm{crit}}} \,. \tag{7}$$

Explicit forms of these wave functions can be found, e.g., in Sokolov and Ternov [13].

It is evident from the form (1) of the wave functions that only the motion along the field direction remains free, and therefore laws of conservation remain only for the energy and the z momentum (the momentum perpendicular to the field is no longer a good quantum number). As a consequence, QED processes that are forbidden in free space on kinematical grounds (conservation of energy and the total three-dimensional momentum) can become allowed in strong magnetic fields. Famous examples are (magnetic) bremsstrahlung of a free electron or one-photon pair creation and annihilation. Results for one-photon pair creation are briefly reviewed in the next section.

2.2 One-Photon Pair Creation in Neutron Star Magnetic Fields

Consider a photon with energy above the pair creation threshold ($\hbar\omega \geq 2m_e c^2$) propagating at an angle $\theta = \pi/2$ with respect to the direction of the magnetic field (all other cases can be reduced to this situation through an appropriate Lorentz transformation in the z-direction, which does not change the magnetic field strength). It is easy to see that the laws of conservation that apply in a magnetic field for the decay of the photon into an electron (with quantum numbers n^-, p_z^-, τ^-) and a positron (with quantum numbers n^+, p_z^+, τ^+),

$$\hbar\omega = E_{n^-,p_z^-,\tau^-} + E_{n^+,p_z^+,\tau^+}, \qquad \hbar k_z (= 0) = p_z^- + p_z^+ \qquad (8)$$

can be fulfilled simultaneously, which proves that this process is kinematically allowed in the presence of a field. Loosely speaking, the magnetic field "absorbs" the transverse momentum of the photon, thus rendering the process possible. However, because of the (in a magnetic field) one-dimensional density of states of the newly created particles, resonances appear in the decay rate whenever the electron and positron are created "at rest" ($p_z^- = p_z^+ = 0$) at the different Landau thresholds, viz.

$$\hbar\omega_{n^+,n^-} = \frac{1}{2}m_e c^2 \left(\sqrt{1 + 2n^+ B/B_{\text{crit}}} + \sqrt{1 + 2n^- B/B_{\text{crit}}} \right). \qquad (9)$$

The possibility of one-photon pair conversion in a magnetic field had been investigated long before the discovery of neutron star magnetic field strengths (e.g. Toll [14], Klepikov [15], Erber [16]), although these authors were primarily concerned with the derivation of expressions in the limit of low fields ($B/B_{\text{crit}} \ll 1$) and photon energies much higher than threshold, where the pair conversion rate becomes a quasi-smooth function of the photon energy on account of the relative smallness of the quantal increments $\Delta E = \hbar\omega_c$. However, at energies near the pair creation threshold, quantum effects due to the discreteness of the electron and positron excitation energies in the magnetic field strongly affect the properties of pair production, in particular for neutron star magnetic fields, and this range was investigated in detail by Daugherty and Harding [17]. Figure 2 shows their results for the attenuation coefficient R (= inverse mean free path ℓ of the photon) due to pair production in a magnetic field of $B = 5 \times 10^8$ T (averaged over both polarizations) for propagation at angle $\theta = \pi/2$. The "sawtooth" behaviour of the attenuation coefficient, which is caused by the singularities at the frequencies (9), is evident from Fig. 2, as is the decrease of the typical spacing between peaks and the overall increase of the size of the coefficient with energy. Shortly above the pair creation threshold (1.022 MeV) the mean free path is on the order of a few centimeters, while at ~ 2 MeV it has decreased to values of a few 10^{-4} cm. This demonstrates that one-photon pair creation in strong magnetic fields is a very efficient mechanism for producing high-energetic electron-positron pairs, which in turn by curvature radiation can generate high-energetic γ-quanta that again decay into further electron-positron pairs. In this way, the process of one-photon pair creation gives rise to *pair cascades* in the vicinity of strongly magnetized neutron stars, and these cascades are believed to play an important rôle in the formation of spectra of pulsars.

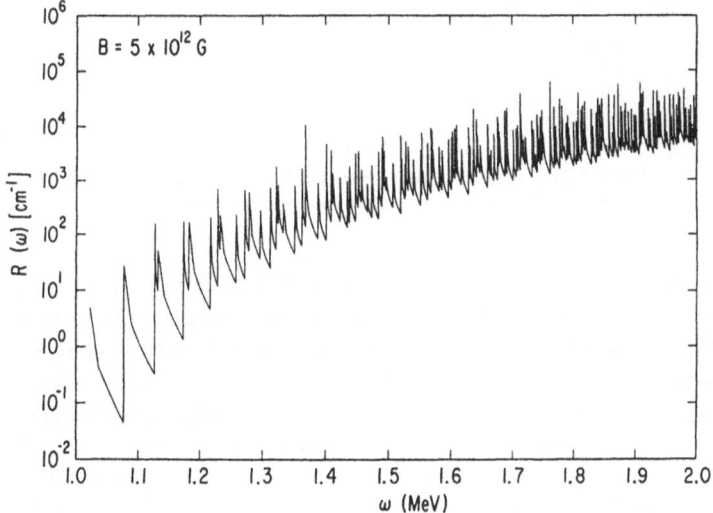

Fig. 2. Attenuation coefficient R (= inverse mean free path of a photon) due to electron-positron pair production by a *single* photon at $B = 5 \times 10^8$ T for propagation of the photon perpendicular to the magnetic field, plotted against photon energy. The threshold energy is $2m_e c^2$ (= 1.022 MeV), below which the process is forbidden. Note the "spikes" (sharp increase of the attenuation coefficient averaged over small energy intervals) at energies where the electron-positron pair is produced in Landau states "at rest" (cf. Equation (9)). (From Daugherty and Harding [17])

Obviously photons with energies below the pair creation threshold possess (in the absence of other scattering particles) infinitely large mean free paths with respect to pair creation. For these sub-pair-creation-threshold photons, however, it is another exotic QED process in a magnetic field which limits their lifetimes, namely the process of *photon splitting*.

2.3 Photon Splitting in Strong Magnetic Fields

This third-order process ($\gamma \rightarrow \gamma' + \gamma''$), whose Feynman diagrams are shown in Fig. 3, is forbidden in free space by the well-known Furry theorem [18], which states that Feynman diagrams containing a fermion loop with an odd number of vertices can be omitted in diagrammatical expansions, since the two amplitudes corresponding to opposite directions of the fermion line exactly cancel each other. On account of the changes in symmetry and the drastically different structure of the electron states, this theorem *does not hold in a strong magnetic field*, and photon splitting becomes an allowed process.

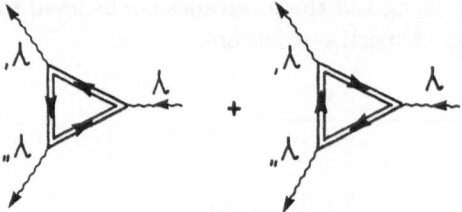

Fig. 3. Feynman diagrams for photon splitting. Note that the internal fermion lines represent virtual electrons and positrons propagating in Landau states.

The actual calculation of the third-order S-matrix for photon splitting is very cumbersome because of the complicated form of the electron propagator in a magnetic field. Stoneham [19] used a representation of the electron propagator derived by the Fock-Schwinger proper-time method and after a long calculation in Fourier space obtained a general expression for the photon splitting rate, which was never evaluated numerically. Only a low-frequency ($\hbar\omega \ll m_e c^2$), low-magnetic-field ($B \ll B_{\mathrm{crit}}$) limit of his expression could be given, which confirmed a result derived earlier by Adler [20] for this limit using an effective-Lagrangian method. The urgent need of quantitative results for photon splitting rates in neutron star magnetic field strengths and photon energies up to 1 MeV had been emphasized by Baring [21], who pointed out the astrophysical importance of photon splitting as a possible cooling mechanism for hot photons below the pair creation threshold in cosmic γ-ray sources.

Quite recently the S-matrix element for photon splitting in a magnetic field has been rederived by Mentzel et al. [22], who expressed the electron propagator in terms of Landau spinors in configuration space and obtained an analytical expression for the photon splitting rate that is exact in the magnetic field strength as well as the photon energy, and lends itself to numerical evaluation. It is of course impossible to repeat their derivation here, and the interested reader is referred to their paper. Once the S-matrix element is derived, the total splitting probability per unit time time, ζ, of a photon in a given initial state i is obtained by summing over all possible final states of the photons

$$\zeta = \frac{1}{2}\left(\frac{V}{(2\pi)^3}\right)^2 \sum_{p',p''=\perp,\|} \int d^3k'\, d^3k''\, \frac{|S_{fi}|^2}{T}\,, \tag{10}$$

where the factor $1/2$ is due to the Bose symmetry of the photons and avoids double counting of final states. It is convenient to break up the S-matrix element into kinematical factors and an "effective" matrix element $C(\gamma \to \gamma'\gamma'')$,

$$S_{fi} = (-2\pi i)\, e^3\, 2\pi e B\, \delta^{(4)}(k^{\mu\prime} + k^{\mu\prime\prime} - k^\mu)\, (2V)^{-\frac{3}{2}}\, \frac{1}{16^3}\, \frac{C(\gamma \to \gamma'\gamma'')}{\sqrt{\omega\omega'\omega''}}\,, \tag{11}$$

in terms of which the photon splitting rate is finally obtained in the form

$$\zeta = \frac{\alpha^3}{8^9 \, \pi^2} \frac{c}{\lambda_c} \left(\frac{B}{B_{\text{crit}}} \right)^2 \sum_{p', p''=\perp, \parallel} \int_0^{\omega/m_e} \frac{d(\omega'/m_e)}{(\omega/m_e)^2} \, |C(\gamma \to \gamma'\gamma'')|^2, \quad (12)$$

where in the integral in the square brackets all quantities are assumed to be expressed in units of m_e. The mean free path ℓ is then given by $\ell = c/\zeta$.

The conservation of four-momentum and the energy-momentum relation of photons imply that the three photons move *collinearly*. Therefore photon splitting does not change the direction of propagation of electromagnetic radiation. As in the case of one-photon pair creation, it is sufficient to consider the case in which the photon travels perpendicular to the field ($k_z = 0$). It immediately follows from the behaviour of photons under Lorentz transformations that the mean free path ℓ' of a photon propagating at an arbitrary angle θ with respect to the magnetic field axis and with frequency ω' is related to the mean free path $\ell(\omega, \pi/2)$ in the $k_z = 0$ frame by

$$\ell'(\omega', \theta) = \ell(\omega' \sin\theta, \pi/2) / \sin\theta. \quad (13)$$

As an example, Fig. 4 shows the mean free paths of photons (with parallel or perpendicular polarization) in the $k_z = 0$ frame with respect to photon splitting for two magnetic field strengths, $B/B_{\text{crit}} = 0.1$ and 1.0, as a function of photon energy below the pair creation threshold. (Above $2\,m_e\,c^2$ photon pair production as a first-order process outweighs the effects of photon splitting.) As a general trend it is seen that the mean free paths decrease rapidly with increasing photon energy, from the range of meters for soft X-ray photons to that of millimeters, or less, for hard quanta, with the mean free paths shortening by a factor of two to three as one goes from $B = 0.1\,B_{\text{crit}}$ to $B = B_{\text{crit}}$. It is noted that photons with the polarization vector parallel to the magnetic field generally possess longer lifetimes against photon splitting than photons with perpendicular polarization. Comparing with Fig. 2 we see that the order of magnitude of the mean free paths shortly below threshold are comparable to those for pair production above the threshold. The results shown in Fig. 4 imply that hard quanta practically have no chance to travel a substantial distance across the field, since they are easily converted into softer photons through photon splitting. It can therefore be predicted that γ-spectra of strongly magnetized cosmic objects should be deficient of such quanta. On the other hand, it must be noted that for directions of propagation not perpendicular to the magnetic field the mean free paths are enlarged by the $\sin\theta$ term as well as the down-scaling of the effective frequency in (13), and, in particular, γ quanta propagating almost parallel to the magnetic field have very large mean free paths ($\ell' \to \infty$ as θ goes to zero) and therefore can escape from the emission regions. In any case we have the result that photon splitting crucially influences the angular radiation characteristics both for hard quanta as well as for X-ray-quanta, and produces a pronounced beaming along the magnetic field direction.

Obviously, a reassessment of the importance of photon splitting to spectral formation in strongly magnetized neutron stars from the hard X-ray to the soft γ-ray regime ($\hbar\omega \lesssim 2\,m_e c^2$) is urgent. The numerical results reported above open the way to such studies.

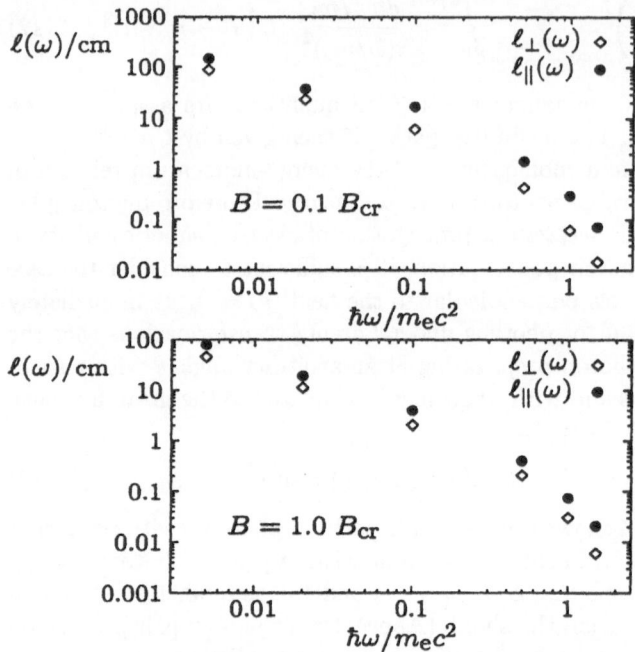

Fig. 4. Total mean free paths for the splitting of photons with parallel or perpendicular polarization into unpolarized photons for energies below the pair creation threshold ($\hbar\omega < 2\,m_e c^2$) and $B = 0.1\,B_{\text{crit}}$ and $B = B_{\text{crit}}$. (From Mentzel et al. [22])

3 Hydrogen Atoms in Magnetic Fields of Arbitrary Strength

3.1 General Considerations

Measuring energies in units of the Rydberg energy E_∞, lengths in units of the Bohr radius a_0, and the magnetic field strength in units of B_0, the Hamiltonian of an electron under the combined action of a static Coulomb potential and a uniform magnetic field reads for spin-down states

$$H = -\Delta - \frac{2}{r} + 2\beta l_z + \beta^2 \varrho^2 - 2\beta \quad , \tag{14}$$

where the magnetic field is assumed to point in the z-direction and $\varrho^2 = x^2 + y^2$. The energies of the corresponding spin-up states are obtained by simply adding 4β. The eigenstates of (14) can be classified according to z-parity π and the z-component l_z of orbital angular momentum, which are exact symmetries of H, but in general no further separation of the two-dimensional problem is possible.

Energies and wave functions calculated from the infinite-nuclear-mass Hamiltonian (14) can be connected to those belonging to finite nuclear mass, m_+, by use of the appropriate mass scaling laws valid in a magnetic field (cf. Ruder et al.

[12]), which in the case where the system has negligible (generalized) momentum perpendicular to the field read:

$$\Phi_m(m_-, m_+, B; \mathbf{r}) = \lambda^{3/2} \Phi_m(m_-, m_+ \to \infty, B/\lambda^2; \lambda \mathbf{r}) \quad , \qquad (15a)$$

$$E_m(m_-, m_+, B) = \lambda E_m(m_-, m_+ \to \infty, B/\lambda^2) - \hbar m e B/m_+ . \qquad (15b)$$

Here, $\lambda = m_+/(m_+ + m_-)$, $m_- = m_e$, and the subscript m denotes the magnetic quantum number of the state. From (15) the scaling laws for electromagnetic dipole transitions can also be derived. It should be noted that for the hydrogen atom the effect of the finite proton mass is not always of the anticipated order of magnitude m_e/m_p. The last term in (15b) causes a shift between states with different magnetic quantum numbers m by (with the proton cyclotron frequency $\Omega_c = eB/m_p$) $\Delta m \, \hbar \Omega_c \approx \Delta m \times 29.6 \, \text{eV} \times B/(4.7 \times 10^8 \, \text{T})$, which, in high fields, is comparable to the Coulomb binding energies, and thus is by no means negligible (cf. Wunner et al. [23]).

The above considerations prove that for hydrogen atoms in arbitrary strong magnetic fields it is sufficient to consider the Hamiltonian (14). For small or extremely large magnetic field strengths, where either the Coulomb forces dominate the Lorentz forces or vice versa, corresponding perturbational approaches to solving the Schrödinger equation for the Hamiltonian (14) are of course appropriate. In the *intermediate* (or *strong-*) *field regime*, however, the electron experiences electric and magnetic forces of comparable strength, and perturbative approaches fail. The mathematical reason for this is that the spherical symmetry of the Coulomb potential, on the one hand, and the cylindrical symmetry of the magnetic field, on the other, prevent a separation of variables so that closed-form analytical solutions are not possible. Thus the Hamiltonian (14) belongs to the class of *nonintegrable* Hamiltonians. The absolute sizes of the field strengths at which one lies in this regime depend on the state of excitation of the electron. By considering the equality of Coulomb and Lorentz forces for an electron in a circular Bohr orbit with principal quantum number n_p one obtains as a rough measure $B_{n_p} \cong B_0/(2n_p^3) \cong (8.7 \, T)(30/n_p)^3$. Therefore for white dwarf and neutron star magnetic fields low-lying states are found to be subjected to a strong-field situation, while at laboratory field strengths studies of the strong-field regime are possible with highly excited states ("Rydberg states").

The nonintegrability of the Hamiltonian (14) represents the basic difficulty of the problem in point. As a consequence of this nonintegrability one is forced, in the quantum theoretical treatment of the system, to resort to numerical methods. Various methods were employed by various authors, and since the calculations have been described in detail in the literature, it is not necessary to repeat these here. Diagonalizations of the Hamiltonian in large complete basis sets have proved most efficient in this problem. Basis sizes as large as ~ 300000 were used in some instances to guarantee convergence of the values for transition strengths. As a result of these multilateral efforts it is possible today to continuously trace the energy values and oscillator strengths of *low-lying states* ($n_p \leq 5$) from zero field up to neutron star magnetic fields, while for Rydberg states in laboratory fields of a few Tesla ($n_p \geq 25$) one can calculate the energies and transition rates of bound-bound and bound-free transitions up to energies well above the ionization threshold. In the following I shall present examples of these results.

3.2 Results for Low-Lying States

3.2.1 Energy Values. Very comprehensive results of the extensive numerical calculations are presented in the book by Ruder et al. [12]. In particular, the energy values (in units of $-E_\infty$) can be found for the 38 lowest-lying states of the static Coulomb problem – corresponding to the H atom with infinite proton mass – and for magnetic field strengths in the range $10^{-4} \leq \beta \leq 10^3$. Here I restrict myself to a graphical rendering of the results and show, in Fig. 5, the continuous dependence of the energy values on the magnetic field strength for all energy levels with $n_p \leq 5$ in the range $10^{-3} \leq \beta \leq 10^{+3}$. It is found that as the principal quantum number n_p increases, the transition from the spherical to the cylindrical expansion occurs more and more at smaller values of β. This effect is easily understood when one compares the level distances in the unperturbed one-electron problem ($\propto 1/n_p^3$) with the perturbation energy associated with the diamagnetic term ($\propto \beta^2 n_p^4$), from which it immediately follows that the transition to the magnetic-field dominated region is shifted to smaller values of β proportional to $n_p^{-7/2}$.

The figure shows, for small values of β, the Rydberg structure of the level scheme corrected by the splitting of the linear and quadratic Zeeman effect, while at values of β around 10^{-2} (for the low-lying states under consideration) a complete rearrangement of the level structure occurs (the breakpoint is shifted left with increasing principal quantum number, in accordance with the $n_p^{-7/2}$ dependence discussed above). It is only for β greater than, say, 10 that the graphs become well-behaved again, and the level order characteristic of the strong-field regime is reached. In this latter case the energies of states with $\nu > 0$ converge towards the Rydberg series $1/\bar{\nu}^2$ again, with $\bar{\nu} = \nu/2$ for ν even, and $\bar{\nu} = (\nu+1)/2$ for ν odd (*hydrogen-like states*), while the energies of the z-nodeless ground states in every $m \leq 0$ band are strongly lowered (*tightly bound states*), and in fact diverge logarithmically in the limit $\beta \to \infty$ (Loudon [24]). The quantity ν denotes the number of nodes of the longitudinal part of the wave function in the $\beta \gg 1$ limit. The perpendicular part is given by Landau states with Landau quantum numbers $n = 0, 1, 2, \ldots$. The energies of states with positive m, which correspond to states in excited Landau levels, become less and less negative with β increasing from zero and eventually merge with the continuum at $\beta < 0.15$, and even sooner for higher principal quantum numbers. Finally, it should be noted that, at intermediate values of β, the monotonous behaviour of the energy graphs of $m \leq 0$ states is more and more destroyed as β is increased; a slight dip is already observed in the state $2s_0/0\ 0\ 2$, a stronger one in $3d_0'/0\ 0\ 4$, and the wiggles become more pronounced in higher states. (Here the states are labeled by their field-free and their $\beta \gg 1$ quantum numbers $n_p lm/n\ m\ \nu$.) This phenomenon is related to the existence of an approximate symmetry in the

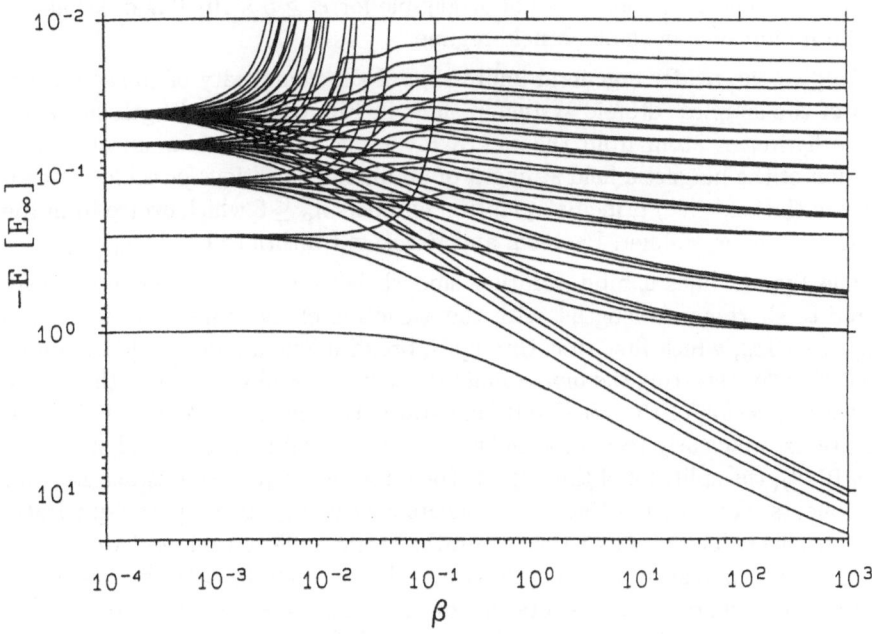

Fig. 5. Energy values (in units of E_∞) of all states of an electron in a static Coulomb potential with principal quantum numbers $n_p \leq 5$ as continuous functions of the magnetic field parameter $\beta = B/(4.701 \times 10^5 \, \text{T})$. The Rydberg series, corrected by the splitting of the linear and quadratic Zeeman effect, is easily recognized for small values of β. At β-values around 10^{-2} a complete rearrangement of the energy levels occurs. States with positive m are shifted to higher Landau levels, and for $\beta \geq 10$ the level structure of the strong-field regime is reached, where states with $\nu > 0$ converge towards the Rydberg series again (*hydrogen-like states*), while the z-nodeless ground states in every $m \leq 0$ band are strongly lowered (*tightly bound states*) and diverge logarithmically in the limit $\beta \to \infty$.

magnetized hydrogenic problem, which manifests itself, for higher n_p-states, by exponentially diminishing level anticrossings (cf. Sect. 6).

3.2.2 Wavelengths of the Hydrogen Atom.

With the accurate energy values for the static Coulomb problem at hand it is an easy task to compute the wavelengths of transitions as functions of the magnetic field strength. The wavelengths of the physical hydrogen atom are obtained from those of the static Coulomb problem by employing the scaling laws from Sect. 3.1 for taking into account the finiteness of the proton mass ($\lambda = m_p/(m_p + m_e) \cong 0.9994557$). While the corrections in the first term on the right-hand side of (15b) are of the expected order of magnitude, the second term (multiples of the proton cyclotron energy $\hbar\omega_p \cong 29.6 \, \text{meV} \times (B/4.7 \times 10^5 \, \text{T})$) can produce sizable shifts if the magnetic field is large enough. For wavelengths below 1000 nm the contributions in $|\Delta m| = 1$ dipole transitions are non-negligible for $B \geq 5 \times 10^4$ T and become of special importance in the high-field regime.

To give the reader a general impression of the complexity of the spectrum, even of the simplest atomic system, I show in Fig. 6 the wavelength spectrum of the hydrogen atom from the soft X-ray range up to the far infrared as a function of the magnetic field strength of all dipole transitions possible between states with (field-free) principal quantum numbers $n_p \leq 5$ which evolve from the field-free Lyman, Balmer, Paschen, and Brackett transitions ($n'_p \neq n_p$).

The Lyman lines exhibit a rather smooth behaviour and are continuously shifted to shorter wavelengths, with the exception of the transition originating in $2p_{-1} \rightarrow 1s_0$, which first runs through a broad maximum of $\lambda = 134.1$ nm at $\beta \approx 0.12$ before starting the monotonous descent essentially caused by the strong energetic lowering of the $1s_0/0\,0\,0$ final state. Turning to Balmer and Paschen transitions, we clearly recognize, in the region of small magnetic field strengths ($\beta \approx 10^{-3}$), the splitting of the unperturbed lines into three equidistant Zeeman components. For larger β these components continue splitting by the quadratic Zeeman effect. The onset of the quadratic Zeeman effect is shifted to smaller β-values with increasing wavelengths. Beyond this region ($\beta \approx 10^{-2}$), where the perturbation theory treatment breaks down, the lines are completely torn apart by the magnetic field within one β-decade and the spectrum becomes totally distorted. Since the energy levels of states with different m and different z-parity are allowed to cross (cf. Fig. 5), the wavelengths of corresponding transitions go to infinity at certain values of β.

Reordering appears only in intense fields, indicative of the fact that, in the limit $B \rightarrow \infty$, the level scheme approaches that of the one-dimensional Coulomb problem, which consists of tightly bound levels and levels whose energies equal those of the field-free hydrogen atom. As a consequence, numerous lines tend toward the wavelengths of the unperturbed hydrogen series on the right-hand side of the figures. Further ordering is evident from the clustering of many lines into two conspicuous bunches (clearly seen in Fig. 6), which decline proportionally to $1/\beta$. The left-hand bunch comprises all electron cyclotron transitions with $\lambda_{cycl}^{(e)} \cong 22.8\,\text{nm}/\beta$. The shortest wavelengths then correspond to cyclotron transitions of the electron from the first-excited to the ground-state Landau level. For neutron star magnetic fields ($\beta = 10^3$), the appropriate photon energies lie at 54 keV, and thus in the X-ray region. For photon energies like these, which amount to more than 10 per cent of the rest energy of the electron, of course relativistic effects of the same order are to be expected. This is indeed confirmed by the rigorous relativistic treatment (cf. Daugherty and Ventura [25]; Herold et al. [26]) for cyclotron transitions in which the transverse state of excitation of the electron changes; for the Coulomb binding energies, however, in which the longitudinal motion plays the essential rôle, relativistic effects remain well below 0.1 per cent even at magnetic fields of 10^9 T (Lindgren and Virtamo [27]), which is understandable because of the smallness of the ratio between the Coulomb binding energy and the rest energy of the electron.

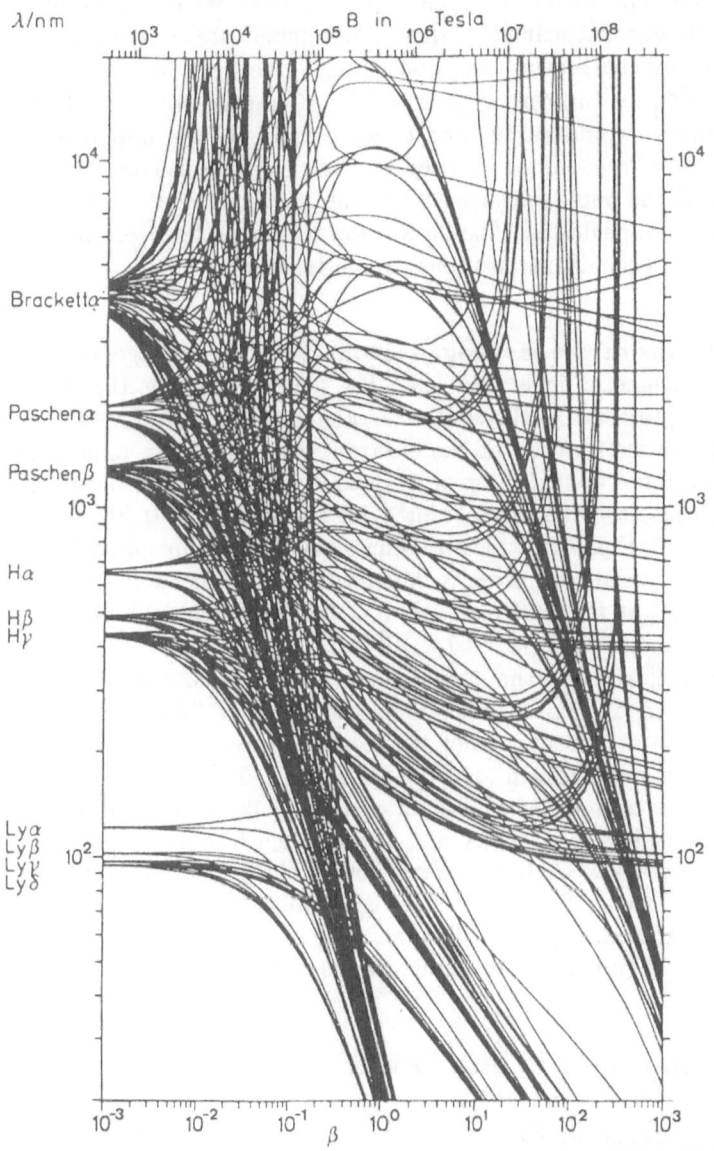

Fig. 6. The wavelength spectrum of the hydrogen atom from the soft X-ray range (30 nm) up to the far infrared (10000 nm) as a function of the magnetic field strength in the interval 470 T to 4.7×10^8 T on a doubly logarithmic scale. All possible transitions between states with (field-free) principal quantum numbers $n_p \leq 5$ and $n'_p \neq n_p$ are included. Effects of the finiteness of the proton mass are taken into account. The two rapidly declining bunches of lines correspond to cyclotron-like transitions of electrons (left-hand bunch) and protons (right-hand bunch), respectively. The *stationary* lines in the intermediate region are particularly well recognizable if the figure is viewed sideways at flat angles.

The bunch on the right-hand side is due to the finiteness of the proton mass, by which levels with adjacent azimuthal quantum numbers are shifted with respect to each other by the proton cyclotron energy, giving rise to transitions with wavelengths $\lambda_{\text{cycl}}^{(p)} = (m_p/m_e)\lambda_{\text{cycl}}^{(e)} \cong 4.18 \times 10^4 \, \text{nm}/\beta$. All lines in the proton cyclotron bunch would either tend to wavelengths of the unperturbed hydrogen atom or even go to infinity (in the case of energy levels coinciding in the $B \to \infty$ limit of the one-dimensional Coulomb problem), if the finiteness of the proton mass were neglected. The lines below $\sim 100 \, \text{nm}$ between the two cyclotron bunches correspond to transitions to tightly bound states.

3.2.3 Wave Functions of the Hydrogen Atom. What do hydrogenic wave functions in strong magnetic fields look like? The starting point is the quantum mechanical interpretation of the squared modulus of the wave function as the spatial probability distribution of the electron. Because of the remaining rotational symmetry about the direction of the field, it suffices to consider intersections of $\psi^*\psi$ with the xz-plane. For four low-lying states, $1s_0/0\ 0\ 0$, $2s_0/0\ 0\ 2$, $2p_0/0\ 0\ 1$, $2p_{-1}/0\ -1\ 0$, Fig. 7 provides contour plots of the probability distribution of the electron (decreasing from maximum to zero in increments of 0.1 of the maximum value) at four different values of the magnetic field strength, $B = 0\,\text{T}$, $1.4 \times 10^5\,\text{T}$, $9.4 \times 10^5\,\text{T}$, and $4.7 \times 10^7\,\text{T}$.

The most striking feature is the strong constriction of the atomic states (roughly proportional to $\beta^{-\frac{1}{2}}$) perpendicular to the magnetic field as the field strength is increased – a consequence of the Lorentz forces becoming larger and larger. Two of the states shown ($1s_0/0\ 0\ 0$ and $2p_{-1}/0\ -1\ 0$) additionally experience a sizable reduction in their linear extension parallel to the field, which causes them to appreciably gain energy and become *tightly bound* states. As demonstrated by the state $2s_0/0\ 0\ 2$, however, other states may even enlarge their extension parallel to the field. This effect is enhanced even more in states with higher numbers of nodes in the z-direction, due to the "lack of space" in the quantized "tubes" of radius a_L prescribed by the magnetic field.

4 Stationary Lines and White Dwarf Spectra

One of the most spectacular applications of the calculations presented in the foregoing section was the identification of absorption features in the optical spectra of magnetized white dwarf stars, which had defied interpretation for almost 50 years, in terms of *stationary* components of hydrogen lines in magnetic fields of several 10^5 T (Angel et al. [28]; Greenstein et al. [29]; Wunner et al. [30]; Schmidt et al. [31], [32]). By *stationary* components we mean those transitions whose wavelengths go through maxima or minima as functions of the magnetic field strength. These lines, between 300 nm and 1000 nm, are particularly well recognized if Fig. 6 is viewed sideways at flat angles. The fact that these transitions can produce sharp absorption features in white dwarf spectra is obvious when one considers that the magnetic field strength varies (in a dipolar

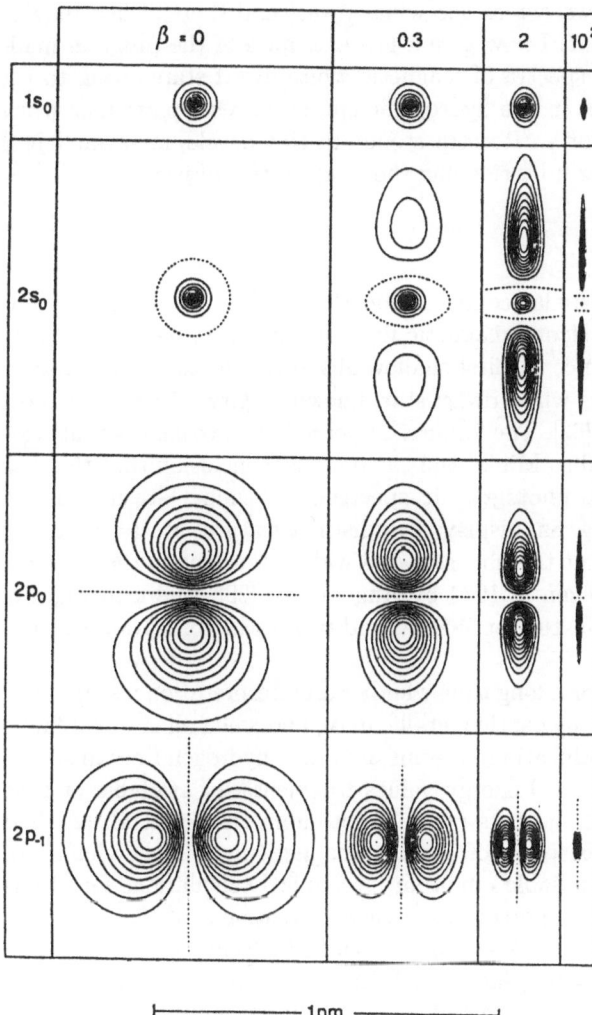

Fig. 7. Contour plots of the spatial probability distribution of the electron in the xz-plane (decreasing from maximum to zero in increments of 0.1 of the maximum value) at four different values of the magnetic field strength. Nodal surfaces are indicated by broken curves. As β increases, the atomic states are squeezed together in the direction transverse to the field; in tightly bound states ($1s_0, 2p_{-1}$) the extension parallel to the field also shrinks, while *hydrogen-like* states become strongly elongated, particularly for higher quantum numbers, due to the necessary orthogonality of the longitudinal wave functions in the high-field limit.

geometry) by a factor of two across the white dwarf and thus all fast moving wavelengths are smeared out. I now give a few examples of the progress made recently in analyses of the spectra of magnetic white dwarf stars owing to the existence of stationary lines in the hydrogenic spectrum. One particular white dwarf has played the rôle of a "Rosetta stone" in this development, and it is therefore appropriate to begin by retelling the story of this object.

4.1 Grw+70°8247

42 light years away in the stellar configuration Draco lies a 13 magnitude star long suspected of having a strong magnetic field. Its spectrum had consistently defied interpretation ever since the first shallow absorption features were discovered in the spectrum of this white dwarf star, known as Grw+70°8247, almost 50 years ago by Minkowski [33]. The features appeared at wavelengths that were completely inexplicable, and so Minkowski came to the conclusion that this was a most unique object. The photographic spectrum taken by Greenstein and Matthews [34], shown in Fig. 8a, displays two distinct features at $\lambda = 3650$ and 4135 Å, with sharp cut-offs at the blue side and wide extensions to the red side, while the CCD spectrum taken in 1974 by Angel et al. [28], shown in Fig. 8b, exhibits a similarly shaped feature at 5850 Å, and two broad absorption features around 7000 Å and 8450 Å.

It had been speculated for a long time whether an interpretation was possible by molecular bands of C_2 or by exotic metallic lines, and when all else had failed Angel [35] proposed an identification in terms of atomic hydrogen lines in strong magnetic fields using variational energy values fragmentarily available at that time. The circular polarization of its optical continuum (Kemp et al. [2]) had in fact given a clue to the existence of a strong magnetic field in the vicinity of this object, but the lack of reliable quantum mechanical calculations for atoms in fields above ~ 30 MG (the perturbation theory treatment of the quadratic Zeeman effect fails beyond 5–10 MG, and extensions by Kemic [4], [5] are valid only up to ~ 30 MG; 1 MG $= 10^6$ Gauss $= 10^2$ Tesla) presented a major obstacle to checking the identification in terms of magnetically strongly shifted atomic lines and pinning down the prevailing field strengths. This obstacle was only removed when the quantum mechanical problem of hydrogen atoms in magnetic fields of arbitrary strength could be solved conclusively.

It is evident from Fig. 6 that any attempt to observe, and resolve, a line spectrum of hydrogen at a *given* magnetic field strength in the intermediate regime is doomed to failure. However, as already mentioned at the beginning of this section, an element of order is brought in even in this domain by the "stationary" transitions whose wavelengths go through maxima and minima in certain intervals of the magnetic field strength and thus approximately produce features around the maximum or minimum values of the wavelengths when the magnetic field varies across these intervals. Since the magnetic field of a white dwarf with a dipolar field configuration has a variation by a factor of two from the pole to the equator, this opens the possibility of well resolved "stationary" line features being produced even in these field strengths. This in fact finally provided

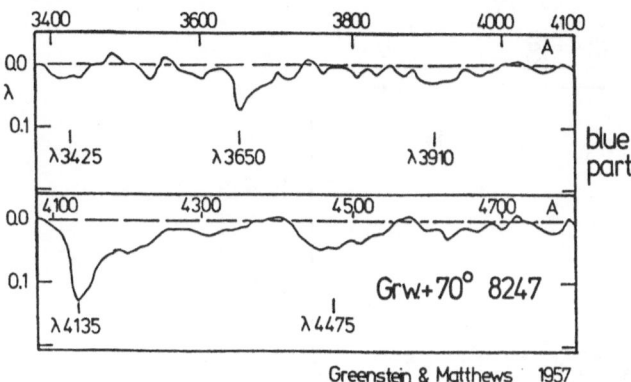

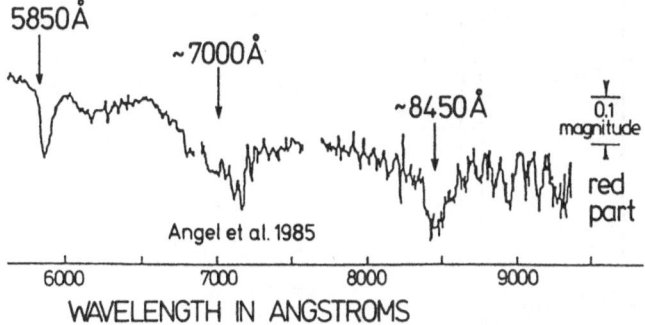

Fig. 8. Optical spectrum of the white dwarf star Grw+70°8247 (a) blue part (Greenstein and Matthews [34]), (b) red part (Angel et al. [28]). The features – occurring at strange wavelengths for a white dwarf – had defied identification ever since their first discovery by Minkowski in 1938. Note in particular the band-head-like features at 3650, 4135 and 5850 Å.

the key to the correct explanation of the shallow absorption features observed in Grw+70°8247, and it opened a new era of stellar atomic spectroscopy, the "spectroscopy of stationary lines".

In Fig. 9 and Fig. 10 I demonstrate the excellent agreement between the wavelength positions of the extrema of stationary components of H_α, H_β, H_γ and absorption features in the red and blue part of the spectrum of Grw+70°8247. In particular, the sharp blue edges and "red-shaded" extensions of the features at 3650 Å, 4135 Å, and 5800 Å are well accounted for by the minimum character of the corresponding stationary components. More complicated structures of other features are produced by blends of stationary components, such as the broad feature around 8500 Å to which a stationary transition of H_α and one of Paschen β contribute. Thus we have the result that — in terms of wavelengths and qualitative line shapes — all spectral features can consistently be explained by *stationary* lines of atomic hydrogen in a dipolar magnetic field with a polar field strength of $\sim$ 320 MG (Angel et al. [28]; Greenstein et al. [29]; Wunner et

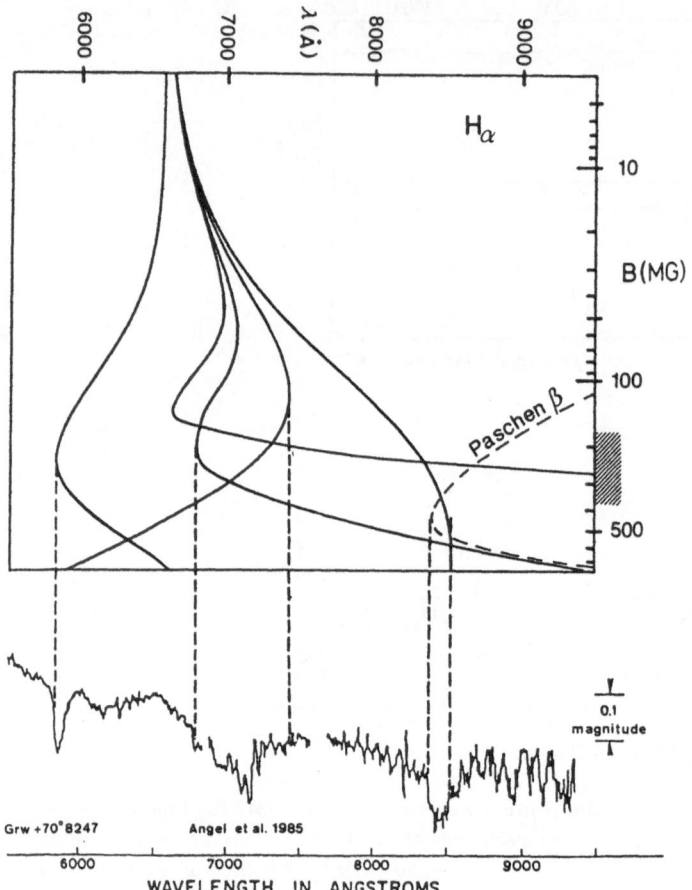

Fig. 9. Stationary H_α transitions of the hydrogen atom in magnetic fields from 4 to 700 MG in comparison with the red part of the optical spectrum of Grw+70°8247. The sharp blue edge of the feature at 5850 Å coincides with the wavelength minimum of a single stationary H_α transition, as indicated by the dashed line, and the red extension of the feature is explained by the variation of the wavelength around the minimum in an extended magnetic field whose strength varies from $\sim$ 160 to $\sim$ 320 MG (see the corresponding hatching along the B coordinate; 1 MG = 10^6 Gauss = 10^2 Tesla). Such a variation is present, e.g., in a dipolar field. The broader feature around 7000 Å is accounted for by a blend of two stationary H_α components in the same range of field, while even a strongly blue-shifted stationary Paschen β component contributes to the feature around 8450 Å.

al. [30]; Henry and O'Connell [36]). No other previously known white dwarf star had a magnetic field even one tenth this value.

Our interpretation so far relied solely on the behaviour of the wavelengths as a function of the field. But astronomers are more ambitious and want to actually calculate synthetic spectra, to be compared with the observed spec-

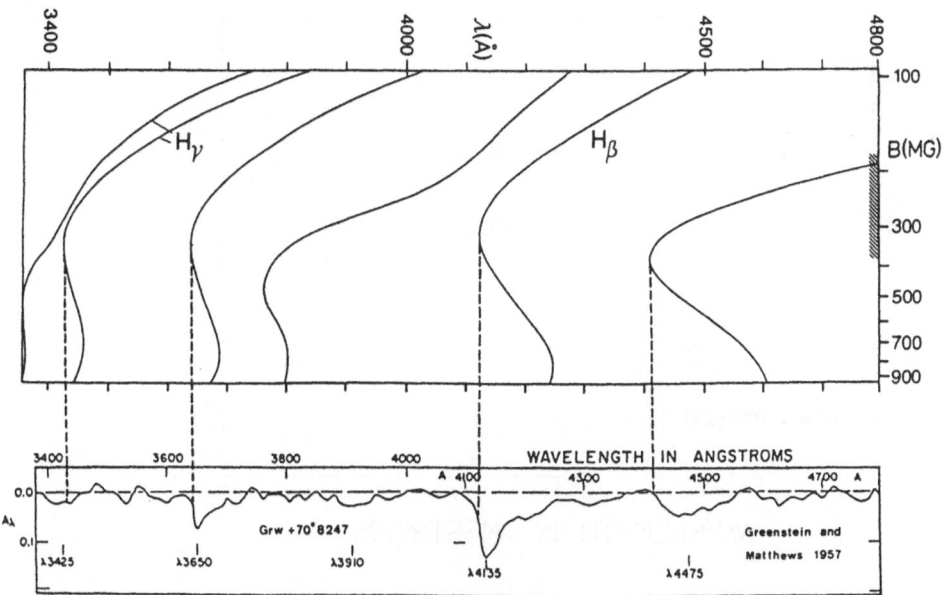

Fig. 10. Stationary H_β and H_γ components in magnetic fields from 100 to 900 MG in comparison with the short-wavelength part of the spectrum of Grw+70°8247. Again all the features are explained in a consistent way in terms of stationary transitions of hydrogen in the range of magnetic field found in Fig. 9 (compare hatching along the B-coordinate).

tra in order to explore the precise emission and absorption conditions in these strongly magnetized objects. Of course, transition rates have to be incorporated in the calculations of synthetic spectra, and it is evident from the strong dependence of the transition rates on the field strength that these calculations provide a very sensitive tool indeed to probe the physical conditions prevailing in these objects. Fig. 11 shows a synthetic spectrum of Grw+70°8247 calculated by Wickramasinghe and Ferrario [37] assuming a centered dipolar field with a polar field strength of 320 MG, in comparison with the observed spectrum. The agreement is excellent and confirms the interpretation given above. However, it must be noted that in the calculations of line shapes parameterized Stark-broadened profiles are assumed, lacking a more detailed theory for broadening of nondegenerate lines. This theory will require the knowledge of the shifts in energy of the lines split by the magnetic field as functions of an arbitrarily oriented electric field caused by the ions and electrons in the atmosphere. This latter problem is still far from being solved.

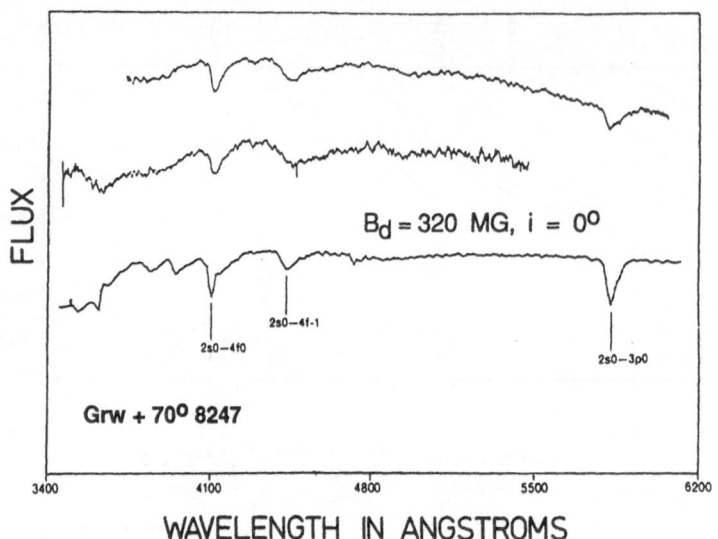

WAVELENGTH IN ANGSTROMS

Fig. 11. Synthetic spectrum of Grw+70°8247 (bottom curve) calculated by Wickramasinghe and Ferrario [37] in a dipole field model using the wavelength and bound-bound transition information of the present work in comparison with observed spectra (two upper curves). A polar field strength of 320 MG and an angle of inclination of 0° between the magnetic axis and the line of sight was assumed in the calculation.

4.2 PG 1031+234

Unlike Grw+70°8247, whose spectrum shows no time variation, the magnetic degenerate PG 1031+234 (Schmidt et al. [31], [32]; Latter et al. [38]) rotates with a period of 3.4 hours and exhibits strongly asymmetric changes in the spectrum and the polarization during one period. Fig. 12 shows the spectrum of this white dwarf star together with the computed behaviour of the wavelengths of the H atom as a function of field strength.

The comparison with Grw+70°8247 shows that the features attributed to the $2s_0 \rightarrow 4f_0$, $2s_0 \rightarrow 4f_{-1}$ and $2s_0 \rightarrow 3p_0$ transitions are slightly red-shifted indicating a somewhat higher field strength. From the positions of the stationary components and the observed phase modulations Schmidt et al. [31], [32] concluded that this object has a nonaxisymmetric field morphology with a maximum field strength of $\sim$ 500 MG. This value currently represents the "world record" for the magnetic field strength of a white dwarf star. At the same time it is the highest magnetic field in which hydrogen lines have ever been seen in nature.

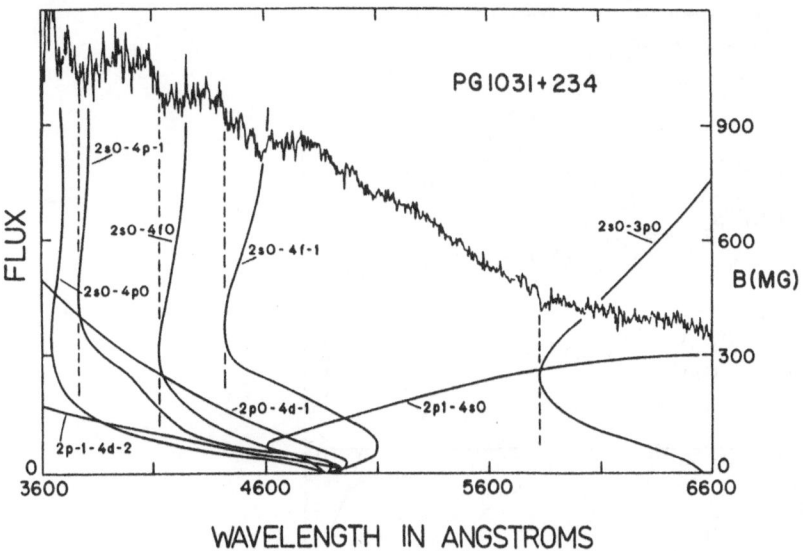

FLUX

PG 1031+234

2s0-4p-1

2s0-4f0

2s0-4f-1

2s0-4p0

2s0-3p0

2p0-4d-1   2p1-4s0

2p-1-4d-2

900

600

B(MG)

300

0

0
3600   4600   5600   6600

WAVELENGTH IN ANGSTROMS

Fig. 12. Observed spectrum of the high field magnetic white dwarf PG 1031+234 (Schmidt et al. [31]) at a given rotational phase, together with the computed behaviour of wavelengths as a function of field strength. A field strength ranging up to ∼ 500 MG is deduced from the data for this object.

4.3 PG 1015+014

The inspection of Fig. 6 for the wavelengths shows that for fields in the range ∼ 50–100 MG, apart from the occurrence of a few stationary components, the optical region is sprayed with slow and moderately fast moving components of Balmer α and β which can have an impact on the spectrum even in the presence of a field spread. PG 1015+014 is an example of a white dwarf with a field in this range.

Spectropolarimetric observations by Wickramasinghe and Cropper [39] revealed that its spectrum and broad-band circular polarization varies with a period of 98.75 minutes. The spectrum at two rotational phases is shown in Fig. 13 together with the suggested identifications. The synthetic spectra shown were obtained by Wickramasinghe and Cropper [39] for a centered dipole field distribution with a polar field strength of 100–120 MG viewed at angle $i = 75°$ to the dipole axis. Again the overall agreement is very good and confirms the presence of a magnetic field of this strength. More refined calculations of synthetic spectra may have to be extended to nonaxisymmetric field structures, for which there are evidences in the observed spectral variations during one period.

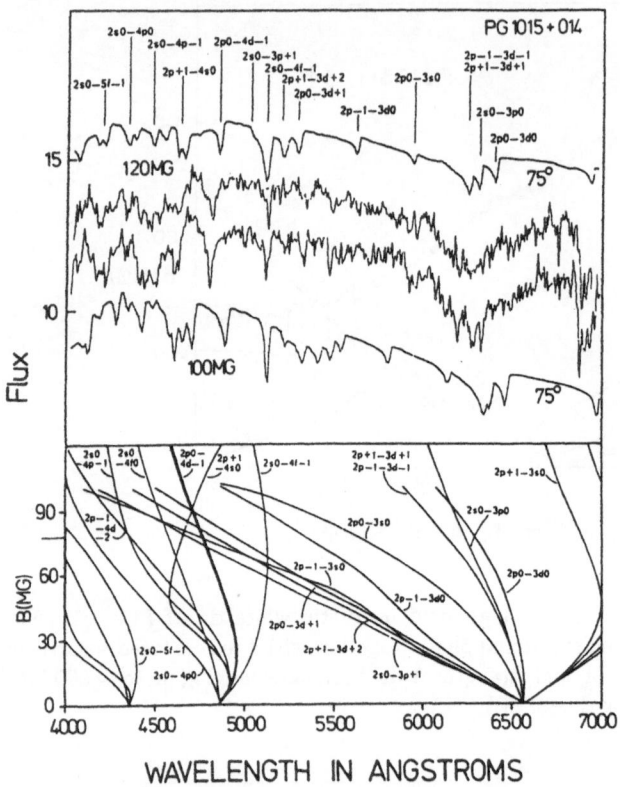

Fig. 13. Spectra of the intermediate field magnetic white dwarf PG 1015+014 at two different rotational phases (two curves in the middle) compared with synthetic spectra (top and bottom curve) using centered dipole models. The magnetic field dependence of wavelengths is shown in the bottom panel. The magnetic field of this object lies in the range 100–120 MG.

4.4 G227–35

This star is an example of a magnetic white dwarf in which the modulations of the flux spectrum are too weak to make possible an identification of stationary lines, but where in polarization spectra features can be detected that are due to stationary lines. Fig. 14 shows the total unpolarized flux F_λ and the normalized Stokes parameter of circular polarization, V, as functions of the wavelength, measured for G227–35 by Cohen et al. [40], in comparison with the behaviour of H_α and H_β components which have stationary points between 20 and 200 MG, or are changing slowly. The position of the prominent feature in the circular polarization near 7450 Å is in excellent agreement with the calculated stationary point of the $2p_{-1} \rightarrow 3d_{-2}$ component of H_α at 7449.4 Å, which occurs at 117 MG.

Further confirmation of this identification is seen, in the expanded view of the feature in Fig. 15, in the existence of an extended blue wing, which is expected because the stationary point is a maximum of wavelength. The wing extends

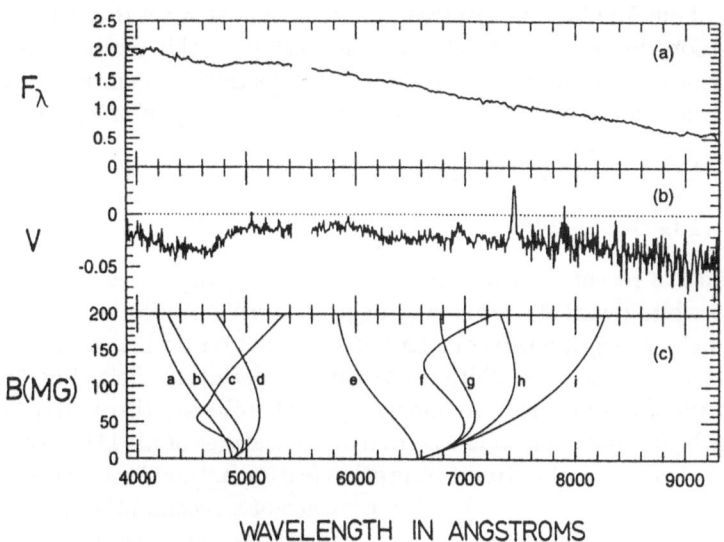

Fig. 14. *(a)* Flux spectrum F_λ for G227–35, *(b)* normalized Stokes parameter V of the circular polarization, and *(c)* wavelength vs. magnetic field in MG for some of the components of H_α and H_β ((a) $2s_0 \rightarrow 4f_0$, (b) $2p_0 \rightarrow 4d_{-1}$, (c) $2p_{+1} \rightarrow 4s_0$, (d) $2s_0 \rightarrow 4f_{-1}$, (e) $2s_0 \rightarrow 3p_0$, (f) $2p_{+1} \rightarrow 3s_0$, (g) $2s_0 \rightarrow 3p_{-1}$, (h) $2p_{-1} \rightarrow 3d_{-2}$, (i) $2p_0 \rightarrow 3d-1$). The feature in the circular polarization near 7450 Å is due to the transition h, which has its stationary point at 7449.4 Å and 117 MG. (From Cohen et al. [40])

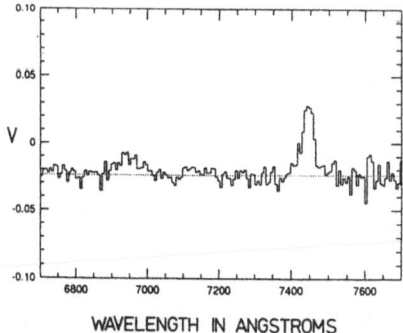

WAVELENGTH IN ANGSTROMS

Fig. 15. Expanded view of the circular polarization spectrum of G227–35 around the feature at 7450 Å. Note the blue-shaded wing of the feature, in accordance with the maximum character of the stationary point. The occurrence of this feature clearly points to the presence of a magnetic field around the stationary point of the transition ($B = 117$ MG). (From Cohen et al. [40])

to about 7400 Å, which corresponds to fields from 80 to 160 MG. A simple centered dipole model for the polarization spectrum leads to a value of the polar field strength of about 130 MG. By contrast, the total flux shows a real, though inconspicuous absorption feature at 7450 Å. It is clear from this example that polarimetry provides an additional powerful tool for detecting stationary atomic lines in strong fields, and thus for measuring the field strengths of magnetic degenerates.

4.5 LB 11146 (PG 0945+245)

Finally I mention the recent discovery (Liebert et al. [41]) of a binary system consisting of *two* white dwarf stars: one a strongly polarized object with an inferred magnetic field strength among the largest yet found on a white dwarf ($\geq$ 300 MG), the other a normal DA white dwarf with no detectable field. Fig. 16 shows the spectrum of the magnetic component (LB 11146b) together with calculated wavelengths for hydrogen. While certain of the features of LB 11146b are found to be due to hydrogen, the strong absorption feature around 5800 Å is not explicable by hydrogen and thus requires the presence of a second atmospheric component. The feature resembles the deep, red-shaded absorption features seen in GD 229 (whose spectrum is also shown in Fig. 16), for which helium in a strong magnetic field was proposed as an explanation. Further progress in the analysis of the spectra will require accurate data for wavelengths of neutral helium in magnetic fields of that strength.

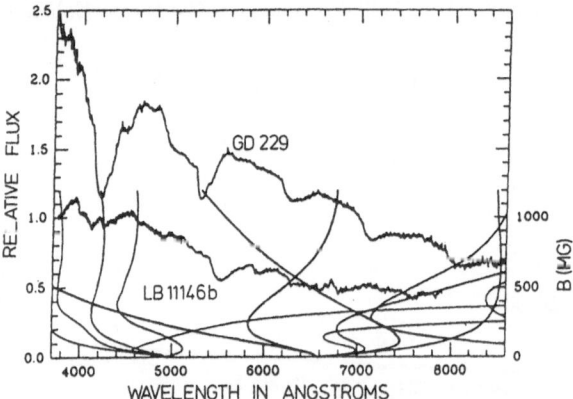

Fig. 16. Spectrum of the magnetic component LB 11146b of the double-degenerate binary system LB 11146 along with the magnetic-field dependence of the wavelengths of transitions in hydrogen in the range up to 1200 MG, compared with the highly polarized magnetic white dwarf GD 229. Predicted relative strengths of the Balmer transitions are indicated by line thickness. Some UV and optical features can be attributed to hydrogen in fields above 300 MG, while the broad feature around 5800 Å may be due to helium. Note the similarity of this feature and the broad features in GD 229, for which helium was proposed as an explanation. (From Liebert et al. [41])

5 Application to Semiconductors in Strong Magnetic Fields

The calculations of the behaviour of energies and wave functions of hydrogen atoms in arbitrary strong magnetic fields find their successful application not only in the interpretation of the spectra of white dwarf stars with strong cosmic magnetic fields but also in the analysis of the properties of doped semiconductors in strong laboratory fields, where the binding of an electron to a positively charged impurity (the donor) is analogous to the binding of an electron to the proton in the hydrogen atom. The characteristic energies and length scales, however, that come into play are altered with respect to those in the "real" hydrogen atom by the dielectric constant ε of the material and the effective mass m^* of the electron. More specifically, one has for the effective Rydberg energy E^*, the effective Bohr radius a^* and the reference magnetic field strength B^*

$$E^* = E_\infty \, \frac{m^*/m_e}{\varepsilon^2} \,, \tag{16a}$$

$$a^* = a_0 \, \varepsilon \, m_e/m^* \,, \tag{16b}$$

$$B^* = B_0 \left(\frac{m^*/m_e}{\varepsilon} \right)^2 = B_0 \left(\frac{a_0}{a^*} \right)^2 . \tag{16c}$$

For a substance like n-type GaAs, with $\varepsilon \sim 18.7$ and $m^*/m_e \sim 0.1$, one obtains the values

$$E^* = 3.9 \, \text{meV} \,, \quad a^* = 10 \, \text{nm} \,, \quad B^* = 13.2 \, \text{T} \,, \tag{17}$$

from which it is seen that a strong-field situation (with Lorentz and Coulomb forces of comparable magnitude), which occurred in white dwarf stars only at field strengths in the range $B \lesssim B_0 = 4.7 \times 10^4 \, \text{T}$, is realized in GaAs in strong laboratory magnetic fields $B \sim 10 \, \text{T}$!

Since the results computed for the hydrogen atom were obtained in dimensionless form (with energies in Rydbergs, lenghts in Bohr radii, and the magnetic field in terms of the applicable reference field) these results can directly be transferred to impurity–electron pairs in strong laboratory magnetic fields. The physics of low-lying states of the hydrogen atom in cosmic magnetic fields can therefore be tested, in these systems, in terrestrial laboratories. In fact it was in the exciton physics of the sixties that the behaviour of hydrogenic systems in strong magnetic fields received its first attention.

As an example I demonstrate that by studying the electrical conductivity of n^+–n^-–n^+ GaAs structures in which the thickness of the n^- layer is comparable with the mean donor separation the effect of a strong laboratory magnetic field on the donor wave function (equivalent to the ground-state wave function of the hydrogen atom in Megatesla fields) can be investigated (Roche et al. [42]). The experimental situation is schematically shown in Fig. 17: two regions of practically metallic doped GaAs which are laid on different electric potentials are separated by a very thin ($\ell \sim 100 - 200 \, \text{nm}$, i.e. $\ell \sim 10 - 20 \, a^*$) layer of n^- doped GaAs. At low temperatures, electrical conduction between the two

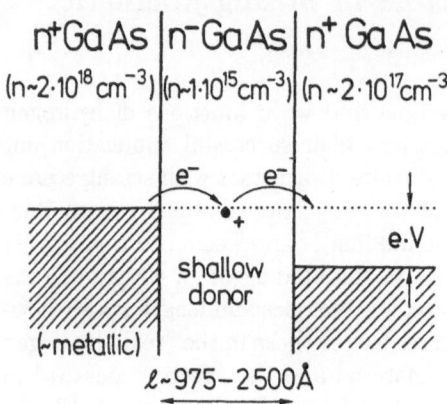

Fig. 17. Schematic representation of the experimental situation (Roche et al. [42]) in impurity-assisted tunnelling in n^+–n^-–n^+ GaAs multilayers. At low temperatures conduction between the n^+ doped regions occurs by tunnelling of electrons via shallow donor states in the very thin intermediate layer.

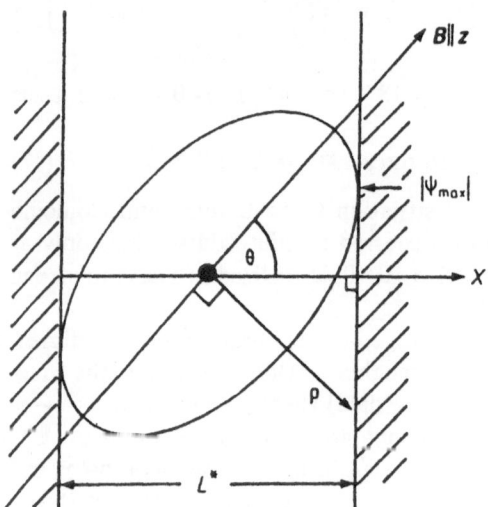

Fig. 18. A schematic illustration of the impurity-assisted tunnelling process in a magnetic field which is oriented at an angle θ with respect to the layer axis. The curve represents a single contour of the wave function amplitude for a donor situated at the centre of the layer; $|\psi_{\max}|$ is the maximum value of the donor wave function at the boundary. This value determines the conductivity, viz. $\sigma \propto 1/|\psi|^2$.

regions is dominated by a process in which electrons tunnel across the barrier via (bound) states of shallow donor impurities close to the centre of the n^- layer. Theory predicts that in this process of impurity-assisted tunnelling the conductivity is proportional to the probability of finding of the electron in the bound state at the boundary of the layer, i.e proportional to the maximum of the modulus squared of the donor wave function at the boundary (cf. Fig. 18).

Since the wave function in a strong magnetic field assumes a cigar-like shape, it follows that the conductivity is *large* when the magnetic field is directed across the barrier, while it is *small* when the field is directed parallel to the barrier. Thus the magnetoresistance directly reflects the anisotropy of the wave function in a strong magnetic field.

Figure 19 shows the ground-state (1s) wave function for magnetic field parameters $\beta = 0.4, 0.6$ and 0.9, corresponding to $B = 5.3, 7.9$ and 11.9 T for GaAs. The full curves represent surfaces of constant $|\psi|^2$ ($|\psi|^2 = 10^{-1}, 10^{-2}, 10^{-3}, \ldots$, down to 10^{-10} from the donor site outwards). As mentioned above, the length scale has units $a^* = 10$ nm for the donor states in GaAs. The diagrams illustrate the stronger and stronger deformation of the impurity wave functions, with the regions in the vicinity of the nucleus (donor site) being influenced to a much lesser extent than those farther out.

In Fig. 20 the experimental results of the measurement of the magnetoresistance $R(\theta)$ of such samples by Roche et al. [42] are compared with the theoretical results derived from the angular shape of the wave function. The theoretical curves show the dependence of $|\psi|^{-2}$ on θ for a fixed sample length. For each value of θ it is necessary to find the maximum value of $|\psi|^2$ at the edge of the active layer. Since the surfaces of constant $|\psi|^2$ are approximately ellipsoidal, the largest value of ψ occurs, except for $\theta = 0°$ and $90°$, at an angle of less than θ with respect to the field direction and a distance $r > L^*/2$ from the donor site (Fig. 18). The agreement between the experimental and the theoretical curves for the anisotropy of the magnetoresistance evident from Fig. 20 is remarkable, in particular, when one realizes that the variation of the resistance as a function of θ extends over two orders magnitude. The excellent agreement proves the correctness of the underlying physical concepts as well as the accuracy of the numerical calculations of the wave functions even at large distances from the nucleus, where the wave function has decayed by many orders of magnitude.

It must be emphasized that this experiment represents one of the rare occasions where the quantum mechanical *wave functions*, as opposed to energy values, are probed by an actual experiment.

As a further example of the application to semiconductor physics in magnetic fields I mention the interpretation of percolative transport in GaAs at 10 T magnetic fields via hydrogen wave functions at Megatesla fields (Buczko et al. [43]). Here one considers an extended, "dilute" semiconductor, with many active impurities separated by more than $\sim 5\,a^*$. Electrical conduction occurs through nearest-neighbour hopping in the network of impurities, with the hopping probability being determined by the donor wave function. It is also in this case that one finds excellent agreement between the measured magnetoresistance of such samples and the theoretical results obtained with the help of hydrogen wave functions originally computed for strong cosmic fields.

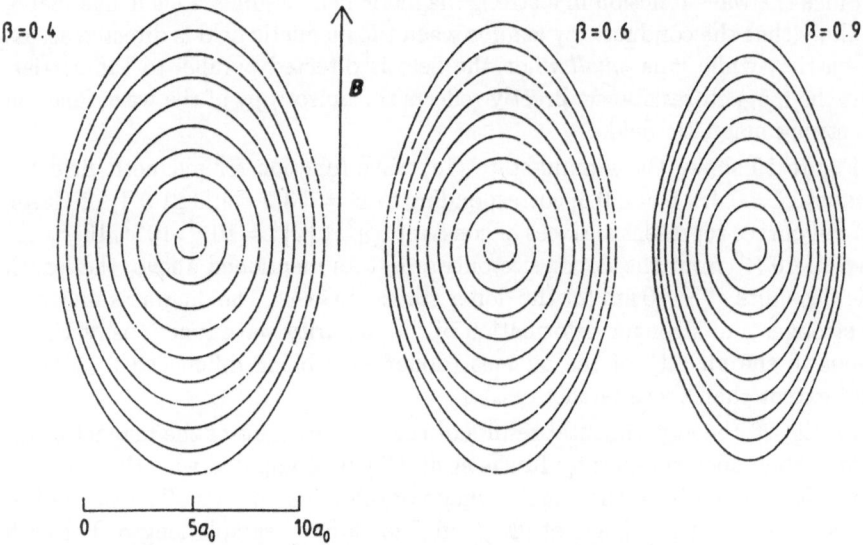

Fig. 19. Contours of constant $|\psi|^2$ of the ground-state hydrogen wave function for different values of the magnetic field parameter β. For GaAs, the corresponding field strengths are $B = 5.3, 7.9$ and 11.9 T.

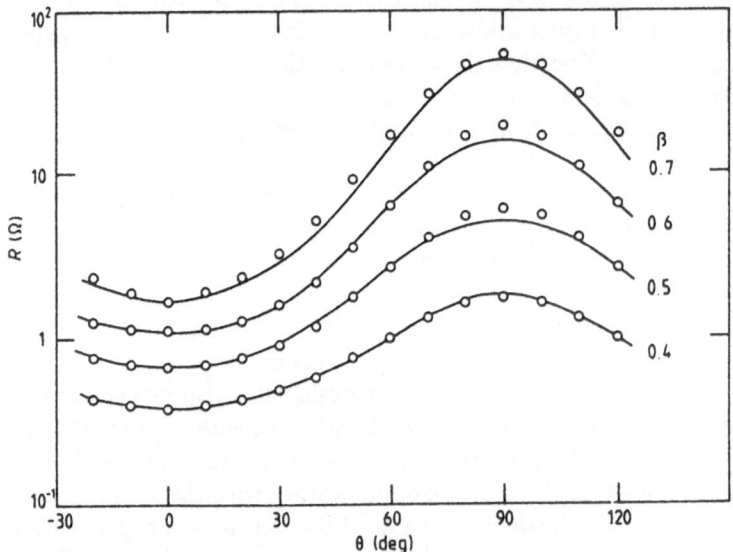

Fig. 20. The resistance at various magnetic fields of a barrier of width $L^* = 97.5$ nm as a function of the angle θ between the current and magnetic field directions. Circles: experimental values, full curves: theoretical predictions using the numerical wave functions. Note that the anisotropy of the wave functions is quantitatively reflected in the anisotropy of the measured magnetoresistance. (From Roche et al. [42])

6 Highly Excited States

In the previous sections I have discussed in much detail the influence of strong magnetic fields on *low-lying* states. The crucial point in the whole discussion was that the systems were allowed to be exposed to magnetic fields of such strengths as to make the effects of the magnetic field of the same order of magnitude as, or even larger than, the Coulomb binding forces acting in low-lying states in the atom. We had already noted in Sect. 3.1 that the reference magnetic field, which is obtained by setting electrons on Bohr orbits and requiring the equality of Lorentz and Coulomb forces, scales with the inverse cube of the principal quantum number n_p

$$B_{n_p} = \frac{B_0}{2n_p^3} \approx \frac{4.70 \times 10^5 \, \text{T}}{2n_p^3} \approx 8.3 \left(\frac{30}{n_p}\right)^3 \text{T} \quad , \tag{18}$$

so that the strong-field situation, which is encountered for low-lying states of the "real" hydrogen atom only in the field strengths of white dwarfs or neutron stars, can be realized for highly excited states ($n_p \geq 30$, e.g.) also in terrestrial laboratory field strengths of several Teslas. The discussion of the influence of strong laboratory magnetic fields on *highly excited states* of the hydrogen atom is the subject of this section.

Although uncovering the behaviour of highly excited states in strong magnetic fields is an interesting problem of physics in its own right, these "Rydberg" states have gained additional importance in recent years in connection with the fundamental question of whether or not there is *chaos* in quantum mechanics. It turns out that the hydrogen atom in a strong magnetic field is a prototype of a nonintegrable system with two degrees of freedom which classically undergoes a transition to chaos so that all the topics that are causing so much excitement in investigations of the relation between classical chaos and quantum mechanical behaviour in nonlinear systems can be studied in this real and simple system both theoretically and experimentally. Therefore a large part of this section will also be devoted to the discussion of the importance of highly excited states of the hydrogen atom in magnetic fields to the topical research area of "quantum chaos".

6.1 Results

6.1.1 Energy Levels. I begin by presenting results for the behaviour of the energies of highly excited states as a function of the magnetic field strength. As we have to deal with hundreds of levels in their magnetic field dependence in this part of the bound spectrum, the results are best presented in graphical form. Because of the large number of levels one wants to describe the diagonalization of the Hamiltonian in large complete basis sets is the method of choice for highly excited states.

As a representative example Fig. 21 shows the behaviour of the energy levels of the hydrogen atom originating from multiplets with principal quantum numbers between 38 and 46 in the magnetic field range 2 to 7 T.

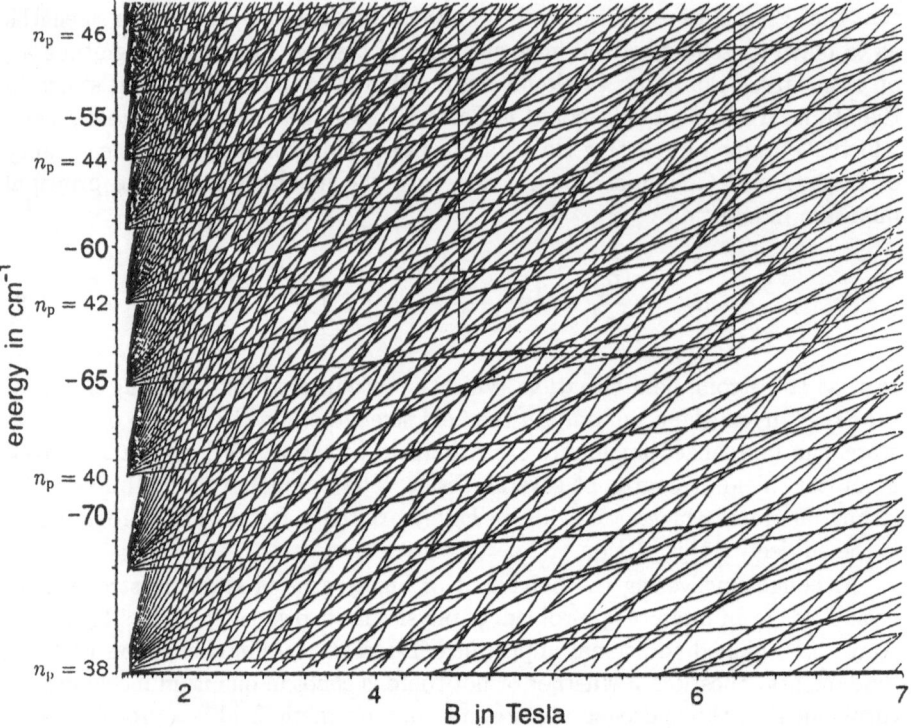

Fig. 21. Level scheme of the hydrogen atom in the energy range $-76\,\mathrm{cm}^{-1}$ to $-50\,\mathrm{cm}^{-1}$ (corresponding to principal quantum numbers $38 \leq n_\mathrm{p} \leq 46$ in the field-free case) as a function of the (square of the) magnetic field strength. It is recognized that the diamagnetic multiplets belonging to different principal quantum numbers are completely interwoven at high field strengths. (By courtesy of D. Wintgen)

It is evident that the diamagnetic interaction causes a mixing of the different multiplets already below 2 T, from where on n_p is no longer a good quantum number (n_p-mixing regime). Nevertheless it is possible to trace the individual levels over a wide range of the field, an indication of the fact that in spite of the competition of the Coulomb and Lorentz forces in the atom the problem is still quasi-integrable (existence of an additional approximate constant of motion, cf. Solov'ev [44]).

The level crossings in Fig. 21 are always avoided crossings, but it is only beyond the breakdown of quasi-integrability that the anti-crossings become sufficiently large to be visible in this figure (see inset). It can also be recognized that in every diamagnetic multiplet the lowest state possesses the minimum rate of variation of its energy as a function of the field: it is the state whose wave function has maximum probability of presence along the magnetic field axis; conversely, the state with the maximum variation of energy with the field is the one with the maximum probability of presence of the electron in the plane perpendicular to the direction of the magnetic field. It is obvious that this distribution in configuration space will react most readily to the diamagnetic term in the Hamiltonian. The general impression conveyed by Fig. 21 is that in a strong magnetic field there is a complete interweaving of levels originating from different principal quantum numbers with the result that the density of states is high and the mean level spacing is small compared to the field-free case.

In spite of the complicated "spaghetti"-like structure of the energy level diagram of hydrogen in a magnetic field, it has been possible to experimentally verify every individual level predicted by the theoretical calculations at a sample magnetic field strength of 6 Tesla. Experiments with highly excited states of hydrogen and deuterium were performed by the Bielefeld group (Holle et al. [45]) in 4 – 6 Tesla fields. The Rydberg states were excited by resonant two-photon absorption: in a first step a (frequency-doubled) Lyman excitation was performed from the ground state to the Paschen-Back resolved $2p$ substates, from where a tunable dye-laser took the electron up to close below or above the ionization threshold. Fig. 22 provides a comparison between the experimental and theoretical spectra in the range -80 cm^{-1} to -20 cm^{-1} at 5.96 T. The agreement between theory and experiment is excellent; moreover, theory reveals where neighbouring lines were no longer resolved in this experiment at high energies. All in all more than a thousand lines were successfully identified in this experiment by the theoretical calculations!

While these comparisons focussed on the bound-state part of the spectrum, progress in the numerical methods has made it possible to compute transition rates also for continuum states, and thus extend the comparison into the positive energy region. Fig. 23 shows, for 6 T and the energy range -30 cm^{-1} to $+30$ cm^{-1}, the comparison between the experimental spectrum obtained by the MIT group (Iu et al. [48]) for odd-parity states of lithium (which are almost hydrogen-like because of the small quantum defects of odd parity states) and the spectrum computed by Delande et al. [49]. They diagonalized the complex-rotated Hamiltonian in large basis sets to calculate photoionization cross sections in the positive energy region. Again the agreement between theory and experiment is excellent. Fig. 23 again testifies to the very complicated, disorderly looking structure and high level density of the level scheme of hydrogen atoms in strong magnetic fields.

It is of course extremely gratifying to have such an excellent agreement between the results of theory and experiment – after all it proves that quantum theory "works" even in this highly nonintegrable regime. The more profound importance, however, of these theoretical and experimental investigations of the Rydberg states of the hydrogen atom in strong laboratory magnetic fields lies in the fact that one has the rare opportunity to look at a real quantum system in a range of parameters where the corresponding classical system exhibits *chaotic* behaviour. For this reason I will now outline the importance of the hydrogen

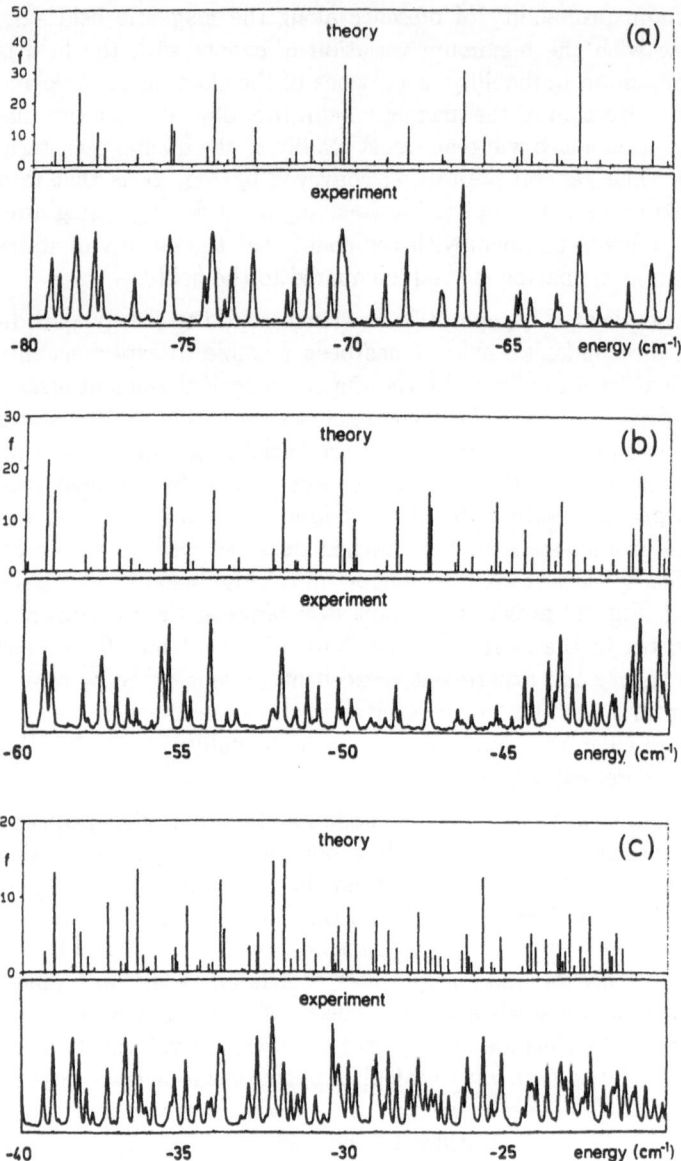

Fig. 22. Deuterium Rydberg atoms in a magnetic field of 5.96 T: comparison between the theoretical oscillator strength spectrum and the experimental photoabsorption spectrum for $\Delta m = 0$ Balmer transitions to $m = 0$, even-parity Rydberg states between $-80\,\mathrm{cm}^{-1}$ and $0\,\mathrm{cm}^{-1}$, the field-free ionization threshold. Note that at the end of the energy range (above ~ -25 cm^{-1}) the corresponding classical system becomes completely chaotic. 639 lines are shown in this spectrum. Oscillator strengths are given in units of 10^{-6}, the experimental intensity scale is in arbitrary units. (From Holle et al. [46] and Zeller [47])

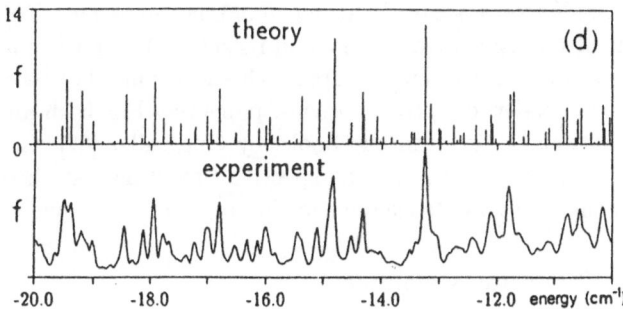

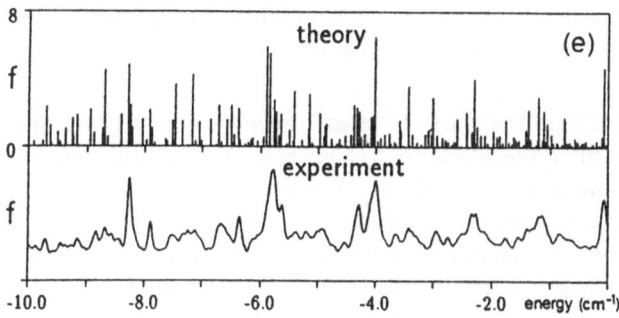

Fig. 22. continued

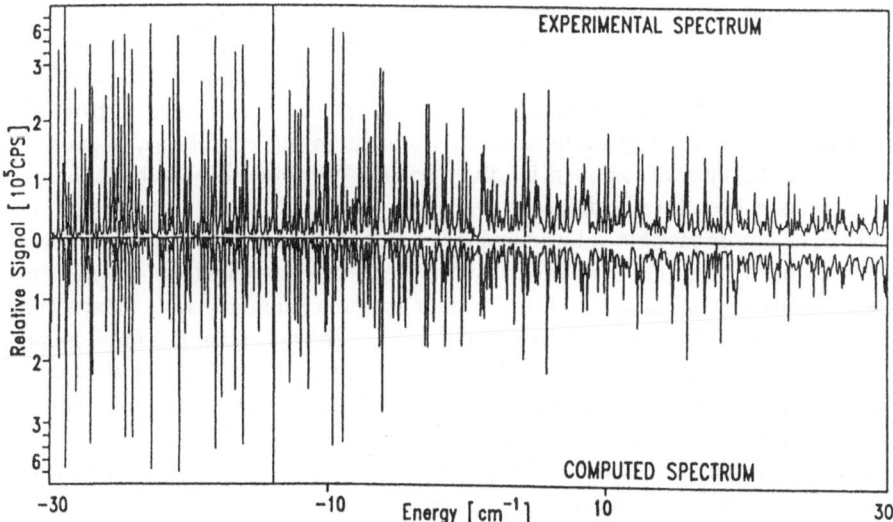

Fig. 23. Comparison of the experimental (Iu et al. [48]) and the theoretical (Delande et al. [49]) spectrum for dipole transitions from the $3s$ state of lithium to odd-parity Rydberg states in a magnetic field of 6 T in the energy range $-30\,\mathrm{cm}^{-1}$ to $+30\,\mathrm{cm}^{-1}$.

atom in strong magnetic fields to studies of the problem of *quantum chaos*. This topic has been the object of intensive investigations in recent years, and I must restrict myself to sketching the essential lines of approach and results. For more detailed accounts I refer the reader, e.g., to the review papers by Friedrich and Wintgen [50] and Hasegawa et al. [51] and the books by Gutzwiller [52], Gay [53] and Haake [54]. To put these studies into perspective I shall discuss them in the broader context of the question "Is there chaos in quantum mechanics?".

6.2 Is There Chaos in Quantum Mechanics?

Can we tell from the properties of a quantum system in the limit of large quantum numbers that the corresponding classical system behaves chaotically? Recent theoretical and experimental studies of hydrogen atoms – the "showpiece" of quantum mechanics – in intensive microwave fields and in strong magnetic fields have produced results that could help answer this question.

6.2.1 Introduction. Studies on the subject "chaos" have become very popular in recent years. One of the main reasons is that chaos is a universal phenomenon that turns up in the problems of the different branches of natural science – biology, chemistry, technical engineering, and physics – and therefore is able to bridge the gap between these branches. As a technical term, chaos describes a behaviour of deterministic systems that are *extremely sensitive to the choice of initial conditions*. *Deterministic* means that the temporal evolution of the system is described by differential equations, in a way that the past and the future are uniquely determined by the initial conditions.

The cause of the extreme sensitivity of the solutions to the choice of initial conditions is a characteristic, local instability of the motion, namely the *exponential* divergence of initially closely neighbouring trajectories. One measure of the "speed" of divergence is the (positive) Liapunov exponent of the trajectory (Fig. 24a). The "chaotic" or statistical aspect can be seen from the fact that, even with only infinitesimally different initial conditions, because of the immediate onset of exponential divergence of the solutions, the behaviour of the systems can be totally different over long periods of time, so that long-term predictions are impossible. The equations of motion that lead to chaos are necessarily nonlinear, and the systems can be either time-dependent, driven systems or conservative, nonintegrable Hamiltonian systems (*nonintegrable* means that no canonical transformation can be found so that the new Hamiltonian depends only on the generalized momenta, i.e., on action variables). Fig. 24 shows the stadium billiard and the Sinai billiard as examples of simple chaotic systems in comparison to the "regular" rectangular billiard.

Chaos is a classical concept. We know, however, that as soon as we study phenomena on the molecular, atomic or nuclear level, classical mechanics must be replaced by quantum mechanics. Because, according to Bohr's correspondence principle, classical physics is contained within the quantum theory in the limit of large quantum numbers, the following question arises: in what way is classical chaos reflected in the characteristics of corresponding quantum systems?

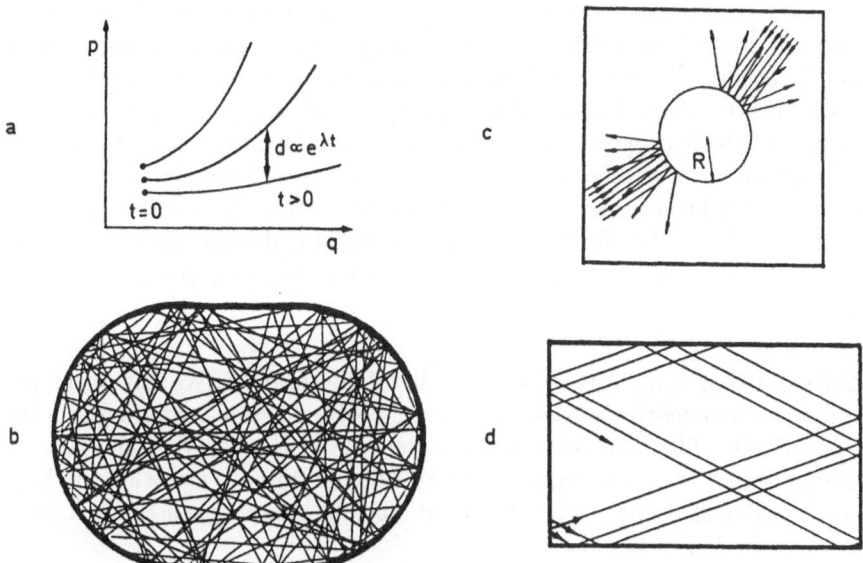

Fig. 24. Chaos and order in classical model systems. (a) Classical chaos: orbits with only slightly different initial conditions diverge exponentially with time. The Liapunov exponent λ of a trajectory in phase space characterizes the average "speed" with which trajectories in the neighbourhood of the given trajectory exponentially separate. The size of the Liapunov exponent determines the degree of instability of an orbit. The mathematically strict definition and calculation of λ is realized via the stability matrix of the trajectory. (In regular systems, the divergence is linear, and the calculation of λ always gives the value zero.) (b) The stadium billiard as an example of a chaotic system with two degrees of freedom. The reflection on the semicircles causes a divergence of neighbouring groups of trajectories. (c) In the "Sinai billiard", the reflection on the disc produces defocusing, and with it divergence, of the trajectories. (d) The rectangular billiard as a standard example of a system with regular orbits only.

To date, no satisfactory definition of "quantum chaos" has been found (cf. Haake [54]), and in this section I will not attempt to find one. One of the main difficulties in finding such a definition lies in the fact that the Schrödinger equation, which determines the temporal evolution of all observables, is linear in the wave function, and therefore the nonlinearity of the classical systems can come into being only in the limiting process $\hbar \to 0$. The current research standpoint is pragmatic: one is looking for *new semiclassical*, but nonclassical, *phenomena* characteristic *of quantum systems* whose classical counterparts exhibit chaos. Berry [55], [56] coined the term "quantum chaology" for this field of study ("chaology" being a theological term borrowed from the 19th century: it describes the study of the condition of the universe at a time in which it was still "chaotic" or "without form and void").

It is obvious that the (numerical) solution to the Schrödinger equation of nonintegrable systems in the range of large quantum numbers is a very complicated problem that can be dealt with only in a few systems. As I have already pointed out, in addition to the chaotic model systems (billiards, nonlinear oscillators), in atomic physics "real" physical quantum systems exist in which one can look for symptoms of chaos, theoretically as well as experimentally. By way of example of the hydrogen atom in a uniform magnetic field – as a prototype of a nonintegrable *conservative* Hamiltonian system – I will explain several of the phenomena that have turned up as candidates for typical behaviour of quantum systems with classically chaotic counterparts. (The hydrogen atom in a strong microwave field as an example of a *driven* chaotic system is reviewed, e.g., in [12].)

6.2.2 Statistical Analysis of Energy-level Sequences. We first focus our attention on the *energy eigenvalues of systems* whose classical counterparts behave chaotically. The motivation for this stems from the fact that energies are associated with stationary states, and a stationary state, in quantum mechanics, is something that goes on forever. Thus a connection to (persisting) classical chaos might be hoped for.

The remarkable result is that the distribution of energy eigenvalues in the semiclassical range points to *universality*. In order to produce comparable conditions, however, the spectra must be transformed in a way that the average level spacing becomes the same. This is accomplished as follows: the average level density $\rho(E)$ is determined, the average number of levels up to energy E is calculated from it,

$$\bar{N}(E) = \int_{-\infty}^{E} \rho(E)\, dE \quad , \tag{19}$$

and then one considers the spectrum $x_j = \bar{N}(E_j)$, which, by construction, has a constant average level spacing that is chosen as the energy unit. The cumulative level density $n(x)$ (number of levels up to x) then is a staircase function that fluctuates around a straight line with slope 1.

Universality is found in the *statistics* of the level sequences. One such statistic, a short range one, is the probability distribution $P(S)$ of neighbouring energy levels, i.e., the distribution of $S_j = x_{j+1} - x_j$. Fig. 25a shows the distribution $P(S)$ (Bohigas et al. [57], [58]) calculated from several hundred levels of the stadium billiard and, above it, the distribution for the Sinai billiard.

The distributions are obviously the same. The solid curve that so exactly approximates the data is the Wigner function

$$P_W(S) = \frac{\pi}{2} S e^{-\pi/4S^2} \quad . \tag{20}$$

The probability of finding two levels in the same location, therefore, tends to zero linearly with S. Another type of statistics, a long-range one, is the spectral rigidity $\Delta_3(L)$. By this one means the mean-square deviation of the staircase function $n(x)$ from the straight line best approximating it,

$$\Delta_3(L) = \left\langle \frac{1}{L} \min_{A,B} \int_{-\frac{L}{2}}^{+\frac{L}{2}} [n(x) - Ax - B]^2\, dx \right\rangle \quad . \tag{21}$$

Fig. 25b shows the rigidity for both these chaotic systems; they are equal and suggest universality.

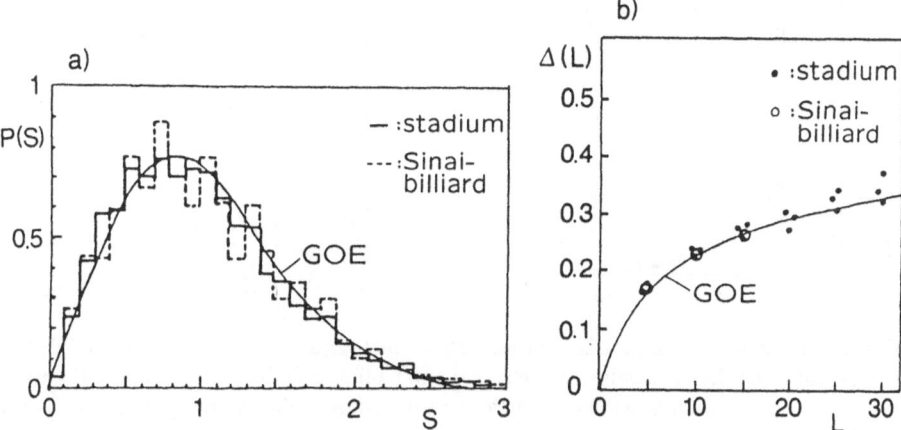

Fig. 25. (a) Histograms $P(S)$ of the distribution of level distances of the quantum energies of the stadium and Sinai billiard (Bohigas et al. [57], [58]). The solid curve is the result for the distribution of nearest-neighbour spacings of symmetric random matrices (Gaussian Orthogonal Ensemble, GOE) and is approximated very well by the Wigner function (20). (b) Spectral rigidity $\Delta_3(L)$ for the eigenvalue sequences of the stadium and Sinai billiard in comparison to the spectral rigidity of the eigenvalue sequences of the real symmetric random matrices (asymptotically $\Delta_3(L) \to (1/\pi^2) \ln L + \text{const for } L \to \infty$).

The solid curve in Fig. 25b, as well as the Wigner function (20), is borrowed from the eigenvalue statistics of infinitely large, real symmetric matrices, whose elements are random numbers (GOE: Gaussian Orthogonal Ensemble). The theory of random matrices was developed in the 1960s to simulate by means of a model the complicated many-body Hamiltonians of atomic nuclei. It is amazing that the same results can also describe exactly the quantum mechanical energy level sequences of classically chaotic systems with few degrees of freedom.

For systems whose classical motion is not chaotic, the corresponding level statistics are much different. Fig. 26 shows the results for the rectangular billiard (Berry [55]).

Here, the statistical results are equivalent to those found in random *numbers*, i.e., of the Poisson type: the distribution of adjacent energy levels reaches its maximum at a level distance of zero, and $\Delta_3(L)$ grows linearly.

A comparison of Figs. 25 and 26 suggests that level repulsion and as a result of this, a maximum probable level distance different from zero is characteristic of quantum systems if the corresponding classical system behaves chaotically; level accumulation, on the other hand, proves to be typical of quantum systems with

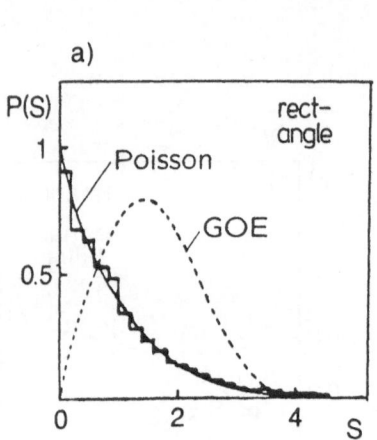

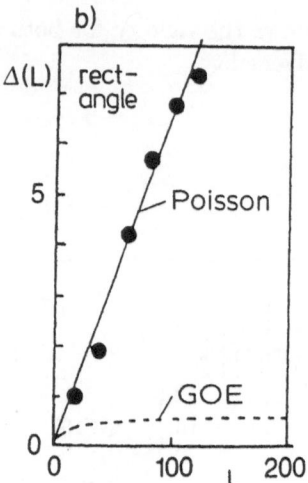

Fig. 26. Nearest-neighbour distributions $P(S)$ (histogram) and spectral rigidity $\Delta_3(L)$ (black dots) for the eigenvalue of the rectangular billiard (Berry [55]). The solid curves depict the corresponding statistics for Poisson-distributed eigenvalues ($P(S) = e^{-S}$ and $\Delta_3(L) = L/15$), the dotted curves show those for real symmetric random matrices.

classically regular behaviour. The concepts developed by aid of these model system have been excellently confirmed through recent theoretical and experimental studies of the hydrogen atom in magnetic fields.

6.2.3 Order and Chaos in the Hydrogen Atom in a Magnetic Field.
In order to find chaos in a hydrogen atom, we must, of course, go beyond the usual perturbation-theoretical Zeeman range to magnetic fields in which the effects of the magnetic fields are comparable to those of the Coulomb field. An estimate of the necessary sizes of the fields in dependence on the principal quantum number was given in (18). At such field strengths, the two different symmetries of the problem – the spherical symmetry of the Coulomb field and the cylindrical symmetry of the magnetic field – prevent even an *approximate* separation of the variables (only the z-component of the angular momentum is conserved), and we have to deal with the prototype of a nonintegrable system with two degrees of freedom.

The appearance of chaotic trajectories in the hydrogen atom in a magnetic field can be illustrated with the method of Poincaré surfaces of section. The equations of motion that result from the diamagnetic Hamiltonian (14) are solved numerically, plane sections are laid in the three-dimensional subspace of four-dimensional phase space defined by the given value of the conserved energy (energy shell), and the points of intersection of the orbits with the plane are marked in that plane. In Fig. 27, Poincaré surfaces of section calculated for a magnetic field of 6 T show that, with decreasing binding energy of the electron, the regions filled with regular orbits gradually break up, until no more regular trajectories are recognizable.

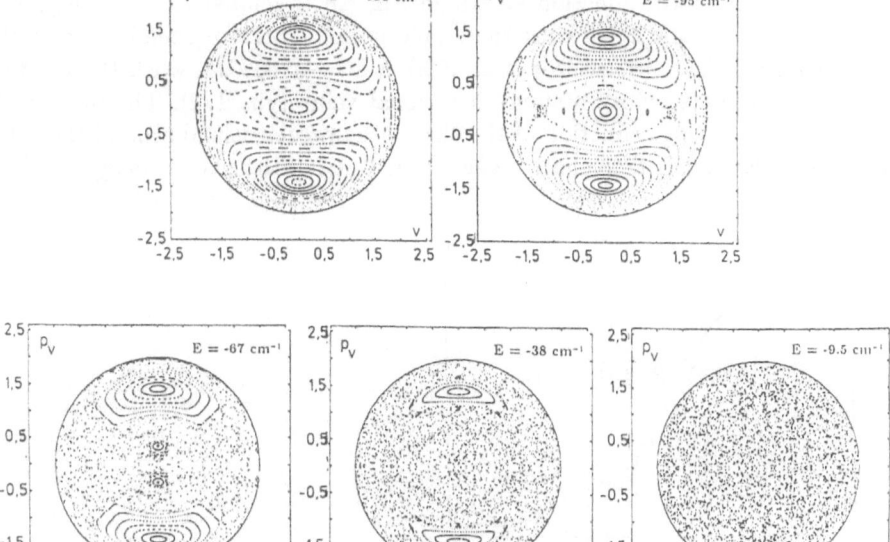

Fig. 27. Poincaré surfaces of section of classical orbits in the hydrogen atom in a magnetic field of 6 T and at different energies. To avoid the singularity at the origin caused by the Coulomb potential, the equations of motion were solved in semiparabolic coordinates u, v ($u^2 = r + z$, $v^2 = r - z$) and the v-p_v-plane ($u = 0$) was chosen as the section. The collapse of regularity with decreasing binding energy is clearly visible. (The Rydberg constant is $R_\infty = 109737$cm^{-1}; i.e. the energy values lie in the range of highly excited states.)

To give a quantitative measure of the increasing collapse of regularity, in Fig. 28 the area fraction of ranges filled with regular orbits is shown as a function of energy for a magnetic field of 6 T. Fig. 28 is actually universal and not limited to this magnetic field strength. This is the result of a remarkable scaling property of the Hamiltonian (14): through the substitutions of the spatial, e.g. cylindrical, coordinates, $(\rho, z) \to \beta^{2/3}(\rho, z)$, $(p_\rho, p_z) \to \beta^{-1/3}(p_\rho, p_z)$ it assumes a form that is no longer explicitly dependent on the magnetic field, but only on the *scaled* energy $\varepsilon = E/(2\beta)^{2/3}$. The intuitive meaning of this transformation is that the *relative* strengths of the Coulomb attraction and the Lorentz forces are held constant at a given scaled energy. This means that the classical dynamics is "frozen" at given ε, apart from the similarity transformation given above. The upper abscissa in Fig. 28 gives the scale for ε. It can be read off Fig. 28 that in the diamagnetic Kepler problem the breakdown of regularity begins around $\varepsilon = -1$, and the last islands of stability visible in the Poincaré surfaces of section disappear around $\varepsilon = -0.25$. The latter turns out to be the case when the periodic orbit perpendicular to the direction of the magnetic field finally becomes unstable (the exact value of ε being -0.2545372). The strong dip in the behaviour of the function shown in Fig. 28 is related to the merging of an unstable periodic orbit with the stable periodic orbit perpendicular to the field, which appears (Schweizer et al. [59]) at $\varepsilon = -0.632372$, when the winding number of the trajectory assumes a rational value (viz. 1/2). The merging is reflected in a strong reduction of the size of the regular island around the orbit perpendicular to the field shortly below this value of the scaled energy.

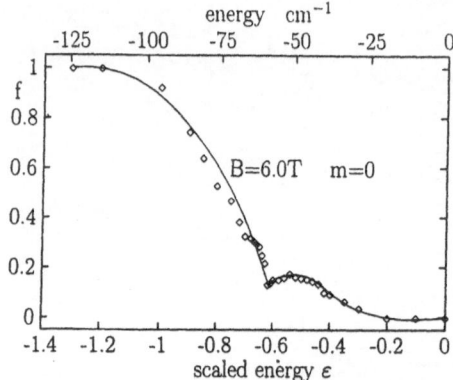

Fig. 28. Area fraction of regular regions in the Poincaré surfaces of section as a function of energy for a magnetic field of 6 T (lower abscissa) and on the scaled energy $\varepsilon = (E/R_\infty)/(B/2.35 \times 10^5 \text{ T})^{2/3}$. The dip in the otherwise more or less smooth behaviour of the function is caused by the confluence of an unstable periodic orbit with the stable periodic orbit perpendicular to the field at a value of $\varepsilon = -0.632372$. (From Schweizer et al. [59])

6.2.4 Level Statistics for the Hydrogen Atom in Magnetic Fields. In Figs. 22,23 I had shown examples of theoretical and experimental quantum spectra of hydrogenic atoms in strong laboratory fields. The figures conveyed a *qualitative* impression of the growing complexity of the quantum spectrum of even the simplest systems, when we push forward to field strengths in which classical movement becomes more and more chaotic.

Inspired by corresponding results in model systems, the energy spectra were subjected to a statistical analysis in order to reach a *quantitative* measure of this complexity. Fig. 29 shows histograms of the distribution of neighbouring energy levels of the hydrogen atom in magnetic fields as a function of the scaled energy (Wintgen and Friedrich [60]). Here, too, the rearrangement from a Poisson-like to a Wigner-like distribution when the classical system changes from regularity to chaos was confirmed. This rearrangement was thus demonstrated for the first time in a *real* physical system. The Δ_3-statistics of the energy levels (Wunner et al. [61]) also demonstrated the expected behaviour as based on the model systems (see Fig. 30).

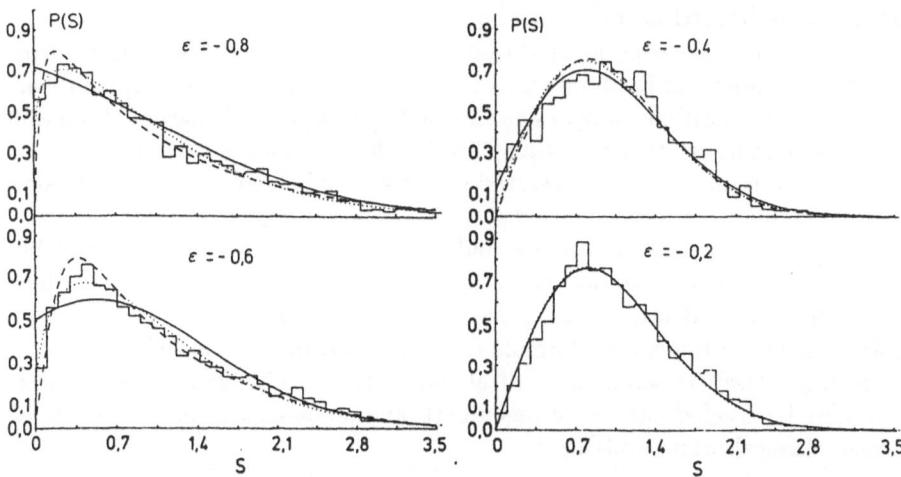

Fig. 29. Histograms $P(S)$ of the level distances of quantum energies in the hydrogen atom in magnetic fields at different values of the scaled energy (Wintgen and Friedrich [60]). The smooth curves are the results of fits to the histograms using various formulas derived in literature that interpolate between the Poisson and the Wigner distributions in the transition range to chaos.

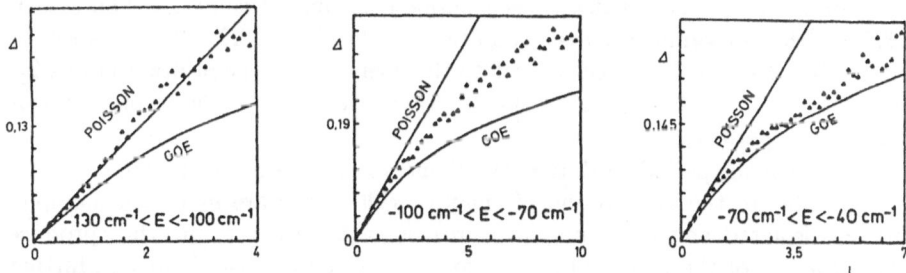

Fig. 30. Spectral rigidity $\Delta_3(L)$ for energy level sequences of the hydrogen atom in a magnetic field of 6 T in three different energy intervals. The swing to the GOE-distribution (cf. Fig. 25b) as soon as the classical motion becomes chaotic is also visible here.

6.2.5 Resonances in Chaos – the Rôle of Periodic Orbits. The Fourier transformation of spectra, such as those of Figs. 22,23, yields a surprising result: one finds distinct peaks, an indication that different "wavelengths" ΔE with periodicities not visible to the human eye are (apparently) present in irregular spectra. More surprising, however, is the fact that these modulations of orbital frequencies can be attributed to *classically periodic*, mostly *unstable* orbits, unstable meaning here that the Liapunov exponent is different from zero. These orbits are embedded in the sea of chaotic orbits.

Fig. 31 shows, as experimental results, the Fourier-transformed spectra from Balmer transitions in Rydberg series taken at a constant scaled energy, in comparison to the positions of peaks as they follow from the calculation of relevant classical periodic orbits (Holle et al. [62]). One finds an almost perfect correlation between the positions of the observed structures and the results of the classical calculations.

How is this correlation to be understood? In the attempt to generalize the semiclassical quantization conditions valid only for stable periodic orbits, Gutzwiller [63] and others, using the Feynman path integral formalism, were able to derive an asymptotic formula that allows expressing the spectral density of quantum levels in terms of classical, closed orbits. The level density can accordingly be divided into an averaged part, and an oscillatory part to which *all* closed, classical orbits contribute:

$$\rho(E) \;=\; \bar{\rho}(E) \;+\; Re\frac{1}{\pi\hbar}\sum_{p}\sum_{r=1}^{\infty}\frac{1}{|\det(M_p^r-1)|^{1/2}}\,e^{i(S_p/\hbar-\mu_p\pi/2)r}\quad. \qquad (22)$$

The sum over p runs over all primitive periodic orbits, the sum over r over all repetitions, S_p is the action and μ_p a corresponding phase shift (Maslov index) of the primitive orbit p. The matrix M_p is the stability matrix of the primitive orbit p, and depends on the the Liapunov exponents. Between a quantum level and a closed orbit, there is therefore generally no longer a one-to-one correspondence; instead, each closed orbit describes a *periodic accumulation* of levels on a scale ΔE, which is determined by the condition $\Delta E/\hbar\, d(rS_p)/dE = 2\pi$. Because the derivative of the action with respect to the energy gives the orbital periods T_p, one has $\Delta E = h/(rT_p)$, and the modulations are given by the orbital periods and their repetitions.

Although the periodic orbits in the Poincaré surfaces of section are no longer recognizable in the background of chaotic orbits (they are infinite in number, but are isolated and have measure zero), it is just these orbits that produce the *structure* of the quantum spectrum in the chaotic range. This is a further phenomenon that is typical of the behaviour of quantum systems with a chaotic counterpart: the quantum support of "regularity" in the chaotic range.

6.2.6 "Scarring" of Wave Functions.
What do wave functions in the chaotic range look like? Because the orbit of the electron in this range comes arbitrarily close to every phase-space point in the course of time, it was very early on suspected that "chaotic" wave functions essentially are delocalized. But Heller [64] discovered that, in the quantum mechanical treatment of the stadium problem, wave functions along classical, unstable closed orbits are "scarred" (cf. Fig. 32), i.e., show an increased probability of presence: the smaller the "discrepancy per period" λT of the classical trajectory, the more pronounced is this behaviour.

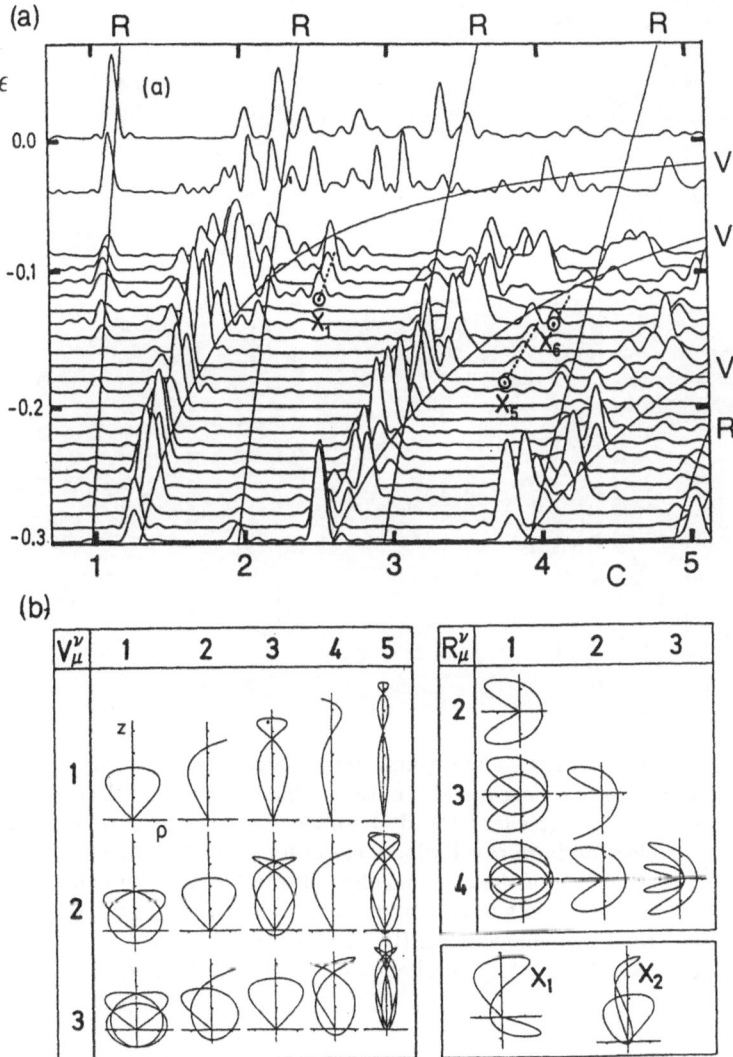

Fig. 31. Resonances in the quantum spectrum of the "chaotic" hydrogen atom ($m = 0$, even parity) (Holle et al. [62]). (a) Fourier-transformed photo-absorption spectra at different scaled energies as functions of the scaled action C ($C = n(2\beta)^{-2/3}$). Each resonance can be attributed to a closed classical orbit (designated, for historical reasons, "rotational" (R), "vibrational" (V) or "exotic" (X) by the authors). The solid curves represent the expected location of the resonances according to classical calculations. Further "vibrational" curves fan out to the left from each particular base curve and coincide with the other observed resonances. The figure shows the quantum support of regularity in the chaotic range. (b) Several closed, mainly unstable ($\lambda \neq 0$), periodic orbits (in cylindrical coordinates ρ, z) of the hydrogen atom in the magnetic field that lead to the observed resonances.

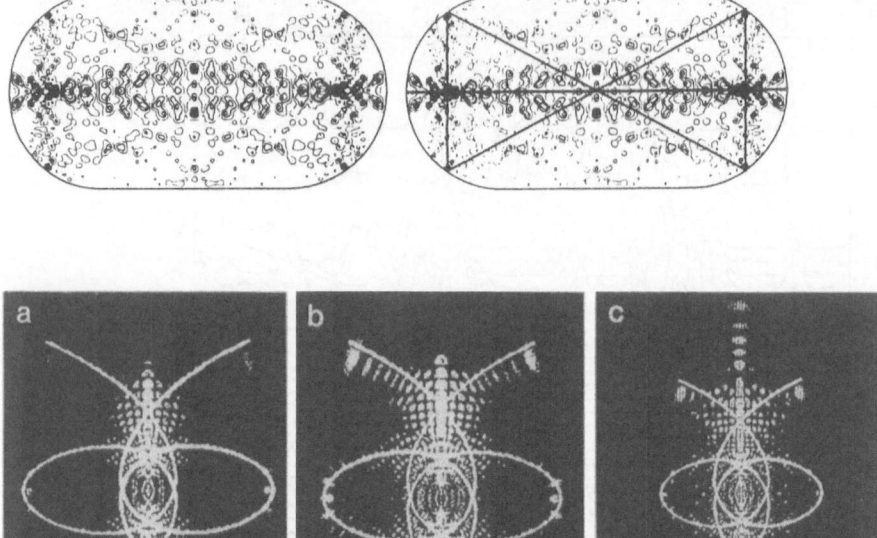

Fig. 32. "Scarring" of wave functions along unstable classical periodic orbits in the chaotic range. Top: Contours of a wave function of the stadium billiard and the classical orbits that contribute most to it (Heller [64]). Bottom: Quantum mechanical probabilities of presence of the electron in $m = 0$ Rydberg states for $B = 5.7$ T together with the projections of selected orbits of the corresponding classical system (in cylindrical coordinates, the magnetic field points upward); (a) 420th excited state ($\varepsilon = -0.3067$), superimposed classical orbit Liapunov-stable, (b) 434th excited state ($\varepsilon = -0.2897$), classical orbit Liapunov-stable, (c) 448th excited state ($\varepsilon = -0.2736$), classical orbit Liapunov-unstable.

Fig. 32b shows examples of wave functions of the hydrogen atom in the chaotic range together with the projection of a selected orbit of the corresponding classical system. Dominant structures in the quantum mechanical system along the (unstable) classical orbit can be clearly recognized. Thus scars are also discovered in the wave functions of highly excited states of the hydrogen atom in strong magnetic fields, which is an indication that scarring of wave functions along unstable, closed orbits is another universal "chaotic" quantum phenomenon.

6.2.7 Outlook. As examples of the characteristic behaviour of quantum systems in the classical chaotic range, I have reported on level statistics, on the appearance of resonances in the quantum spectrum caused by unstable periodic orbits, and on the scarring of wave functions. This is of course not the complete list. Further characteristics that are still being analyzed are, e.g., the parameter sensitivity of spectra, the distribution of wave function values, the statistics of transition probabilities in the chaotic range, and the scarring of quantum states in phase space (e.g. of Wigner or Husimi distribution functions).

The question of "chaos in quantum mechanics" has certainly not been handled conclusively to date. But the results reported on here can give an impression of the studies being done around the world on nonintegrable quantum systems – in mathematical theory, with numerical calculations and in laboratory experiments. The particular fascination of these studies can certainly be attributed to the fact that, more than 60 years after the development of quantum mechanics, they still allow new and unexpected insights into the connection between the world of classical physics and that of quantum theory.

Acknowledgments

Many of the results reported here on atoms in strong magnetic fields are the outcome of a cooperation with my former colleagues and students at the University of Tübingen. This cooperation has been going on for years, and I am grateful in particular to Hanns Ruder, Heinz Herold and Wolfgang Schweizer for many fruitful discussions and suggestions on the subject. This work was supported by Deutsche Forschungsgemeinschaft.

References

1. J. Rafelski, B. Müller: Phys. Rev. Lett. **36** (1976) L517
2. J.C. Kemp, J.B. Swedlund, J.D. Landstreet, J.R.P. Angel: Astrophys. J. **161** (1970) L77
3. J.R.P. Angel, E.F. Borra, J.D. Landstreet: Astrophys. J. Suppl. **45** (1981) 457
4. S.B. Kemic: Astrophys. J. **193** (1974) 213
5. S.B. Kemic: Joint Institute Laboratory Astrophysics Report **113** (1974)
6. R.H. Garstang, S.B. Kemic: Astrophys. Space Sci. **31** (1974) 103
7. R.H. Garstang: J. Phys. (Paris), Colloque C2 **43** (1982) 19
8. S.L. Shapiro, S.A. Teukolsky: "Black Holes, White Dwarfs, and Neutron Stars", Wiley and Sons, New York (1983)
9. D.J. Helfand, J.-H. Huang (eds.): "The Origin and Evolution of Neutron Stars", Reidel Publishing Company, Dordrecht (1987)
10. F. Nagase: Publ. Astron. Soc. Japan **41** (1989) 1
11. J. Trümper, W. Pietsch, C. Reppin, B. Sacco, E. Kendziorra, R. Staubert: Ann. N.Y. Acad. Sci. **302** (1977) 538
12. H. Ruder, G. Wunner, H. Herold, F. Geyer: "Atoms in Strong Magnetic Fields", Astronomy and Astrophysics Library, in press, Springer-Verlag Heidelberg (1994)
13. A. A. Sokolov, I. M. Ternov: "Synchrotron Radiation", Pergamon, New York (1968)
14. J. S. Toll, Ph. D. thesis, Princeton University (1952)
15. N. P. Klepikov: Zh. Eksp. Teor. Fiz. **26** (1954) 19
16. T. Erber: Rev. Mod. Phys. **38** (1966) 636

17. J. K. Daugherty, A. K. Harding: Astrophys. J. **273** (1983) 761

18. W. H. Furry: Phys. Rev. **51** (1937) 125

19. R. J. Stoneham: J. Phys. A **12** (1979) 2187

20. S. L. Adler: Ann. Phys. NY **67** (1971) 599

21. M. B. Baring: Astron. Astrophysics **249** (1991) 581

22. M. Mentzel, D, Berg, G. Wunner: Phys. Rev. D **50** (1994)

23. G. Wunner, H. Ruder, H. Herold: Phys. Lett. **79A** (1980) 159

24. R. Loudon: Am. J. Phys. **27** (1959) 649

25. J.K. Daugherty, J. Ventura: Astron. Astrophys. **61** (1977) 723

26. H. Herold, H. Ruder, G. Wunner: Phys. Lett. **91** (1982) 272

27. K.A.U. Lindgren, J.T. Virtamo: J. Phys. B: At. Mol. Phys. **12** (1979) 3465

28. J.R.P. Angel, J. Liebert, H.S. Stockman: Astrophys. J. **292** (1985) 260

29. J.L. Greenstein, R.J.W. Henry, R.F. O'Connell: Astrophys. J. **289** (1985) L25

30. G. Wunner, W. Rösner, H. Herold, H. Ruder: Astron. Astrophys. **149** (1985) 102

31. G.D. Schmidt, H.S. Stockman, S.A. Grandi: Astrophys. J. **300** (1986) 804

32. G.D. Schmidt, S.C. West, J. Liebert, R.F. Green, H.S. Stockman: Astrophys. J. **309** (1986) 218

33. R. Minkowski: Ann. Rept. Dir. Mt. Wilson Obs. **28** (1938)

34. J.L. Greenstein, M.S. Matthews: Astrophys. J. **126** (1957) 14

35. J.R.P. Angel: Ann. Rev. Astron. Astrophys. **16** (1978) 487

36. R.J.W. Henry, R.F. O'Connell: Pub. of the Astron. Soc. of the Pacific, **97** (1985) 333

37. D.T. Wickramasinghe, L. Ferrario: Astrophys. J. **327** (1988) 222

38. W.B. Latter, G.D. Schmidt, R.F. Green: Astrophys. J. **320** (1987) 308

39. D.T. Wickramasinghe, M. Cropper: Mon. Not. R. Astr. Soc. **235** (1988) 1451

40. M.H. Cohen, A. Putney, R.W. Goodrich: Astrophys. J. **405** (1993) L67

41. J. Liebert, P. Bergeron, G.D. Schmidt, R.A. Saffer: Astrophys. J. **418** (1993) 426

42. I.P. Roche, G.P. Whittington, P.C. Main, L. Eaves, F.W. Sheard, G. Wunner, K.E. Singer: J. Phys.: Condensed Matter **2** (1990) 4439

43. R. Buczko, J.A. Chroboczek, G. Wunner: Phil. Mag. Lett. **56** (1987) 251

44. E.A. Solov'ev: Sov. Phys. JETP Lett. **34** (1982) 265; Sov. Phys. JETP **55** (1982) 1017

45. A. Holle, G. Wiebusch, J. Main, B. Hager, H. Rottke, K.H. Welge: Phys. Rev. Lett. **56** (1986) 2594

46. A. Holle, G. Wiebusch, J. Main, K.H. Welge, G. Zeller, G. Wunner, T. Ertl, H. Ruder: Z. Phys. D **5** (1987) 279

47. G. Zeller: PhD thesis , Univ. Tübingen (1990)

48. C. Iu, G.R. Welch, M.M. Kash, D. Kleppner, D. Delande, J.C. Gay: Phys. Rev. Lett. **66** (1991) 145

49. D. Delande, A. Bommier, J.C. Gay: Phys. Rev. Lett. **66** (1991) 141

50. H. Friedrich, D. Wintgen: Phys. Rep. **183** (1989) 37

51. H. Hasegawa, M. Robnik, G. Wunner: Progress of Theoretical Physics Supplement **98** (1989) 198

52. M.C. Gutzwiller: "Chaos in Classical and Quantum Mechanics", Springer-Verlag New York, Berlin, Heidelberg (1990).

53. J.-C. Gay (ed.): Comments At. Mol. Phys **25** (1990) 1; also appeared as a book (1990), Gordon & Breach, Science Publishers S. A.

54. F. Haake: "Quantum Signatures of Chaos", Springer-Verlag Berlin, Heidelberg, New York (1991)

55. M. Berry: Proc. R. Soc. London A **413** (1987) 183

56. M. Berry: Physica Scripta **40** (1989) 335
57. O. Bohigas, M.J. Giannoni, C. Schmit: Phys. Rev. Lett. **52** (1984) 1
58. O. Bohigas, M.J. Giannoni, C. Schmit: J. Physique Lett. **45** (1984) L1015
59. W. Schweizer, R. Niemeier,G. Wunner, H. Ruder: Z. Phys. D **24** (1993) 95
60. D. Wintgen, H. Friedrich: Phys. Rev. A **35** (1987) 1464
61. G. Wunner, U. Woelk, I. Zech, G. Zeller, T. Ertl, F. Geyer, W. Schweizer, H. Ruder: Phys. Rev. Lett. **57** (1986) 3261
62. A. Holle, J. Main, G. Wiebusch, H. Rottke, K.H. Welge: Phys. Rev. Lett. **61** (1988) 161
63. M.C. Gutzwiller: Phys. Rev. Lett. **45** (1980) 150
64. E.J. Heller: Phys. Rev. Lett. **53** (1984) 1515

Theoretical Aspects of Quantum Electrodynamics in Strong Fields

Joachim Reinhardt and Walter Greiner

Institut für Theoretische Physik, Johann Wolfgang Goethe-Universität,
D-60054 Frankfurt am Main, Robert-Mayer-Str. 8-10, Germany

Abstract: In strong electric fields the neutral electron-positron vacuum state becomes unstable and decays through pair creation. This phenomenon is related to the radiation from cosmic strings and to the Hawking radiation near black holes. In a hypothetical atom with a charge exceeding the critical value $Z_{cr} \simeq 173$ a charged vacuum will be formed, signalled by the spontaneous emission of positrons.

In the laboratory sufficiently strong electric fields can be created transiently in heavy-ion collisions with total nuclear charge $Z_1 + Z_2 > Z_{cr}$. Detailed coupled channel calculations for the time dependent two-center Dirac equation have been performed to describe the dynamics of the electron shell in superheavy quasimolecules. K-hole production as well as δ-electron and positron emission are found to be sensitively dependent on the strong field in a collision. The emission of positrons is strongly enhanced in high-Z collisions. Clear experimental signals for supercritical positron creation are expected in collisions above the nuclear Coulomb barrier if a 'giant' dinuclear system is formed with a life time larger than about 10^{-20} sec.

In the spectra of positrons produced in heavy ion collisions narrow structures with unexpected properties have been found in several experiments performed at GSI (Darmstadt). The subsequent observation of correlated e^+e^- emission compatible with the two-body decay of a neutral particle state with mass around 1.8 MeV has led to widespread theoretical speculations on the existence and the nature of these objects. However, despite intense experimental and theoretical research no satisfactory explanation has been found. New elementary particles are clearly ruled out and also models of composite extended objects are beset with problems.

1 Introduction

1.1 The Charged Vacuum in Supercritical Fields

When atomic structure is extrapolated from the known boundary of chemical elements (nuclear charge $Z = 109$ into the region $Z = 170 \ldots 190$ one finds that the $1s_{1/2}$-state, the atomic K-shell, gains tremendously in binding energy. As shown in Fig. 1 the $1s_{1/2}$-state – and also the next higher state, the $2p_{1/2}$-level – traverses the gap between the positive and negative energy continuum solutions of the Dirac equation, and is predicted to reach a binding energy of $2m_e=1.022$ MeV at the *critical nuclear charge* $Z_{cr}=173\pm1$. The uncertainty derives from our lack of precise knowledge of the extrapolated nuclear charge distribution and from possible radiative corrections of higher order that are not accounted for in the calculations.

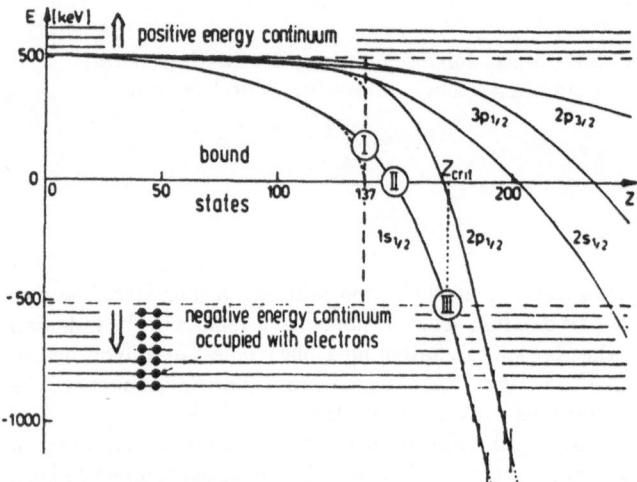

Fig. 1. Atomic binding energies as function of nuclear charge

What happens at and beyond this critical charge was clarified in the early 1970's by our group at Frankfurt [1, 2] and by another group in Moscow [3]. The transition from a just subcritical 1s-state to the supercritical state is most easily understood in the framework of Fano's theory of configuration interaction and autoionizing states. We start with the reduced Hilbert space of a just critical atom spanned by the 1s-state $|\phi_0\rangle$ and the negative energy continuum of s-wave states $|\phi_E\rangle$, as shown in Fig. 2 (left part):

$$H_C|\phi_0\rangle \approx -m_e|\phi_0\rangle , \tag{1}$$

$$H_C|\phi_E\rangle \approx E|\phi_E\rangle , \quad E < -m_e . \tag{2}$$

When a few, say Z', protons are added to the nucleus to render the potential supercritical, the 1s-state is drawn into the continuum and only continuum solutions exist:

$$(H_C + Z'U(r))|\Psi_E\rangle = E|\Psi_E\rangle , \quad E < -m_e . \tag{3}$$

The solutions $|\phi_E\rangle$ of the supercritical Hamiltonian are expanded in terms of those of (1,2):

$$|\Psi_E\rangle = a(E)|\phi_0\rangle + \int_{-\infty}^{-m_e} dE' \, b_{E'}(E)|\phi_{E'}\rangle . \tag{4}$$

Elementary methods for solving integral equations yield the following result for the probability $|a(E)|^2$ of admixture of the critical $1s$-state to the supercritical continuum state $|\Psi_E\rangle$:

$$|a(E)|^2 = \frac{1}{2\pi}\Gamma_E \left((E - E_\phi - F(E))^2 + \frac{1}{4}\Gamma_E^2 \right)^{-1}, \qquad (5)$$

where

$$E_\phi = -m_e + Z'\langle\phi_0|U(r)|\phi_0\rangle, \qquad (6)$$

$$V_E = Z'\langle\phi_E|U(r)|\phi_0\rangle, \quad \Gamma_E = 2\pi|V_E|^2, \qquad (7)$$

$$F(E) = P \int_{-\infty}^{-m_e} dE' \frac{|V_{E'}|^2}{E - E'}. \qquad (8)$$

Obviously, the bound $1s$-state turns into a resonance in the negative energy continuum located at $E_{res} = E_\phi + F(E_{res})$. The width of the resonance, Γ_E, typically is of the order 1 keV, corresponding to a lifetime in the range $10^{-18} - 10^{-19}$s. The supercritical situation is shown schematically in Fig. 2 (right part).

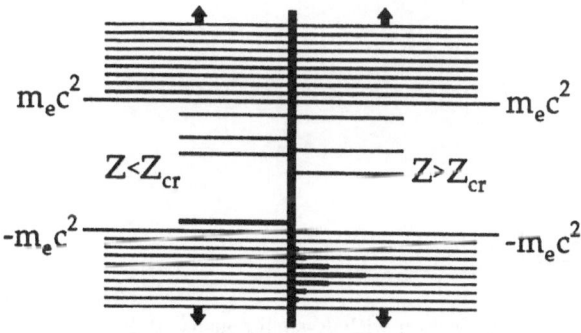

Fig. 2. Transition from Z_{cr} (left) to $Z_{cr}+Z'$ (right).

The reason why the $1s$ bound state turns into a resonance is intuitively clear: the vacant K-shell is unstable against pair-decay when the binding energy E_K exceeds twice the electron rest mass. A pair is created, the electron occupying the $1s$-state while the positron is emitted freely with kinetic energy $E_p = |E_K|-2m_e$. When the K-shell is fully occupied by two electrons the spontaneous decay process is stopped by the action of the Pauli principle. The intuitive picture is easily corroborated by arguments based on second quantized field theory [4, 5].

As the supercritical K-shell resonance is part of the negative energy continuum, i. e. of the Dirac sea, it is customary to consider it as part of the vacuum.

Consequently one speaks of the *charged vacuum* state in supercritical quantum electrodynamics, and the spontaneously occurring process of pair creation involving the vacant 1s-state is known as *decay of the (neutral) vacuum*. That the supercritical vacuum, indeed, contains a nonvanishing charge is illustrated in Fig. 3, showing the vacuum polarization charge density

$$\rho^{\mathrm{VP}}(r) \; = \; -|e| \left(\frac{1}{2} \sum_{E<-m_e} \Psi_E^+(r)\Psi_E(r) - \frac{1}{2} \sum_{E>-m_e} \Psi_E^+(r)\Psi_E(r) \right) \qquad (9)$$

for the supercritical nuclear charge Z=184. It is very similar to the charge distribution contained in the just subcritical occupied 1s-state at Z=172, which is also shown. Note that the space integral over ρ^{VP} for Z=184 does not vanish, indicating that the vacuum charge is real and not only a displacement charge as for $Z < Z_{\mathrm{cr}}$.

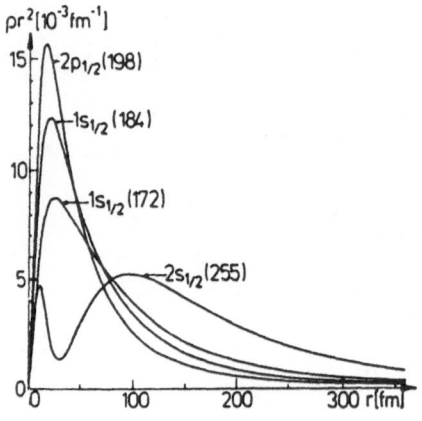

Fig. 3. Real vacuum polarization in comparison with K-shell density at Z_{cr}.

So far we have restricted our considerations to the single-particle picture. It is a relevant question whether these conclusions survive when higher order processes of quantum field theory are taken into account. The basic corrections are shown in the Feynman diagrams of Fig. 4. It is clear that the usual perturbation expansion in powers of $Z\alpha$ cannot be trusted at the critical Z, hence all orders must be summed by use of exact propagators in the external field (indicated by thick lines). This is done in the Furry bound state interaction picture. Denoting the exact electron propagator in the external field by $G(x,y)$ and the free photon propagator by $D(x,y)$, the corrections to the binding energy of the 1s-state can be written as:

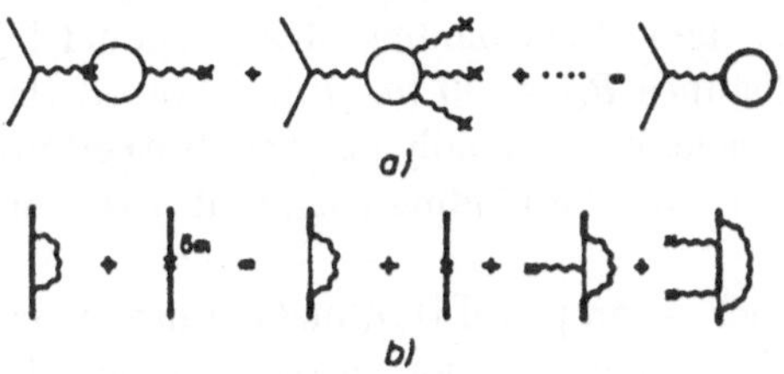

Fig. 4. Vacuum polarization (above) and self-energy corrections (below) to all orders in the nuclear charge.

a) vacuum polarization (Fig. 4a):

$$\Delta E_{1s}^{\mathrm{VP}} = -\mathrm{i}e^2 \int d^3x\, d^3y\, \bar{\Psi}(x)\gamma^0\Psi(x)D(x-y) \int \frac{d\omega}{2\pi} \mathrm{Tr}\big[\gamma^0 G(y,y';\omega)\big]_{y\to y'} ; \tag{10}$$

b) self-energy and vertex corrections (Fig. 4b):

$$\Delta E_{1s}^{\mathrm{SE}} = +\mathrm{i}e^2 \int dt\, d^3x\, d^3y\, \bar{\Psi}(x)\gamma^\mu G(x,y)\gamma_\mu\Psi(y)D(x-y)$$

$$+\delta m \int d^3x\, \bar{\Psi}(x)\Psi(x) . \tag{11}$$

Here $\Psi(x)$ denotes the $1s$ wavefunction. Eq (10) was evaluated [6, 7] to give a shift:

$$\Delta E_{1s}^{\mathrm{VP}}(Z_{\mathrm{cr}}) = -10.68 \text{ keV} \tag{12}$$

at the critical point, increasing the binding energy. Expression (11), which is more difficult to evaluate, was calculated as well [8] to result in a repulsive contribution

$$\Delta E_{1s}^{\mathrm{SE}}(Z_{\mathrm{cr}}) = +10.99 \text{ keV} \tag{13}$$

at the diving point. The almost complete cancellation between the two contributions means that the total shift in the K-shell energy due to field theoretic corrections of order α is only $+0.31$ keV, less than 10^{-3} of the total binding energy. It would be highly surprising, if higher orders in α (not $Z\alpha$) would change this picture. It therefore seems clear that the transition to a charged vacuum state must occur at a critical nuclear charge $Z \approx 173$.

It is legitimate to ask the question: How far do we have experimental proof that binding energies comparable to the rest mass of the electron, or even twice the rest mass, can actually be achieved? This question actually has two aspects: First, how can the strong binding be set up experimentally and second, how can its presence be observed? The answer to the first question was given around 1970 by the Moscow as well as by the Frankfurt group (see [9] for a historical perspective). In collisions of two very heavy nuclei the electric field of a nucleus with charge $Z_1 + Z_2$ is simulated temporarily. The solution of the Dirac equation with two Coulomb centers showed that the critical binding of $2m_e$ should be reached in, e.g., U+U collisions at a distance $R_{cr} \approx 30$ fm [10]. However, the binding energy varies with time as the nuclei move rapidly on their Rutherford trajectories and it is not so easy to determine the binding energy at a certain internuclear distance experimentally.

An approximate method was, nonetheless, proposed [11], making use of the generalization of Bang and Hansteen's [12] scaling law for direct ionization. In the case of superheavy collision systems, ionization occurs predominantly at the point of closest approach of the nuclei, R_0. One can then show [13, 14, 15], that the ionization probability of the $1s$-state on a given scattering trajectory depends – to a good approximation – only on the binding energy at distance R_0:

$$P = D \exp\left(-\frac{\gamma R_0}{v} E_B^{1s}(R_0)\right) , \qquad (14)$$

where γ is a numerical constant, v the beam velocity, and D is a function of $Z_1 + Z_2$ only which can be obtained by comparison with full-scale numerical calculations. In Fig. 5 we have shown binding energies extracted from measurements of $1s$-vacancy production in the Pb+Cm system by Liesen et al. [16], in comparison with results of a two-center Dirac calculation carried out by W. Betz [17]. Although we cannot conclude that critical binding has been achieved, the existence of binding energies in the range between 500 and 800 keV appears to be fairly well established.

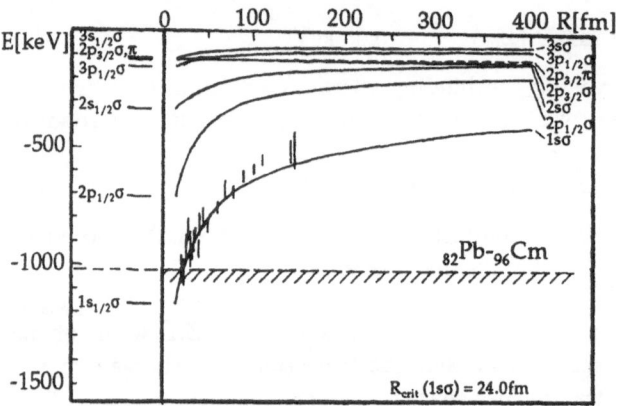

Fig. 5. Binding energy of the $1s\sigma$-state in Pb+Cm, determined from measured Cm K-hole probabilities with help of the scaling law.

For a detailed overview of vacuum properties in the presence of supercritical fields see [18].

2 Dynamics of the Electron Field

2.1 Supercritical Heavy Ion Collisions

We now address the question of the dynamics of the electron field in collision systems where the combined charge of both nuclei is sufficiently large to let the quasimolecular 1s-state enter the Dirac sea at a critical distance R_{cr}. The electronic wavefunction $\Psi_i(\mathbf{r}, t)$, which satisfies the usual boundary conditions as $t \to -\infty$ and in addition solves the time dependent problem, may be expanded in a basis $\{\psi_j\}$ containing bound states and two sets of continuum states. Specifically we choose this basis to coincide with the set of adiabatic, quasimolecular eigenstates of the two-center Dirac Hamiltonian [19], i. e.

$$\left(E_j(\mathbf{R}) - \hat{H}_{TC}(\mathbf{R}, \mathbf{r})\right)\psi_j(\mathbf{R}, \mathbf{r}) = 0 , \tag{15}$$

$$\Psi_i(\mathbf{r}, t) = \sum_j a_{ij}(t)\psi_j(\mathbf{R}(t), \mathbf{r})e^{-i\chi_j(t)} , \quad \chi_j(t) = \int_{t_0}^{t} dt \langle \psi_j | \hat{H}_{TC} | \psi_j \rangle . \tag{16}$$

The time dependent expansion coefficients $a_{ij}(t)$ are determined by solving a truncated set of coupled ordinary differential equations, the coupled channel equations [19].

In supercritical collisions the situation is complicated by the fact that the 1s state vanishes from the bound spectrum and becomes admixed to the lower continuum. In order to obtain numerically reliable results for the evolution of the occupation amplitudes in continuum states, describing the resonance of width $\Gamma \approx 1$keV, it would be necessary to include continuum states spaced by much less than 1 keV at several points per 1 fm on the nuclear distance grid. This clearly would be a very inefficient way to solve the problem.

The difficulty can be avoided by employing an improved version of the auto-ionization picture. One artificially constructs a normalizable resonance wavefunction $\tilde{\varphi}_r$ for the supercritical 1s-state, e.g. by cutting off a continuum wavefunction in the center of the resonance at its first zero (more sophisticated procedures have been devised and are routinely used [19]). In the next step a set of orthogonal states $\tilde{\varphi}_E$ are constructed in the negative energy continuum with the help of a projection technique. Those states are solutions of the projected Dirac equation

$$(H_{TC} - E)\tilde{\varphi}_E = \langle \tilde{\varphi}_r | H_{TC} | \tilde{\varphi}_E \rangle \, \tilde{\varphi}_r \quad (E < -m_e) . \tag{17}$$

Since $\tilde{\varphi}_r$ and $\tilde{\varphi}_E$ do not diagonalize the two-center Hamiltonian, there exists a nonvanishing static coupling between the truncated 1s-resonance state and the modified negative energy continuum which describes the spontaneous decay of a vacancy in the supercritical 1s-state. The spontaneous decay width is given by the expression

$$\Gamma_E = 2\pi |\tilde{V}_E|^2, \quad \tilde{V}_E = \langle \tilde{\varphi}_E | H_{TC} | \tilde{\varphi}_r \rangle . \tag{18}$$

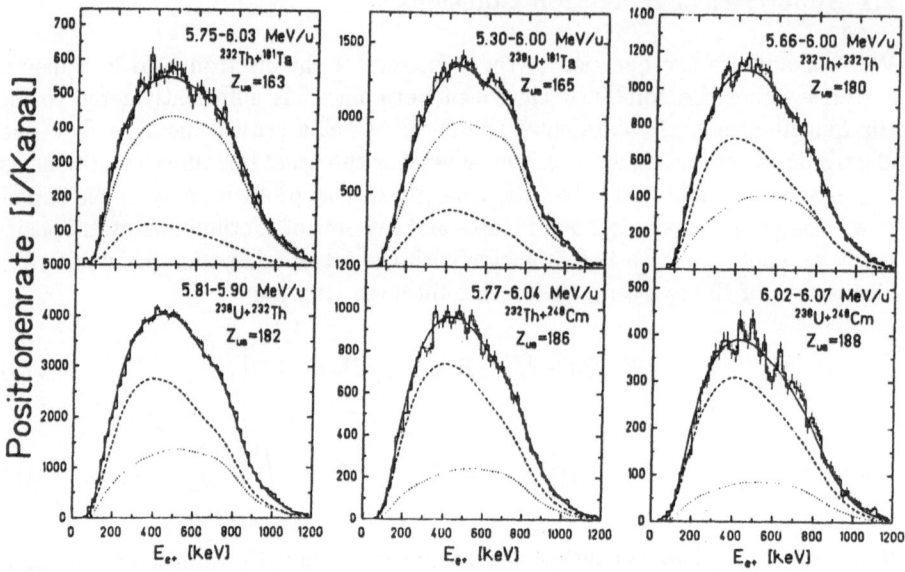

Fig. 6. Total positron spectra for various collision systems. Data: EPOS collaboration; dashed lines: QED pair production; dotted lines: nuclear pair conversion; solid lines: sum of both.

Similarly the coupled channel equations for the amplitudes $a_{ik}(t)$ are amended by a "spontaneous" matrix element that does not vanish in the limit where the nuclei do not move:

$$\dot{a}_{ik} = -\sum_{j\neq k} a_{ij}\left(\langle\tilde{\varphi}_k|\frac{\partial}{\partial t}|\tilde{\varphi}_j\rangle + \mathrm{i}\langle\tilde{\varphi}_k|H_{\mathrm{TC}}|\tilde{\varphi}_j\rangle\right)\exp(\mathrm{i}\chi_k - \mathrm{i}\chi_j)\,. \qquad (19)$$

Careful investigations have shown that the asymptotic amplitudes $a_{ik}(\infty)$ are insensitive to the precise way of constructing the supercritical $1s$-state, although the individual matrix elements may differ somewhat for the various procedures. This means that the concept of "spontaneous" pair production has no unique definition in the dynamical environment of a heavy ion collision, except in the limiting case when the nuclei fuse (at least for some time) into a single compound nucleus. For collisions without nuclear contact, when the two nuclei move on hyperbolic Rutherford trajectories, the contribution from the "spontaneous" coupling constitutes only a small fraction, in any event. Accordingly, the calculations do not yield any perceptible change in the shape of the predicted positron spectra for such collisions when one goes from subcritical to supercritical systems. The total positron yield at fixed beam energy per nucleon for $Z_u > 137$ is predicted to grow at a very rapid rate that can be roughly parametrized by the effective power law

$$\sigma_{e^+}(Z) \propto Z^n, \qquad n \approx 20, \tag{20}$$

showing no discontinuity at the transition to supercriticality. The large value of the exponent demonstrates the entirely nonperturbative nature of pair production in collisions of heavy ions. The fact that this prediction has been verified in the experiments at GSI is a major confirmation of our ability to accurately treat quantum electrodynamics in strong Coulomb fields by the theoretical methods described above, based on the adiabatic quasimolecular basis and the monopole approximation [19].

2.2 The "Atomic Clock" Phenomenon

When in the course of a heavy ion collision the two nuclei come into contact, a nuclear reaction occurs that lasts a certain time T. The length of this contact or *delay time* depends on the nuclei involved in the reaction and on the beam energy. For light and medium heavy nuclei the nuclear attraction is greater than, or comparable with, the repulsive Coulomb force, thus allowing for rather long reaction times of the order of 10^{-20}s or even more. For very heavy nuclei, in or beyond the Pb region, however, the Coulomb interaction is the dominant force between the nuclei, so that delay times are typically much shorter and in the mean probably do not exceed $1 - 2 \times 10^{-21}$s. Nuclear reaction models predict that the delay time increases with the violence of the collision, as measured by inelasticity (negative Q-value), and mass or angular momentum transfer.

A delay in the collision due to a nuclear reaction can lead to observable modifications in atomic excitation processes. The two main observable effects in such collisions are: (a) interference patterns in the spectrum of δ-electrons [20], and (b) a change in the probability for K-vacancy formation [21]. These effects have become known as the *atomic clock* for deep-inelastic nuclear reactions.

The origin of the atomic clock effect is most easily understood in a simple semiclassical model for the nuclear motion, where the nuclear trajectory is described by the classical function $R(t)$ and the only effect of the nuclear reaction is to introduce a time delay T between approach and separation of the nuclei, i.e. $\dot{R}(t) = 0$ for $0 \leq t \leq T$. To retain lucidity of the argument we make use of first-order perturbation theory for the excitation amplitudes a_{ik}:

$$a_{ik}(\infty) = -\int_{-\infty}^{\infty} dt \dot{R}(t) \langle \varphi_k | \frac{\partial}{\partial R} | \varphi_i \rangle \exp\left[i \int_0^t dt'(E_k - E_i) \right]. \tag{21}$$

The range of the main integral splits into three parts: (a) $t < 0$, (b) $0 \leq t \leq T$, and (c) $t > T$. Because $\dot{R}$ enters as a factor in (21), the median part does not contribute. For the last part one can rewrite $t \to t + T$ so that the integral runs from 0 to ∞. Because this exit part of the nuclear trajectory is just the time-reverse of the entrance part, i.e. $\dot{R}(-t) = -\dot{R}(t)$, the amplitude from part (c) can be expressed as the complex conjugate of the amplitude $a_{ik}(0)$ from part (a) of the integral, except for a phase factor resulting from the variable substitution in the phase integral in (21). Thus we find the relation:

$$a_{ik}^T(\infty) = a_{ik}(0) - a_{ik}^*(0) \exp\left[iT(E_k - E_i)\right] , \qquad (22)$$

where the energies have to be taken at the distance of nuclear contact. If we write $a_{ik}(0)$ in the symbolic form $a_0 e^{i\alpha}$, the final probability for excitation between states i and k becomes

$$P_{ik}(T) = |a_{ik}^T(\infty)|^2 = 4a_0^2 \sin^2\left(\frac{1}{2}T\Delta E - \alpha\right) , \qquad (23)$$

where $\Delta E = (E_k - E_i)$ is the transition energy. The excitation probability is obviously an oscillating function, either of T for a given transition $i \to k$, or a function of transition energy ΔE for fixed delay time T.

In reality, of course, things are more complicated. Except in truly elastic collisions the outgoing trajectory is not a precise mirror image of the approaching trajectory. Furthermore, multi-step excitations play an important role in very heavy systems. The total excitation amplitude therefore contains a mixture of contributions from different intermediate states. The simple expression, (22), must then be replaced by the formula

$$a_{ik}^T(\infty) = \sum_j a_{ij}^{in} e^{-iE_j T} a_{jk}^{out} . \qquad (24)$$

Finally, the nuclear delay time T is usually not sharply defined, so that an average over a distribution $f(T)$ of delay times has to be taken in (23). The common result of these refinements is a dilution of the interference patterns, i.e. the oscillations become less pronounced. For short delay times and a large uncertainty of T all that remains is a partially destructive interference between the incoming and outgoing branches of the trajectory, observable as a decrease in the K-vacancy yield or a steepening of the slope of the low-energy part of the δ-electron spectrum.

It was demonstrated by O. Graf et al. [22] that collisions of *fully stripped ions* with a small nuclear time delay could serve to trigger supercriticality. As a unique signal for spontaneous pair creation a change in the *angular correlation* between electron and positron is expected that occurs only in supercritical systems: For nuclear delay times of 2×10^{-21}s the normal forward correlation should turn into a backward correlation.

2.3 Positron Production in Delayed Collisions

In principle, the positron spectra contain the same information about nuclear time delay as the electron spectra. However, because of their low emission probability, positrons are not as useful from a practical point of view, at least in subcritical collision systems. Nevertheless, positron spectra emerging from deep-inelastic heavy ion collisions have been measured [23], and the first experimental observation of the atomic clock phenomenon in heavy ion collisions came in fact from positrons [24]. The yield argument does not necessarily apply to supercritical collision systems for two reasons. Firstly, a reaction-induced nuclear time delay may allow for the detection of spontaneous pair-creation in these systems, as will be discussed below. Secondly, a tiny component of very long reaction times ($T > 10^{20}$s) might become visible in the positron spectrum, because the spontaneous emission mechanism acts as a kind of "magnifying glass" for long delay times [25].

In order to see why this is so, we return to (19) for the amplitudes a_{ik} in a supercritical system, which contained the additional time-independent couplings $\tilde{V}_E$ between the resonant bound state and the (modified) positron continuum states. The presence of this coupling has the effect that the contribution to the integral in (21) from the central time interval $0 \leq t \leq T$ does not vanish any longer. The total amplitude a_E^T for emission of a positron from the supercritical bound state contains therefore an additional term compared with (22):

$$a_E^T(\infty) = a_E(0) - a_E^*(0) \exp\left[iT(E - E_r)\right] - \tilde{V}_E \frac{e^{iT(E-E_r)} - 1}{E - E_r}, \qquad (25)$$

where E_r is the energy of the supercritical state when the nuclei are in contact. For $T = 0$ the new term vanishes, but grows rapidly with increasing T. For delay times considerably greater than 10^{-21}s the additional term in (25) begins to dominate over the first two terms , causing the emergence of a peak in the positron spectrum at the energy of the supercritical bound state:

$$|a_E^T(\infty)|^2 = \frac{\Gamma_E}{2\pi} T^2 \frac{\sin^2[(E - E_r)T/2]}{[(E - E_r)T/2]^2}, \qquad (T \ll \Gamma_E^{-1}). \qquad (26)$$

The energy distribution has a width $\Gamma(T) = 2\pi/T$ as would be expected on grounds of the uncertainty relation, and the total probability for positron emission grows proportional to T. For extremely long delay times, $T > \Gamma_E^{-1}$, the energy distribution instead of (26) goes over into a Breit-Wigner curve centered at E_r with width equal to the spontaneous decay width Γ_E.

The expected emergence of a peak in the positron spectrum for strongly delayed supercritical collisions is strikingly demonstrated in Fig. 7. In the subcritical case the delay causes interference patterns like those already known from electron spectra, effectively reducing the positron yield. In the supercritical case the positron yield increases dramatically when the delay time exceeds about 3×10^{-21}s. Unfortunately, this is beyond the range accessible for the average time delay in deep-inelastic reactions where not much more than 10^{-21}s has been observed. Still, the situation may not be entirely hopeless, because the intensity of the line structure grows with T and simultaneously becomes more localized at the resonance energy. In principle, even a very small tail of the delay time distribution $f(T)$ could acquire sufficient weight to be visible in the positron spectrum [26].

Such long delay times could occur if an attractive pocket is present in the internuclear potential for supercritical collision systems, for which no conclusive theoretical or experimental evidence exists at present. But even if a pocket were

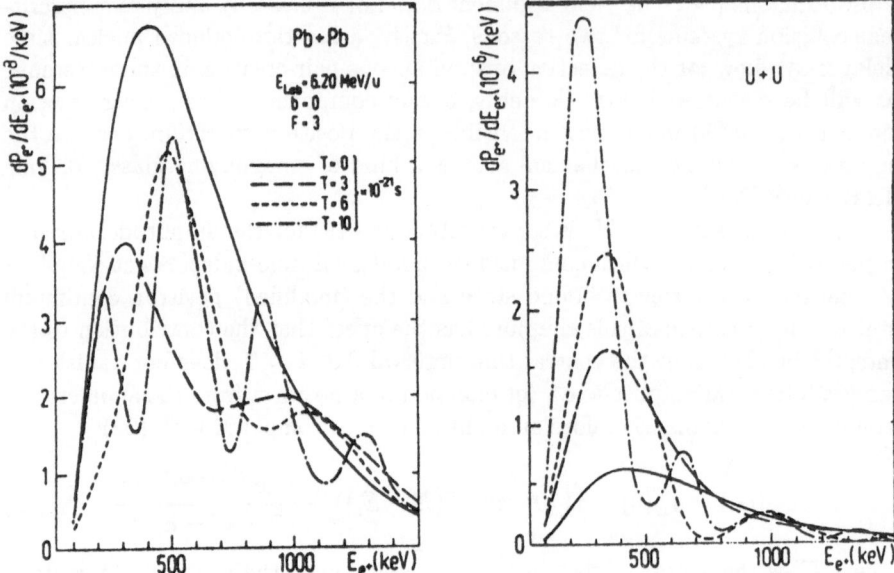

Fig. 7. Effect of a nuclear time delay on the positron spectrum. Left: subcritical system (Pb+Pb). Right: supercritical system (U+U).

there, the existence of a sufficiently large tail of long delay times in the distribution $f(T)$ is not ensured [27]. Although very narrow positron peaks were obtained for beam energies in a small window around the Coulomb barrier, their intensity was much too low to allow for observation. But again it must be emphasized that these models are too simple to permit definite conclusions.

2.4 Structures in the Positron Singles Spectrum

When line structures in the positron spectra were first detected at GSI, they were associated with the spontaneous positron emission line that was predicted by theory for supercritical collision systems with long nuclear time delay. This was quite natural, because that had been the aim and inspiration of the experiments from the beginning. For the first two systems that were investigated, U+Cm and U+U [31, 28], this explanation worked quite nicely; the position of the line agreed rather well with the expected spontaneous emission peak. The measured spectra could be described in the framework of schematic models involving the intermediate formation of a long-lived ($T \approx 5 \times 10^{-20}$s) "giant" di-nuclear system [26, 29].

Of course, the experimentalists were cautious to make sure that the lines would not be a trivial artifact caused by pair decay of some excited nuclear state. This can be checked experimentally, since a pair-decaying nuclear state can always decay through alternative channels, either by photon emission (if the transition multipolarity is not $L = 0$) or by internal conversion to a K-shell electron. The latter process works for any multipolarity. The branching ratios for the various decays can be calculated essentially model independently, because the nucleus is small compared to the wavelength of the emitted real or virtual photon. The photon and electron spectra were measured simultaneously in the relevant energy range (beyond 1 MeV), but no associated structure was found [31, 28].

The observed line width of 70-80 keV provided a second argument against nuclear pair decay. If the structures were emitted from the scattered nuclei they would have to be Doppler broadened due to the motion of the source. At 45° (lab) scattering angle the broadening would amount to 100 keV, i.e. more than the observed line width even for an intrinsically monochromatic structure. However, the positron spectrum emerging from a normal nuclear pair decay is not monochromatic at all! The energy of the transition is shared between the electron and the positron, and a broad peak develops at the upper end of the positron spectrum only for heavy nuclei due to Coulomb effects.

An intrinsically monochromatic positron line could, in principle, be caused by a process called monoenergetic pair conversion, which can occur if an inner atomic shell is not fully occupied. The electron can then be captured in this bound state, and the positron carries away the full remaining energy. The sharply defined energy is characteristic of a two-body decay $A^* \rightarrow (A + e^-) + e^+$, whereas the normal pair decay into a free electron-positron pair is a three-body decay $A^* \rightarrow A + e^- + e^+$. Although a large number of inner-shell vacancies are created in the heavy ion collision, monoenergetic pair conversion is expected to be strongly suppressed, because the vacancies are filled by transitions from outer shells within about 10^{-17}s. This filling time is at least two orders of magnitude shorter than the lifetime of nuclear excited states. Therefore, a possible origin of the line structures by monoenergetic pair conversion was ruled out, too, on experimental [31] and theoretical grounds [30].

The dependence of the line structures as function of combined nuclear charge $Z_u = Z_1 + Z_2$ afforded a crucial test of the hypothesis that they could be attributed to spontaneous positron production. The line must then occur at the positron kinetic energy corresponding to the energy E_{1s} of the $1s$-resonance that is embedded in the Dirac sea: $E_{peak} = |E_{1s}(Z_u)| - m_e$. The surprising result of such a study by the EPOS collaboration [32] is shown in the left-hand part of Fig. 8: The position of the peak was always in the range 350 ± 30 keV essentially independent of Z_u! For comparison, the expectation for a peak caused by spontaneous pair creation in the strong Coulomb field of a long-lived dinuclear system is also shown in part (b) of the figure. Starting at about 320 keV in the U+Cm system the line should move to lower energies and decrease in intensity, assuming similar nuclear delay times for all systems. Fig. 9 demonstrates that the observed structures fall into three groups, at the positron energies of about 250, 330, and 400 keV, respectively.

The main experimental results on structures in the positron singles spectrum can be summarized as follows (cf. also [33, 34])

- Lines have been observed for a large variety of collision systems, ranging from $Z_u = 163$ (Th+Ta) up to $Z_u = 188$ (U+Cm) and involving nuclei with

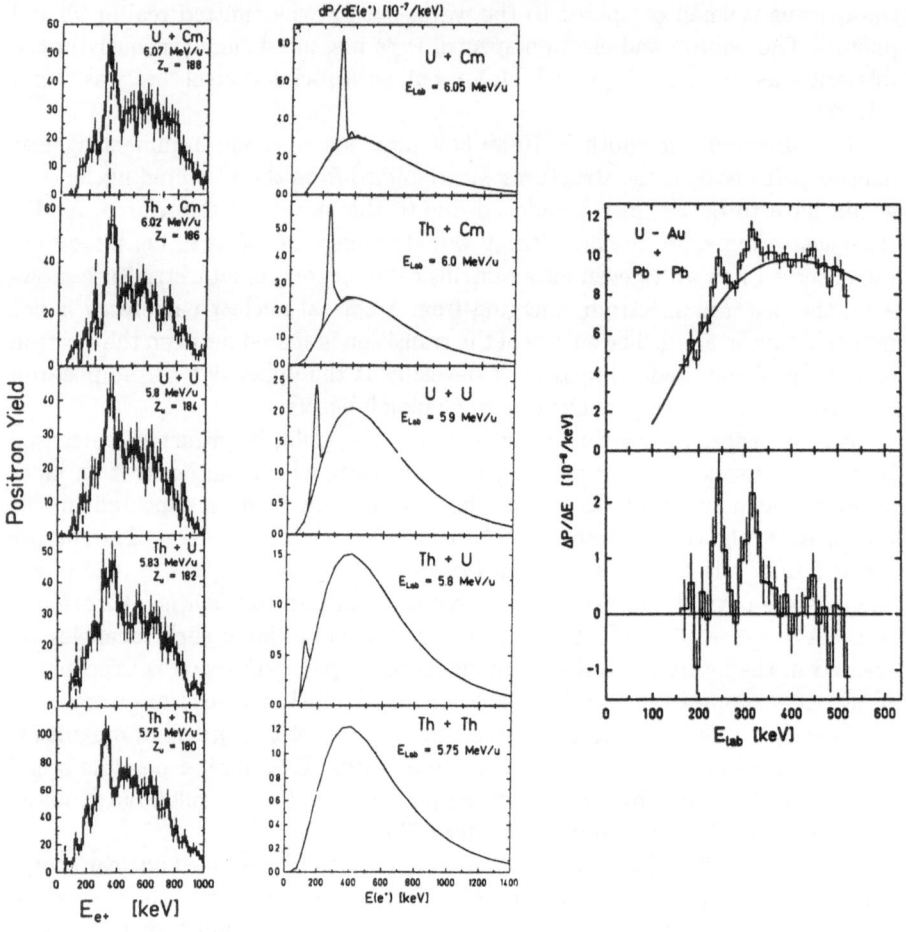

Fig. 8. Left: (a) Positron spectra measured by the EPOS group; (b) expectations for spontaneous positron creation. Right: Combined positron spectra from the systems Pb+Pb and U+Au measured by the ORANGE group.

widely different structure.

- The line positions appear to fall into several groups between 250 and 440 keV; their width is about 70 keV, if all positron emission angles are covered. This value corresponds to the Doppler width of a sharp line emitted by a source moving with center-of-mass velocity.
- A number of lines are common to different collision systems and to both experiments (ORANGE and EPOS).
- Nuclear pair conversion processes $(A^* \rightarrow A + e^- + e^+)$ appear to be excluded from γ-ray and electron spectra, linewidth, and A-independence.

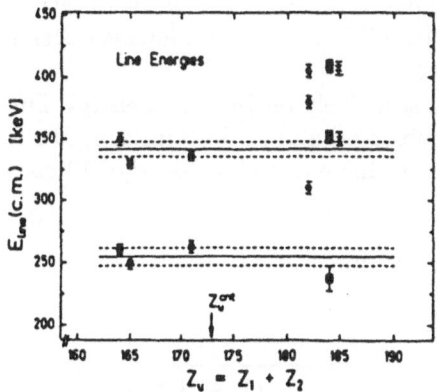

Fig. 9. Peak positions measured as function of Z_u by the ORANGE group.

2.5 Correlated Electron-Positron Lines

The A- and Z-invariance of the line energies strongly hint at a common source that in itself is not related to the nuclei or the strong electric field, although the strong Coulomb field might play a role in the production of this source. Since no systematic Z-dependence is seen, the most natural candidate for such a source would be some (neutral) object that moves with the velocity of the center of mass and eventually decays into a positron and a single other particle. (A two-body decay must be invoked to explain the narrow linewidth, as mentioned before.) Could the second decay product simply be a second electron, i.e. could it be that one sees the pair decay of a neutral particle, $X^0 \rightarrow e^+ + e^-$, with a mass somewhat below 2 MeV [35]?

The search for a correlated electron peak, as performed first by the EPOS collaboration, was successful [36, 37]. Later the existence of this correlated line structure could be confirmed by the ORANGE collaboration as well [38, 39]. We restrict ourselves to a summary of the experimental results that have been accumulated concerning correlated line structures in electron-positron coincidence spectra:

- Lines at 620 and 810 keV sum energy have been observed by the EPOS and the ORANGE collaboration. A third line at 750 keV was only seen by the EPOS group. The peaks seem to occur at the same positions in various systems, e.g. U+Ta ($Z_u = 165$), U+Th ($Z_u = 182$) and U+U ($Z_u = 184$).
- The width of the sum energy peaks lie in the range 10-40 keV; they are much narrower than the positron singles peaks. The source must move slowly ($\beta_s \leq 0.05$).
- The 810 keV line appears to be consistent with back-to-back emission. Several other lines, e.g. 750 keV in U+Ta, are either forward correlated or isotropic.

- The $\Delta E = E_{e^+} - E_{e^-}$ difference-energy spectra exhibit a broad peak near zero energy, indicating that the lepton pair is not produced inside the strong Coulomb field. For the 750 keV and 810 keV lines in U+Ta, however, a shift of roughly $\Delta E = +200$ keV is observed,
- The line intensities appear to depend sensitively on the beam energy. Furthermore, the 640 keV line has been observed also in deep inelastic U+Ta collisions at 6.3 MeV/u with an intensity increased by a factor of 10 compared to elastic collisions [39].

Finally, it should be mentioned that the line intensity appears to depend very sensitively on the beam energy.

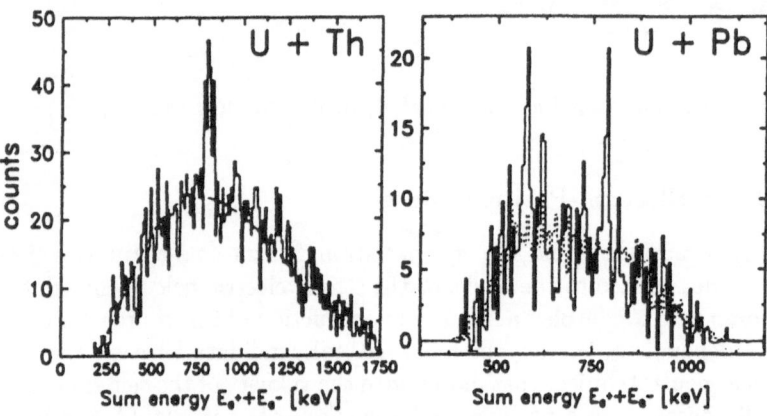

Fig. 10. Two typical examples of coincident electron-positron spectra measured by the EPOS group in the system U+Th (left) and by the ORANGE group in U+Pb collisions (right). When plotted as a function of the sum energy of electron and positron very narrow line structures are observed.

Many features appear to be compatible with the assumption that one observes the pair decay of at least three neutral particle states in the mass range between 1 and 2 MeV. These states must have a lifetime of more than 10^{-19}s (because of the narrow linewidth) and less than about 10^{-9}s (because the vertex of the lepton pair is within 1 cm of the target). On the other hand, some pieces of data do not really fit into this picture, e.g. the characteristics of the 750 keV line observed in U+Ta.

The very idea that a whole family of neutral particle states in the MeV mass range should have remained undetected through more than 50 years of nuclear physics research is hard to accept for the conservative mind. Most physicists, when first confronted with the GSI data, have therefore tried to explain the data in terms of known nuclear or atomic physics. As mentioned before, nuclear pair decay would be the most natural explanation. However, none of the proposed scenarios has yet stood up against a detailed comparison with the experiments. Similar remarks apply to attempts to explain the GSI peaks in terms of atomic physics. None of the ideas that were studied quantitatively have been successful, even if they were based on plausible, but unfounded, ad hoc assumptions.

A topic still under investigation is the polarization dependence of nuclear pair conversion. The excited nuclei emerging from a close heavy ion collision can be in a polarized or aligned state while the pair conversion coefficients up to now employed in the analysis of experiments were calculated for unpolarized nuclei. It might be conceivable that strange angular correlations of the emitted pair taken together with the restricted acceptance intervals of the detectors might fake the characteristics of a two-body decay. According to recent calculations of Ch. Hofmann, however, the angular distribution of pair conversion is only moderately affected by polarization effects [40].

2.6 Limits on Light Neutral Bosons from Precision Experiments

Even when the hypothesis of a new neutral particle was first seriously discussed [35, 41], it was recognized that the precision experiments of quantum electro-dynamics provide stringent limits on the coupling of such light particles to the electron-positron field and to the electromagnetic field. The strength of this argument lies in the fact that any particle X^0 which decays into an e^+ - e^- pair must couple to the electron-positron field. At least in the low-energy limit, the coupling can be expressed by an effective interaction of the form:

$$L_X = g_i(\bar{\psi}\Gamma_i\psi)\phi \,, \tag{27}$$

where ψ denotes the electron-positron field, ϕ the X^0 field, and Γ_i with $i =$ S,P,V,A stands for the vertex operator associated with the various possible values of spin and parity of the X^0 particle. Given the interaction Lagrangian (27) one can calculate the lifetime of the X^0 particle against pair decay as well as the contributions to QED processes by virtual exchange of an X^0. The most sensitive of these is the anomalous magnetic moment of the electron a_e, because of the high experimental and theoretical accuracy [42], $\Delta a_e < 2 \times 10^{-10}$.

As illustrated in Fig. 11, the contribution of a hypothetical X^0 particle to the value of a_e involves two vertices between an electron or positron and an X^0, and is thus proportional to the effective coupling constant $\alpha_{Xe} = g_i^2/4\pi$. The same applies to the decay rate τ_X^{-1}, which involves the square of an amplitude with a single vertex, and to the contribution of an X^0 particle to the hyperfine splitting of the positronium ground states.

The limits derived from these considerations [43] on the X^0-coupling constant and its lifetime are listed in Table 1. Particles with lifetime $\tau_X > 10^{-13}$s cannot be ruled out by this argument. Considering that the experimental conditions only require a lifetime below about 1 ns, there remained an unexplored range of four orders of magnitude in τ_X.

Similar upper limits can be derived for the coupling of an X^0 boson to other known particles [43]. A limit on the product of the coupling constants to the electron and to nucleons is obtained from the Lamb shift in hydrogen and from the K-shell binding energy in heavy elements, one finds $\alpha_{Xe}\alpha_{XN} < 10^{-14}$. For

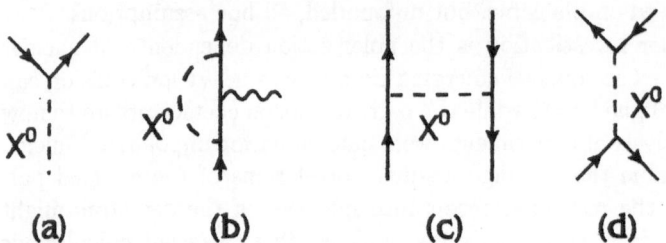

Fig. 11. Feynman diagrams for (a) pair decay of X^0, (b) contribution to the electron anomalous magnetic moment, (c,d) positronium hyperfine splitting.

Table 1. Limits on the coupling constant, lifetime, and pair decay width of neutral elementary bosons with mass $M_X = 1.8$ MeV derived from the anomalous magnetic moment of the electron.

Particle type	Spin J^π	Vertex Γ_i	Max. coupling $\alpha_{Xe} = g_i^2/4\pi$	Min. lifetime τ_X (s)	Max. width $\Gamma_X^{e^+e^-}$ (meV)
Scalar (S)	0^+	1	7×10^{-9}	2×10^{-13}	3.0
Pseudoscalar (P)	0^-	$i\gamma_5$	1×10^{-8}	1×10^{-13}	6.8
Vector (V)	1^-	γ_μ	3×10^{-8}	4×10^{-14}	16
Axial vector (A)	1^+	$\gamma_\mu\gamma_5$	5×10^{-9}	5×10^{-13}	1.4

scalar particles an extremely stringent bound on the coupling to nucleons can be derived from low-energy neutron scattering: $\alpha_{XN} < 10^{-9}$ [44]. Finally, measurements of nuclear Delbrück scattering yield an upper limit on the coupling of a spinless X^0 boson to the electromagnetic field through an effective interaction of the type

$$L_{X\gamma\gamma} = g_S(E^2 - H^2)\phi_X \quad \text{(scalar)}$$
$$L_{X\gamma\gamma} = g_P(\mathbf{E} \cdot \mathbf{H})\phi_X \quad \text{(pseudoscalar).} \qquad (28)$$

The limits are: $g_S < 0.02$ GeV^{-1} and $g_P < 0.5$ GeV^{-1}. They provide lower limits for the lifetime against decay into two photons: $\tau_{\gamma\gamma}(X^0) > 6 \times 10^{-11}$s for a scalar particle and $\tau_{\gamma\gamma}(X^0) > 4 \times 10^{-13}$s for a pseudoscalar particle [45].

2.7 Inadequacy of Perturbative Production Mechanisms

One consequence of these results was that the elementary particle hypothesis could be rejected off-hand. On the other hand, the condition that the coupling constant between the hypothetical X^0 boson and the particles involved in the heavy ion collision, i.e. electrons and nucleons, must be very small creates severe problems for any attempt to explain the measured line intensities by a perturbative interaction of the type shown in (27) [35, 41, 43]. Also the cross section for production by the strong electromagnetic fields present in the heavy ion collision falls short by several orders of magnitude, if it is based on the Lagrangian (28) or similar perturbative interactions [46, 47, 48, 49, 50, 51].

A second serious difficulty with the interactions (27) and (28) is that they favour the production of particles with high momenta due to phase space enhancement. For collisions with nuclei moving on Rutherford trajectories the calculated spectra typically are very broad, peaking at velocities $\beta_X > 0.5$, while the experiments require an average particle velocity $\beta_X < 0.05$! The energy spectra of the X^0 and e^+ assuming various perturbative electromagnetic production mechanisms of Spin-0 and Spin-1 particles have been studied systematically in [52]. In all cases the velocity was found to be too high by several orders of magnitude.

Both these problems could, in principle, be circumvented by the assumption that a very long-lived, excited giant compound nucleus is formed [46, 50], but only at the price of violating other boundary conditions set by the experimental data, e.g. the absence of a much larger peak in the positron spectrum caused by spontaneous pair production [53]. One might also consider the possibility that the X^0 particles are somehow slowed down after production, but this cannot be achieved with the interactions discussed above.

Two mechanisms remain, which can conceivably ensure the survival of the particle hypothesis: (1) a form factor that cuts off production at large momenta [54]; and (2) a non-perturbative production mechanism, e.g. production in a bound state around the two nuclei [55, 56]. Both mechanisms require particles with internal structure.

2.8 Axion Searches

At first the axion [57], i.e. the light pseudoscalar Goldstone boson associated with the breaking of the Peccei-Quinn symmetry required to enforce time-reversal invariance in quantum chromodynamics, seemed like a plausible candidate for the suspected X^0 boson. The interest in an axion was revived when it was realized that there was, indeed, a gap left by previous axion search experiments for a short-lived axion in the mass range around 1 MeV [58]. However, new experimental studies of J/Ψ and Υ decays [59, 60, 61] quickly ruled out the standard axion.

2.9 Beam Dump Experiments

Beam dump experiments, in particular those with a high-energy electron beam, are an excellent source of rather model-independent bounds on the properties of hypothetical light neutral particles [62]. The initial electron might undergo a bremsstrahlung process. Of course, instead of radiating a photon, the electron can emit some other light neutral particle X^0, if any exists. Except for effects from the particle mass and spin, the expected cross section is given by the cross section for photon radiation, multiplied by the ratio of the coupling constant of the emitted particle to the electron and the electromagnetic coupling constant:

$$\frac{d\sigma_X}{d\Omega dE} \simeq \frac{\alpha_{Xe}}{\alpha} \frac{d\sigma_\gamma}{d\Omega dE} . \tag{29}$$

An upper limit for the measured cross section for the X^0-particle cross section hence yields an upper limit for the coupling constant α_{Xe}.

However, for every beam dump experiment there is not only a lower bound for the range of excluded coupling constants but also on upper bound, for the following reason. The lifetime of the hypothetical particle against pair decay is inversely proportional to the α_{Xe}. For sufficiently large values of the coupling constant almost all produced particles therefore decay inside the beam dump, and the e^+e^- pair produced in the decay is absorbed in the target. It is clear that a good value for this other limit requires a short beam dump, whereas high cross-section and low background require a thick target.

Hence, the result of a beam dump experiment is a region of excluded values of α_{Xe}, i.e. the coupling cannot be in the range $\alpha_{Xe}^{\min} < \alpha_{Xe} < \alpha_{Xe}^{\max}$. In the analysis one assumes that the neutral particles interact so weakly that they pass essentially undisturbed through the target.

The conditions and results of various beam dump experiments are listed in Table 2, where the excluded ranges of the coupling constant are given for pseudoscalar particles of mass 1.8 MeV. For a scalar particle the bounds would be similar, but for spin-one particles about one order of magnitude better lower limits would be obtained. Also listed in Table 2 are two proton beam dump experiments. Due to the production of secondary electrons and positrons in the target, a limit is obtained also for the coupling to electrons.

Table 2. Excluded ranges of the coupling constant α_{Xe} of a pseudoscalar particle of mass 1.8 MeV derived from beam dump experiments.

Experiment	Beam	Target	$\alpha_{Xe}^{\min}$	$\alpha_{Xe}^{\max}$
Konaka et al. (KEK) [63]	e^- (2.5 GeV)	W + Fe(2m)	10^{-14}	4×10^{-8}
Davier et al. (Orsay) [64]	e^- (1.5 GeV)	W (10cm)	10^{-11}	10^{-8}
Riordan et al. (SLAC) [65]	e^- (9 GeV)	W (10-12cm)	10^{-12}	10^{-7}
Bjorken et al. (SLAC) [66]	e^- (20 GeV)	(200m)	3×10^{-16}	
Bechis et al. (Bethesda) [67]	e^- (45 MeV)	Ta (1cm)	10^{-13}	10^{-10}
Brown et al. (FNAL) [68]	p (800 GeV)	Cu (5.5m)	10^{-10}	10^{-7}
Faissner et al. (PSI) [69]	p (590 GeV)	Cu (8m)	2×10^{-20}	

Together, the experiments exclude the range of coupling constants α_{Xe} between 10^{-14} and 10^{-7}, corresponding to lifetimes against pair decay in the range

10^{-14}s $< \tau_X < 10^{-7}$s. When combined with the bounds derived from the electron anomalous magnetic moment a_e and by experimental conditions, the beam dump results conclusively rule out any elementary neutral particle as source of the GSI e^+e^- events.

However there is still a "loop-hole" for neutral particles left, as was revealed by an analysis of A. Schäfer, who calculated the bremsstrahlung production cross section for extended particles [62]. He showed that a finite *form factor* can invalidate the experimental bounds, if the emitted particle has a radius of more than about 100 fm (10^{-11}cm). Basically this comes along with an effective suppression of the e^--X^0-vertex due to the X^0 form factor, when the relevant electronic de Broglie wavelength is small compared to the spatial extension of the X^0.

2.10 Bhabha Scattering at MeV Energies

All the limits on the possible existence of a light neutral X^0-boson discussed so far were derived assuming that the particle has no internal structure. When one allows for a particle with finite size, they become model dependent, as was mentioned in the previous section. Similar considerations apply to the bounds from the anomalous magnetic moment, Delbrück scattering, positronium hyperfine structure, and so on. The reason for this model dependence lies in the fact that all these processes involve particles off their mass shell, either the electron or the X^0-boson, whereas all particles are on mass shell in the pair decay $X^0 \rightarrow e^+ + e^-$. A form factor, therefore, enters in different ways into these processes.

In order to obtain model-independent bounds it is necessary to consider the process, in which the boson is produced on shell by electrons and positrons that are also on mass shell. This can be achieved by Bhabha scattering on resonance [71] which is represented by the Feynman diagram (a) in Fig. 12. The cross section from this diagram alone is narrowly peaked around the beam energy E_R corresponding in the center-of-mass system to the rest mass of the X^0-boson:

$$\sigma_X(E) = \frac{\pi \alpha_{Xe}^2 f_{J^\pi}(m_X/m_e)}{4(E - E_R)^2 + (m_X \Gamma_X/m_e)^2} \tag{30}$$

where $E_R = (m_X^2/2m_e) - 2m_e$, and $f_{J^\pi}(x)$ is a dimensionless function of order unity that depends on spin and parity of the X^0-boson [43]. Right on resonance, i.e. for $E = E_R$, the cross section exhausts the unitarity limit for a single partial wave

$$\sigma_X(E_R) = \frac{\pi \alpha_{Xe}^2 m_e^2 f_{J^\pi}}{m_X^2 \Gamma_X^2} \approx \frac{\pi}{m_X^2} \cdot \frac{\Gamma_X^{e^+e^-}}{\Gamma_X} , \tag{31}$$

unless other final states, such as $\gamma\gamma$, $\gamma\gamma\gamma$ or $\bar{\nu}\nu$, contribute significantly so that the partial e^+e^- width $\Gamma_X^{e^+e^-}$ is smaller than the total width Γ_X. On the other hand, the QED Bhabha scattering cross section, described by the Feynman diagrams (c) and (d) in Fig. 12, is of the order

$$\sigma_{QED}(E_R) \approx \frac{\alpha^2}{m_X^2} \approx 10^{-4} \sigma_X(E_R) . \tag{32}$$

At first glance, therefore, it appears as if the resonance caused by the new particle would give a tremendous signal in Bhabha scattering. In reality, the cross section must be averaged over a finite energy interval ΔE, which depends on the experimental conditions. The QED cross section (32) must then be compared with the energy averaged resonance cross section

$$\frac{1}{\Delta E} \int dE \sigma_X(E) = \frac{2\pi^2 (2J+1)}{m_X^2 - 4m_e^2} \cdot \frac{\Gamma_X^{e^+e^-}}{\Delta E} . \tag{33}$$

In practice, the energy resolution is not determined by the uncertainty in the positron beam energy, but by the Fermi motion of the electrons in the target [70]. In the limiting case of a free electron gas with Fermi momentum k_F, the energy resolution is given by

$$\Delta E = 2k_F \sqrt{2E_R/m_e} = 2k_F \sqrt{\frac{m_X^2}{m_e^2} - 4} . \tag{34}$$

At fixed beam intensity this value can be reduced only at the expense of scattering rate, because the Fermi momentum is related to the electron density n_e in the target, viz. $k_F = (3\pi n_e)^{1/3}$. Using light target materials such as Be, the effective resonance width ΔE is about 30 keV as can be deduced from the Compton profile [71].

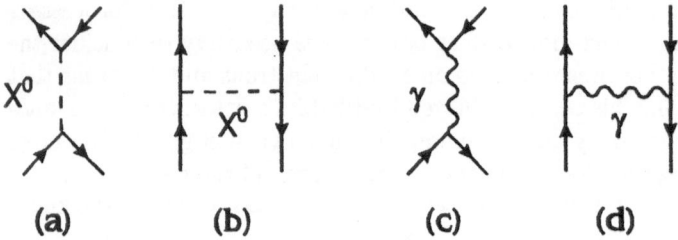

Fig. 12. Feynman diagrams contributing to Bhabha scattering. Only the s-channel diagram (a) is resonant and can compete with the QED process represented by the diagrams (c) and (d).

Several experimental searches for resonances in low-energy Bhabha scattering have been performed [72]. Their negative outcome has narrowed down the maximum allowable resonance width to about $\Gamma_X^{e^+e^-} < 10^{-3}$ eV. This left open an unexplored window of about 3 or 4 orders of magnitude in the lifetime. Fortunately the experimental sensitivity can be vastly increased for long-lived resonances by separating the delayed pair decay of the resonance from the "background" of prompt Bhabha scattering events. This has been done using an active shadow [73, 74] or an energy loss technique [75]. Presently, the highest sensitivity has been achieved by a group working at ILL Grenoble [73] and by a Yale-Brookhaven collaboration [74]. The upper limit for the e^+e^- width and lifetime of a hypothetical spin-1 X^0-boson at rest masses around $m_X \simeq 1.8\,\text{MeV}$ are reported to be [74]:

$$\Gamma_X^{e^+e^-} < 7 \times 10^{-7}\text{eV} \qquad \text{or} \qquad \tau_X^{e^+e^-} > 9 \times 10^{-10}\text{s}. \tag{35}$$

A synopsis [74] of the lifetime regions excluded by various experiments is shown in Fig. 13.

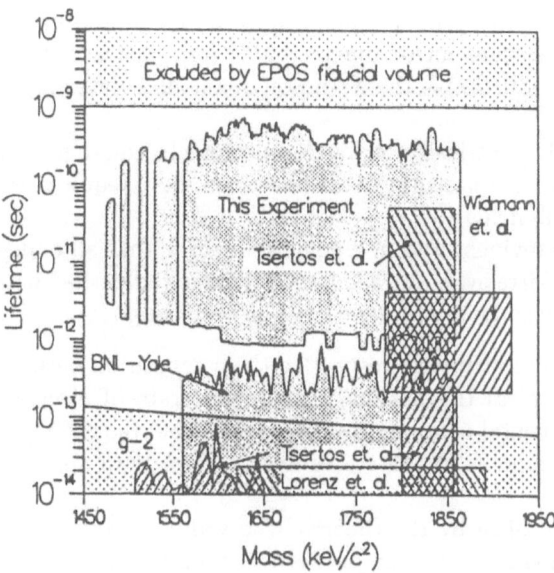

Fig. 13. Excluded regions for a resonance in Bhabha scattering drawn in the mass-lifetime plane assuming spin $J = 0$. Also shown are the lifetime constraints imposed by the fiducial volume of the EPOS detector and by the electron g factor (assuming a structureless point particle).

We can conclude that the Bhabha scattering experiments have now reached a sensitivity that allows to establish limits on new particles which are much more stringent than those derived from other QED precision experiments. The crucial advantage of the new limits is that they are measured on mass shell and, therefore, independent of assumptions about the structure of the X^0-boson.

The latest limit (35) has reached the lifetime bound set by the GSI experiments, i.e. $\tau_X < 10^{-9}$s, which means that there is no room left to explain the positron lines by the decay of an object $X^0 \rightarrow e^+ + e^-$. There are only two caveats: Conceivably the decay might not take place in free space but in a surrounding that is present only in the heavy-ion experiments but not in Bhabha scattering. One scenario for this possibility will be discussed in Sect. 5.4. Furthermore the limit from Bhabha scattering is no longer valid if X^0 were to decay predominantly into other channels, presumably into photons [76].

3 Models of New Extended Neutral Particles

3.1 General Considerations

The postulate of new neutral particles with finite size, or substructure, can simultaneously solve several general difficulties of any explanation of the GSI data in terms of particle decay. These are:

- The fact that several line structures have been seen is naturally explained as the decay of internally excited states of the same particle.
- The small velocity of the pair-decaying source may be explained in two ways: either as a high-momentum cut-off due to the X^0 form factor, if $R_X > 20m_X^{-1} \approx 2000$ fm; or by production of the X^0-boson in a bound state around both nuclei.
- A composite particle with electrically charged constituents could be efficiently produced by some non-perturbative mechanism that requires the presence of strong Coulomb fields.
- As already argued in the previous section, a general bonus is that all experimental limits are rendered irrelevant for a sufficiently large radius R_X, with the exception of those derived from resonant Bhabha scattering.

A general conclusion can be drawn with respect to the competition between two-photon and pair decay. Unless the particle is a bound state of electron-positron pairs, or has a fundamental coupling to the electron field (as the axion would!), the photon decay dominates for all states except those with spin one and negative parity.

Two general routes can be taken by the theorist who wants to construct a model of extended particles in the mass range between 1 and 2 MeV:

- One can speculate that there exists an undiscovered, "hidden" sector of low-energy phenomena within the framework of the standard model of particle physics, i.e. within the $SU(3) \times SU(2) \times U(1)$ gauge theory. This might be a non-perturbative, strongly coupled phase of quantum electrodynamics, low-energy phenomena associated with the Higgs sector of the Glashow-Salam-Weinberg model, or some unknown long-range properties of QCD. It has even been speculated that the standard electromagnetic interaction between charged particles with spin behaves quite differently at short distances than normally assumed in perturbation theory.
- One can invoke new interactions which, for some reason that remains to be explained, do not normally show up in experiments. Examples are many-body forces between electrons and positrons that do not contribute in positronium, or new light fermions that are confined by equally new, medium ranged interaction.

Both roads have been extensively explored, overall with little success. One must be aware that any attempt to fit a scheme of new low-energy composite particles into the standard model faces awesome obstacles, viz. the wealth of experimental data and precision measurements accumulated over fifty or more years. For the first class of models, i.e. those models based on some obscure aspect of the standard model itself, the problem is that there is essentially no free parameter. Every conjectured phenomenon can be calculated reliably, at least in principle.

Another intensively discussed hypothesis, first put forward by Celenza et al. [77] assumes that QED may possess a second strongly coupled phase resembling in its properties the normal vacuum of QCD and that this new vacuum may be formed in heavy ion collisions. Then the GSI peaks are interpreted as being caused by the decay of 'abnormal QED mesons' [78, 79, 80, 81]. However, no mechanism is known [82, 83] by which the field of the colliding heavy ions can trigger a transition to the new phase (which requires a large value of the fine structure constant α [84]). Furthermore it seems doubtful that the new phase is metastable, i.e. it can exist in the absence of the catalyzing nuclear charges as would be required by the characteristics of a two-body decay.

Although free parameters can be introduced in abundance, the second route is no less treacherous. A new force active in the MeV energy range can potentially show up in every atomic, nuclear, or particle physics experiment. This has led, for instance, to the rejection of speculations about many-body forces between electrons.

In the following we do not attempt to give a complete review of the theoretical attempts to construct models for extended X^0-objects, rather we will concentrate on a few selected models.

3.2 New Bound States of the Electron-Positron System

Various attempts have been undertaken to explain the GSI resonances as objects composed of interacting electrons and positrons. In the simplest case one searches for resonant or quasibound states consisting of a single e^+e^- pair. Alternatively bound states having the structure of $(e^+e^-)^2$ or even $(e^+e^-)^n$ have been discussed.

Electron-Positron Resonances. Early speculations on the existence of "unconventional" resonances in the electron-positron system had been put forward by A.O. Barut [90] based on the idea that the strong magnetic interaction caused by the anomalous magnetic moment of the leptons could become dominant at small separations and lead to a new type of highly localized bound states. Numerical calculations based on this [91] or related [92] concepts indeed appeared to show several resonances in the e^+e^- scattering cross section. However, Geiger [93] did not find the effect using Barut's equations and furthermore criticized the underlying concept. More recently in [94] it was demonstrated in detail that the calculation of [92] was in error. In any case, it is obvious from the outcome of the Bhabha scattering experiments described in the last section, that calculations of this type which lead to e^+e^- resonances having a width of several keV are off the mark by at least ten orders of magnitude!

A further group of calculations is less easily discounted. B.A. Arbuzov and collaborators [95] treated the relativistic two-body problem in the framework of the quasipotential method. This amounts to the numerical solution of an involved integral equation for the scattering wave function. The result was a large number of very narrow S-wave resonance states covering the region between the threshold $2m_e$ and several MeV kinetic energy. The authors claimed that their resonances resemble the well-known von Neumann-Wigner bound states embedded in a continuum. Such states can be caused by the constructive interference of the wave function reflected by the ridges of a periodically oscillating potential of the type $V(r) \propto \cos(cr)/r$. However, Walet et al. [96] have demonstrated that the two-body continuum of spinless QED (Wick-Cutkosky model), which was considered by Arbuzov for simplicity, does *not* show any resonances when treated properly.

The single remaining claim for e^+e^- resonances within the framework of QED is that of Spence and Vary [97]. These authors solved several versions of relativistic two-body wave equations in momentum space derived as approximations to QED: A Tamm-Dancoff equation, a Breit equation with positive-energy projection operators, and a Blankenbecler-Sugar equation derived from an instantaneous approximation to the Bethe-Salpeter equation. The numerical solution of these equations consistently produced six electron-positron resonances (dubbed "photonium" states by the authors) at total kinetic energies 329, 476, 637, 808, 987 and 1173 keV. These states appear to have zero width and therefore potentially do not clash with the negative results from the Bhabha scattering experiments. (Note, however, that the annihilation interaction is missing in the calculations which makes this statement doubtful). However, a calculation similar to that of Spence and Vary was performed by Horbatsch [98] who did not find resonances and points out that it is easy to produce resonance-type behaviour through numerical artifacts when solving the integral equations.

Summarizing the previous discussion, at present there is no firm theoretical evidence for the existence of e^+e^- resonances at MeV energies. Claims to the contrary either have been disproved or at least are disputed. The result of Spence and Vary deserves further investigation, a major drawback lies in the fact that the physical mechanism which is supposed to cause the resonance has not been identified.

Let us also note that a general proof has been given by Grabiak [99], which states that within the framework of QED the existence of narrow resonances at MeV energies is impossible since it contradicts the relativistic virial theorem. This argument is valid provided that the state consists mainly of a single e^+e^- pair.

Poly-Positronium, Quadronium. Instead of looking for resonances in the continuum of the e^+e^- two-particle system one also could think of truely bound states of several electron-positron pairs. However, no mechanism is known within the framework of QED for the strong binding required to bring such states several hundred keV below the threshold of at least (for $(e^+e^-)^2$) $4m_e$. (The $(e^+e^-)^2$ system has a very weakly bound state with binding energy of a few eV [86], having the structure of an ordinary positronium molecule.) Nevertheless

J. Griffin [85] conjectured that strongly bound states of two electron-positron pairs ("quadronium") hold the key to the solution of the GSI positron puzzle. He attempted to explain various properties of the GSI experiments in terms of quadronium production and decay. The central question concerning the internal structure of the bound state and its justification from QED was not addressed. A Hartree-Fock calculation failed to give any indication for the existence of $(e^+e^-)^2$ bound states [87].

Strongly bound "poly-positronium" states $(e^+e^-)^n$ probably would require the assumption that some new, non-QED force exists between electrons and positrons. On this basis, a rather satisfactory phenomenological explanation of the GSI events could be constructed [89], if the poly-positronium system would have a size of several 100 fm. The states would be expected to be produced in the heavy ion collision by the action of the strong electric fields with a cross section and kinematic characteristics similar to that of the QED pairs [88].

Is the required new interaction between electrons and positrons compatible with our knowledge of e^+e^- physics? E.g., one might postulate the existence of a short range attractive many-body force that does not act between a single e^+e^--pair, thus avoiding problems in electron-positron scattering at high energy and in the normal positronium system that is well described by QED. The question was systematically studied by Ionescu et al. [89], who considered the limits set by spectroscopic data from heavy atoms on nonlinear interactions of the form

$$L_{\text{int}} = \lambda(\bar{\psi}\psi)^n , \tag{36}$$

where n is some integer greater than one. Such forces would contribute measurably to the K-shell binding energy in heavy atoms, if the effective coupling constant λ is too large. The following limits were obtained in this way: $\lambda(n = 2) < 5 \times 10^{-4}$ and $\lambda(n = 3) < 2 \times 10^{-3}$. On the other hand, the values of λ required to support a poly-positronium bound state are at least $8(n = 2)$ or $130(n = 3)$, respectively [89]. Thus, poly-positronium states based on a new e^+e^--interaction of type (36) can be excluded.

4 A Meson-Type Model for the Composite X^0-State

We refer now to the "second road" and introduce a phenomenologically motivated explicit model [100, 101] for a composite object X^0. The model describes X^0 as a meson-like object built up by a *pair of electrically charged fermionic constituents* f^+f^-, interacting via yet unknown forces which are treated in terms of an effective potential $V(r)$. If the rest masses of the constituents account for the major fraction of the total X^0-mass, one may employ a non-relativistic Schrödinger equation in the relative coordinate $\mathbf{r}$:

$$\left(\frac{\hat{\mathbf{p}}^2}{2m_{\text{red}}} + V(r)\right)\phi(\mathbf{r}) = E\phi(\mathbf{r}) , \tag{37}$$

$m_{\text{red}} = \frac{1}{2}m_{\text{f}}$ being the reduced mass of the $f^{\pm}$. $V(r)$ is the potential between the constituents which is chosen as follows

$$V(r) = \begin{cases} a_X\, r & , \quad r > r_0 \\ V_0 = \text{const.} & , \quad r \leq r_0 \end{cases} . \tag{38}$$

Here V_0 is a positive constant simulating a repulsive interaction of range r_0 between f^+ and f^- ("hard core") the need for which will be explained later. a_X denotes the string tension of the long-range, confining part of the potential which is the analogue of the vacuum pressure B_X in a bag model approach [100]. The parameters m_{f} and a_X can be used to fit the ground state energy and level spacing of X^0. Invoking the flux tube model it is possible to deduce a correspondence between the bag constant B_X and the string tension a_X which has the form [102]

$$a_X = \sqrt{8\pi B_X \alpha_X} \tag{39}$$

with α_X being the coupling constant of the gauge interaction responsible for confinement. Its precise value is of course unknown, but in analogy with the experience from QCD it may be assumed to be of the order of 1. The values of a_X derived from a bag model calculation fixing B_X and subsequently applying (39) and compared to a calculation within the potential model shows fair agreement. For masses $m_{\text{f}} \approx 800\,\text{keV}$ the resulting string tension is of the order $a_X \approx 0.1\,\text{keV/fm}$.

5. Production in Heavy Ion Collisions

5.1 Production of f^+f^- Pairs in Heavy Ion Collisions

The most attractive feature of the f^+f^- model, which was the reason for its conception, is the interaction of the electrically charged constituents with the strong electromagnetic field created by the colliding heavy nuclei. The f^- will be deeply bound by the attractive Coulomb potential while the f^+ is pushed to a larger distance without much change of its energy. Thus the X^0 state can be bound in quasimolecular orbitals around the nuclei and may lose much of its rest mass, allowing for copious production. In order to describe such a configuration, we first assume that the particle is centered at the center-of-mass of the two ions. The negative constituent f^- is strongly attracted to the charge center by the Coulomb interaction of the nuclei. On the other hand, the positively charged constituent f^+ is repelled to the outer boundary of the confinement region, enlarging its radius. Neglecting the contribution of the hard core interaction the total energy of the object can be written as

$$E_{X^0}(Z) = E_{f^-}(Z) + E_{f^+}(Z) . \tag{40}$$

In order to calculate the energy of the negatively charged constituent E_{f^-} in a first approximation one may neglect the influence of the f^+ on the wavefunction of the strongly bound f^- and solve the Dirac equation in the Coulomb field of the two nuclei. Taking the nuclear Coulomb potential $V_C(r)$ in the monopole approximation [103], the Dirac equation

$$\left(\hat{p}\cdot\gamma - m_f - \gamma^0 V_C(r)\right)\psi_{f^-} = 0 \tag{41}$$

was numerically solved. The solution yields an energy $E_{f^-}(Z)$ depending on the total charge of the collision system and the internuclear distance R.

In the case of the positively charged constituent, the interaction with the strongly localized f^- has been taken into account by solving the Dirac equation for the f^+ in the Coulomb field of the nuclei including the scalar potential $V(r)$ of (38) as an additional central potential. Thus the equation of motion reads

$$\left(\hat{p}\cdot\gamma - m_f + \tilde{V}(r)\right)\psi_{f^+} = 0 \tag{42}$$

with the potential

$$\tilde{V}(r) = \gamma^0 V_C(r) + V(r) . \tag{43}$$

By solving eqs.(41,42) for the lowest energy eigenvalues one determines the energy of the X^0-particle in the Coulomb field of the two nuclei. The result in the case of two uranium nuclei can be seen in Fig. 14. The particle is deeply bound in the strong electric field, approaching zero total energy for small internuclear distances R. If the energy were to decrease below zero, spontaneous supercritical production of the neutral particle could occur. Even if this does not happen, the small energy gap favours the pair production of bound f^+f^--pairs in the collision.

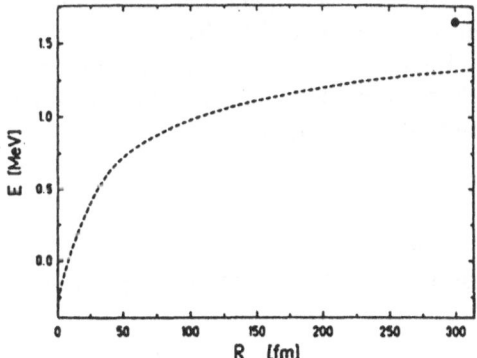

Fig. 14. Energy of the extended 1.8 MeV particle in the Coulomb field of two Uranium nuclei for m_f=850 keV as function of nuclear separation R. The energy of the particle bound to a single U-nucleus is marked to be 1.65 MeV.

The f^--state of lowest energy is strongly bound with a binding energy which – depending on the total charge of the collision system – may exceed $2m_f$. The f^+ states very much resemble positron states, but their energy levels are discretized due to the confinement in the bag.

As in the case of electrons and positrons the wavefunctions of the constituent particles vary strongly with the nuclear distance due to the rapid change of the Coulomb fields of the nuclei. This yields large dynamical transition matrix elements

$$M_{ij} \sim \langle \varphi_{i,\mathfrak{f}-} | \partial/\partial t | \varphi_{j,\mathfrak{f}+} \rangle \tag{44}$$

which mediate the creation of $(\mathfrak{f}^+\mathfrak{f}^-)$-pairs. These pairs are confined due to the potential (38) giving rise to neutral states. For an exact treatment of the electromagnetic production process in principle one has to take into account the confinement interaction (38) between the particles. However, since the extension of the particle is large, $R_X \sim 1000$ fm, an estimate of the total production probability of the $(\mathfrak{f}^+\mathfrak{f}^-)$ states can be obtained by treating the $\mathfrak{f}^\pm$ like ordinary electrons and positrons with the mass replaced by $m_\mathfrak{f}$. The production of $\mathfrak{f}^+\mathfrak{f}^-$ pairs then can be calculated with the same coupled channel code which was successfully applied to dynamical electron-positron production [104].

A numerical calculation in a U+U collision for a constituent mass $m_\mathfrak{f} = 900$ keV leads to a value of $P_{\mathfrak{f}+\mathfrak{f}-}$ which is ten times larger than the calculated total e^+-production in the same heavy-ion collision and about three orders of magnitude larger than the corresponding cross section of the observed line structure. The large production probability originates from the fact that in contrast to the case of electrons there are no occupied $\mathfrak{f}^-$-states in the beginning of the collision and therefore no Pauli suppression for production of $\mathfrak{f}^-$ particles in bound states occur. Note that the given numbers do not yet include the effect of the hard core repulsion.

The intensity of the correlated e^+e^- line structures in heavy-ion collisions thus can be explained within the $\mathfrak{f}^+\mathfrak{f}^-$ model. At this point one should further mention that the similarity of the production process of X^0 with dynamical and spontaneous e^+e^- production in the collision suggests that its production cross section should scale with the total nuclear charge $Z_u = Z_1 + Z_2$ of the collision system like the positron cross section in (20), i.e. roughly

$$\sigma_X \sim (Z_1 + Z_2)^{20} . \tag{45}$$

The experiental evidence on this point is ambiguous. The strong increase with nuclear charge reported earlier no longer seems to be supported by experiment [106].

5.2 Final State Effects

When the extended particle has been created as a bound state in the center-of-mass frame of the nuclei one has to consider the break-up of the collision system. Although a detailed study of the dynamics has not been performed, one may look at the energy of the particle in the Coulomb field of a single ion in order to get an insight into the strength of binding of the neutral state to target or projectile. One solves (41,42) with the Coulomb potential of a single ion. The energy of the total state, given by (40), is about 1.65 MeV as marked in Fig. 14. The X^0 in the Coulomb field of a single nucleus still has a binding energy larger than 100 keV for constituent masses $m_\mathfrak{f} > 800$ keV.

Combining these results one cannot definitely answer the question what happens to the produced particle after the collision. Although adiabatically an X^0 produced in its lowest energy state should be dragged along with a single target or projectile ion, the influence of the dynamics can change this behaviour. It seems to be plausible to expect that a fraction of the produced particles is getting bound by a single ion and another fraction is set free with small velocity with respect to the CM system. This may explain the experimental finding that the difference energy of some of the correlated e^+e^- lines is not centered at $E_{e^+} - E_{e^-} = 0$ but is shifted to positive values. In some measurements there seem to be indications for two 'peaks' in the difference energy spectrum, one at approximately zero difference energy and one several hundred keV off zero. In addition the decay into e^+e^- pairs from such a bound state would of course not necessarily exhibit a back-to-back correlation. We will come back to this point at the end of the next section.

5.3 Influence of Short Range Repulsion

Decay of the X^0. The neutral X^0 state can decay in a similar way as the lowest states of the charmonium system [107]. Since the constituents are electrically charged the object can decay into photons or electrons and positrons depending on the quantum numbers of the specific state. In the case of an 0^{-+}- ('para'-) state with opposite spins of the constituents the f^+f^- may annihilate into two photons.

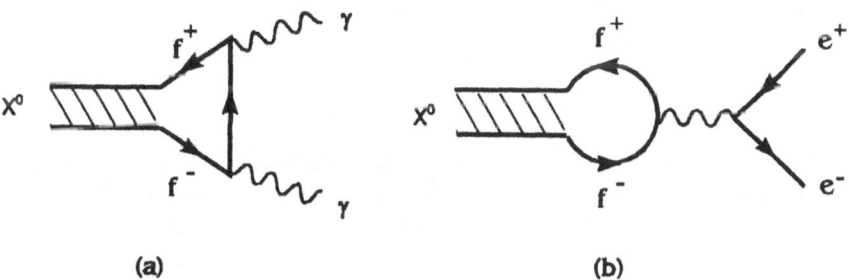

(a)   (b)

Fig. 15. Feynman diagrams for X^0 decay in a) two photons and b) into an e^+e^- pair.

The 1^{--}- ('ortho'-) state can annihilate into a virtual photon which subsequently decays into a correlated e^+e^--pair. A calculation of the diagrams in Fig. 15 for non-relativistic bound states yields the decay widths

$$\Gamma_{0^{-+}\to\gamma\gamma} = \alpha^2 \frac{4\pi}{2\left(\frac{M_X}{2} - m_f - \sqrt{\frac{M_X^2}{4} + m_f^2}\right)^2} \frac{M_X^2}{m_f^2} |\phi_{f\bar{f}}(0)|^2 \qquad (46)$$

and

$$\Gamma_{1^{--}\rightarrow e^+e^-} = \alpha^2 \frac{16\pi}{3}\left(4 + \frac{16}{9}\frac{M_X^2}{m_e^2}\right)\frac{m_e^2}{M_X^5}\left(\frac{M_X^2}{4} - m_e^2\right)^{\frac{1}{2}}|\phi_{f\bar{f}}(0)|^2 \qquad (47)$$

which depend on the probability to find the constituents at the same place, $|\phi_{f\bar{f}}(0)|^2$. The resulting decay widths and corresponding lifetimes for e^+e^- or two γ-decay are shown in Table 3 for two sets of parameters for the potential (38) which differ in the strength of the hard core. The upper limit for the lifetime of a decaying neutral object set by the heavy-ion collisions is given by the condition that the particle should decay inside of the experimental set-up which yields a value $\tau < 10^{-9}$s which is satisfied by the results shown in Table 3.

Table 3. Expectation values and decay widths of f^+f^- states for two different sets of parameters.

m_f	0.85 MeV	
a_X	$9.05 \cdot 10^{-3}$ MeV^2	
R_0	4.64 fm	$4.64 \cdot 10^{-3}$ MeV
V_0	340 MeV	$234 \cdot 10^3$ MeV
E_{1s}	1.8 MeV	
E_{2s}	1.88 MeV	
E_{2p}	1.847 MeV	
$\langle r\rangle_{1s}$	1560 fm	
$\langle r\rangle_{2s}$	2725 fm	
$\Gamma_{e^+e^-}(1^{--})$	$7.0 \cdot 10^{-6}$ eV	$2.56 \cdot 10^{-4}$ eV
$\Gamma_{\gamma\gamma}(0^{-+})$	$3.1 \cdot 10^{-5}$ eV	$1.14 \cdot 10^{-3}$ eV
$\Gamma_{\rm rad}(2p \rightarrow 1s)$	2.4 eV	2.4 eV

However, (46) points to a serious problem: Assuming the radial wave functions to be independent of the spin configuration, the two-photon decay of the $J = 0$ states is predicted to be faster than the e^+e^- pair decay of the 1^{--}-states! Since in a heavy-ion collision both sets of states should be populated with comparable strength this is in conflict with the non-observation of correlated photon pairs in Ref. [109].

As already discussed, a lower limit for the lifetime of the particle state can be found from recent Bhabha scattering experiments [73, 74] to give $\tau > 9 \cdot 10^{-10}$s. This calls for a very strong hard core which will suppress the amplitude for annihilation and therefore increase the lifetime. But simultaneously this approach decreases the production cross section in heavy ion collisions and will run into conflict when attempting to explain the measured line intensities.

QED Precision Tests. The contribution of X^0 to QED precision tests would be unacceptably large without the suppression brought about by the postulated

hard core repulsion. Using dispersion relation techniques the contribution to the anomalous magnetic moment can be expressed [101] in terms of the production cross section $\sigma_{e+e-\to X^0}$ which is easily calculated within the model. The anomalous magnetic moment of the muon turns out to impose the most stringent bounds on the model's parameters [101]. The results are governed by the influence of the hard core, which provides an energy dependent suppression factor at each f^+f^--vertex.

The range of parameters V_0 and r_0 characterizing the repulsive core compatible with the corresponding measurements and experimental requirements on the lifetime τ_{e+e-} (not including, however, the latest results [73, 74] from Bhabha scattering which have closed the remaining gap of allowed parameters) is depicted in Fig. 16. The model clearly needs the hard core.

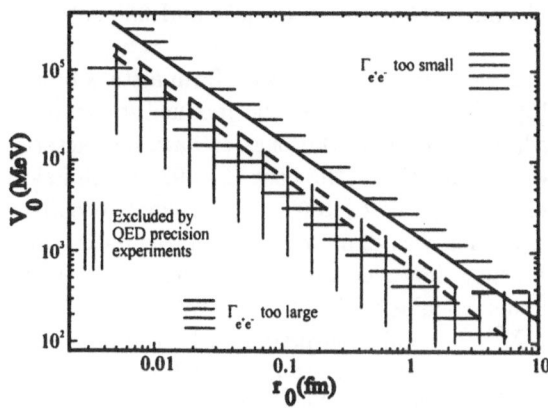

Fig. 16. The range of parameters V_0 and r_0, compatible with QED precision measurements and experimental constraints on the decay width Γ_{e+e-}.

f^+f^--Production in Heavy Ion Collisions. To estimate the total f^+f^--production the coupled channel equations for the production of free fermion pairs of mass m_f were solved. To account for the hard core the result was multiplied by an energy-dependent Gamov factor $\gamma(E)$, which depends on its height V_0 and width r_0. As an example, the parameters: $V_0 = 234$ GeV, $r_0 = 4.64 \cdot 10^{-3}$ fm, $m_f = 850$ keV yield a low-energy Gamow factor $\gamma(2m_0)^2 = 1.7 \cdot 10^{-3}$. The resulting g-2-contributions, as well as the results for the Lamb shift in transitions of muonic lead, are compatible with experimental uncertainties [101]. The total rate of dynamical f^+f^--pair production now yields a fraction of about 1.7% of the dynamically induced e^+e^- pairs, which compares reasonably well with the experimental value.

After the collision, a fraction of the created composite particles may remain bound to one of the separating nuclei. The X^0 ground state and possibly various excited bound states will be populated according to the collision dynamics. Annihilation of a bound X^0 could in principle give rise to e^+e^- coincident lines at discrete energies below M_{X^0}. According to Fig. 14 the model predicts that the energy shift is of the order of -150 keV for the ground state (f^- in a $1s$ orbit around the nucleus) and perhaps -50 keV for a possible excited state (f^- in a $2s$ orbit). (Precise numbers for the mass shifts require a three-body approach. Such calculations recently have been performed by Ehrnsperger [108], who used a variational method to solve the nonrelativistic problem of two charged particles in a fixed Coulomb potential.) Whether the excited states show up in the e^+e^--sum-energy spectrum depends on the competition between annihilation and radiative de-excitation. Since the bound X^0 decay proceeds in the vicinity of the nucleus, its two-body characteristics (angular correlation, e^+e^- energy difference) will be disturbed. The angular correlation of e^+e^- pairs has been most closely studied in the U+Ta system. Pair emission which is clearly not back-to-back has been reported for the lines at 750 keV and 620 keV by EPOS [37] and 635 keV and 805 keV by ORANGE [39].

However, actual calculations for the decay $X^0 \rightarrow e^+ + e^-$ from a state bound to a heavy nucleus by Stein [110] have revealed that the momentum transfer to the nucleus is not sufficiently large to destroy the back-to-back characteristics of the emitted pair. As shown in Fig. 17 for the case $Z = 92$ the emission occurs predominantly at an opening angle $\Theta \simeq 180^0$, with a small broadening of about $\Delta\Theta \simeq 25^0$. (The calculation was done in Born approximation and does not include Coulomb effects which, however, are not expected to alter the conclusions.) A further result of this calculation [110] is the rate for the competing *single-photon annihilation* $X^0 \rightarrow \gamma$ which turns out to be the dominant decay channel of the bound X^0 in heavy atoms. For uranium the branching ratio $\Gamma_\gamma/\Gamma_{e^+e^-}$ is of the order 10. Both results argue against the hypothesis that some of the e^+e^- lines can be interpreted as being caused by the decay of an X^0 state bound to one of the nuclei emerging from the collision.

5.4 Induced X^0 Decay

As we have seen in Sect. 2.10 the last generation of Bhabha scattering experiments largely rules out the existence of any particle states X^0 decaying into e^+e^- pairs in vacuo in the lifetime region relevant for the GSI experiments. Apart from postulating competing but unobserved decay branches there remains one way to avoid the Bhabha constraint: The creation and/or decay of X^0 might depend on external conditions, in particular the presence of strong fields, which are only realized in the heavy-ion experiments.

The f^+f^- model has the potential to fit into this framework [111]: Assume that the ground state of X^0 has the spin and parity assignment 1^{++}. Such a state will be metastable since it can not decay into 2γ or 3γ; the decay $X^0 \rightarrow e^+ + e^-$ can only proceed via at least two virtual photons and thus is suppressed by a factor α^2. This observation suggests the following scenario: In the heavy-ion experiments the X^0 is produced in various excited states from which it rapidly cascades to the postulated 1^{++} groundstate. In a *secondary collision* with an

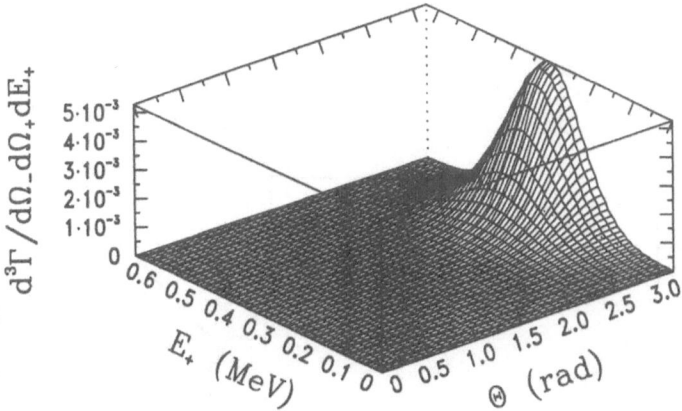

Fig. 17. The triple differential pair decay width $d^3\Gamma/d\Omega_{e^-}d\Omega_{e^+}dE_{e^+}$ assuming that X^0 is bound to a uranium nucleus, drawn as a function of positron kinetic energy and opening angle.

atom of the high-Z target the 1^{++} subsequently undergoes an induced decay to the e^+e^- pair observed in the GSI measurements. In the Bhabha experiments the X^0 may be produced in a 1^{--} state via a single photon and undergo a radiative decay to the 1^{++} state. We then expect the field of the low-Z target atoms used in these experiments not to be strong enough to induce the e^+e^- decay, furthermore the 1^{--} state may be shifted outside the tested energy range.

The induced-decay scenario described above agrees with suggestions from the ORANGE group. According to Koenig [106] some of the e^+e^- lines with narrow sum energy and small opening angle might be explained by a two-step process involving a third heavy partner which takes up the recoil but no energy.

E. Stein [111] has calculated the $X^0(1^{++})$ annihilation cross sections into an e^+e^- pair and into a single photon in the field of a nucleus with charge Z, cf. the Feynman graphs of Fig. 18. The Bhabha experiments differ in two aspects from the GSI setup. They are characterized by the use of low-Z targets ($Z = 4$) and high particle velocity ($v = 0.83c$). The hypothetical particle in heavy-ion collisions is expected to travel with low velocity ($v \simeq 0.05 \ldots 0.1c$) whereas the target-$Z$ is very high (e.g., $Z = 92$). Both differences tend to suppress the induced annihilation process in Bhabha scattering. For the parameters of a typical Bhabha experiment Stein finds total cross sections $\sigma_{X^0 \to e^+e^-} = 6.3 \cdot 10^{-30} \text{cm}^2$ and $\sigma_{X^0 \to \gamma} = 1.4 \cdot 10^{-29} \text{cm}^2$. This is to be compared with the corresponding values for the GSI experiments which are $\sigma_{X^0 \to e^+e^-} = 5.7 \cdot 10^{-24} \text{cm}^2$ and $\sigma_{X^0 \to \gamma} = 1.2 \cdot 10^{-26} \text{cm}^2$. Obviously the envisaged scenario appears to work: the induced decay of $X^0(1^{++})$ would not be visible in the Bhabha experiments since its cross section is suppressed by six orders of magnitude. (There are additional suppression effects: the single-photon decay is favoured over pair decay, and the two-body kinematics is severely disturbed, with a mean momentum transfer to the recoiling nucleus $\langle \mathbf{q}^2 \rangle e^{1/2} \simeq 800$ keV, hiding the resonance from detection).

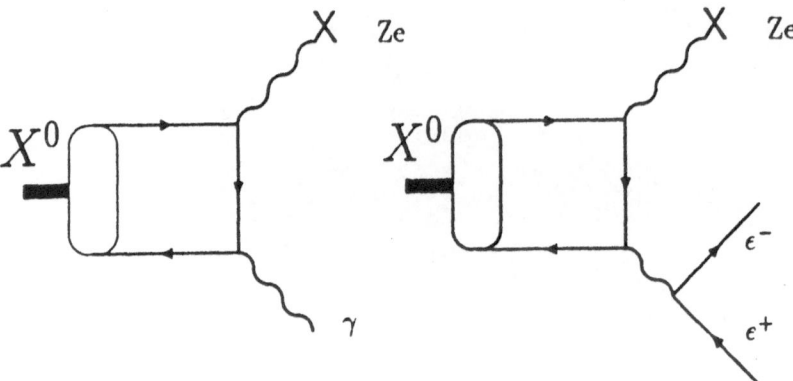

Fig. 18. Feynman diagrams for the induced X^0 annihilation in the field of a target nucleus.

However, unfortunately the story does not end on this optimistic note. The proposed two-step mechanism only can work if the cross section for the induced pair decay is large enough so that a significant fraction of the particles can decay within the target foil. For the $380\mu g/cm^2$ U target employed at GSI this would call for a cross section $\sigma_{X^0 \rightarrow e^+e^-} > 10^{-20} cm^2$, five to six orders of magnitude larger than the value calculated from the model. Thus we have to conclude that the idea of a metastable particle undergoing induced decay shows some promising potential to reconcile the controversial outcome of heavy-ion and Bhabha experiments but it fails the test when it comes to quantitative calculations.

6 Conclusions

The subject of Quantum Electrodynamics of strong fields presents a fascinating area of research. It offers a "clean" laboratory where a fundamental quantum field theory can be studied theoretically and tested through experiment. One facet, upon which we did not touch in these lectures, are high-precision checks of QED radiative corrections, e.g. the Lamb shift, which with the advent of highly stripped heavy ion beams now can be studied in the nonperturbative high-Z regime. Apart from these topics, which are an extension of the traditional investigations of QED pursued for nearly half a century, a qualitatively new phenomenon is expected to occur under the influence of strong external fields: the decay of the neutral to a charged vacuum.

This phenomenon is intimately connected with the relativistic orbitals and the deep binding acquired by the inner shell electrons when the nuclear charge gets of the order $Z \simeq 1/\alpha$. Collisions of very heavy ions are the only means to realize this situation in an experiment, although impeded by the short duration of these collisions. Elaborate coupled-channel calculations based on the quasi-molecular picture had to be performed to understand the dynamics of the electron shell during such a collision. Inner-shell hole production, δ-electron emission and positron creation all are sensitive to the strong electric field. These processes have been studied experimentally in great detail and are well described by theory. Although an accurate "spectroscopy of superheavy quasimolecules" still has not been realized there is clear evidence for the rapid increase in binding energy and strong localization of inner shell orbitals in high-Z systems

The ultimate goal to detect the process of spontaneous positron creation and thus the instability of the QED vacuum in the presence of a supercritical electromagnetic field, however, remains elusive. To overcome the problem posed by the short time scale of supercriticality ($\tau \simeq 10^{-21}$s) hopes have rested on the idea to select collisions in which a nuclear reaction with sufficient time delay occurs. Whether such a situation can be realized in an experiment still is an open question which should be addressed in future experiments laying particular emphasis on the nuclear-physics side of the collision.

During the last decade the development of this field was overshadowed by the spectacular narrow lines in the positron spectrum and later the monoenergetic electron-positron pairs discovered by the EPOS and ORANGE groups at GSI. Attempts to link this effect to the spontaneous positron production process looked promising at the beginning. However, the effect was found to be largely independent of the nuclear charge and has been seen also in subcritical systems like Th+Ta ($Z_u = 163$) and perhaps even in Xe+Au collisions ($Z_u = 133$).

This apparent universality of the positron lines has created much excitement and led to the belief that some fundamental new process had been discovered. A large variety of speculations, most of them based on very shaky ground, were put forward to explain the observations. The most natural explanation for a constant line energy and two-body decay characteristics would be the creation and subsequent decay of a new elementary particle, e.g. the axion. This, however, soon could be ruled out by various arguments, in particular by many control experiments (high-energy beam dump searches, pair production in nuclear transitions). The discovery of several e^+e^- lines at different energies and the fact that their angular correlation, which in the first experiments pointed to a two-body decay, in several cases no longer is back-to-back has complicated the picture considerably.

The particle hypothesis had to be modified to describe more complicated objects with internal structure. These might live within QED, e.g. as polypositronium states or as resonances caused by the two-body interaction of electron and positron. No convincing calculations have been put forward to support these ideas. A more radical phenomenological approach, the f^+f^- model, treats the constituents of the object under considerations as new particles with an equally new confining interaction. We have discussed the properties and problems of this model in some detail in these lectures. Employing an additional ad hoc assumption (a repulsive short range interaction between the constituents) it was possible to explain quite a number of features of the GSI measurements without running into blatant conflicts with established facts in other areas of physics.

However, a probably fatal blow was dealt at the hypothesis of a new particle, be it elementary or composite, by a set of experiments looking for resonances in electron-positron scattering in the mass region around 1.8 MeV. The outcome of these experiments (which are sensitive to resonances with a width down to the μeV level and have fully covered the relevant region of life times) has been completely negative.

Thus one has to conclude that the GSI positron lines are only observable in experiments which involve heavy ions. The complexity of the phenomenon (threshold-type impact energy dependence near the Coulomb barrier, a collection of lines with varying angular dependence, ...) feeds the suspicion that some kind of nuclear process is responsible. No consistent "conventional" explanation for the whole set of experimental data has been found so far but we feel that all aspects of unusual nuclear pair conversion processes should be reanalyzed with priority.

Hopefully the ongoing independent experimental investigation by the APEX group (Argonne National Lab.) will shed more light on the phenomenon of the positron lines. Whatever the outcome, the interest should not be detracted from the subject of QED of strong fields.

References

1. B. Müller, H. Peitz, J. Rafelski, W. Greiner: Phys. Rev. Lett. **28** (1972) 1235
2. B. Müller, J. Rafelski, W. Greiner: Z. Phys. **257** (1972) 62 and 183
3. Ya. Zel'dovich, V. S. Popov: Sov. Phys. Usp. **14** (1972) 673
4. J. Rafelski, B. Müller, W. Greiner: Nucl. Phys. **B68** (1974) 585
5. L. P. Fulcher, A. Klein: Ann. Phys. (N. Y.) **84** (1974) 335
6. M. Gyulassy: Nucl. Phys. **A244** (1975) 497
7. G. A. Rinker, L. Wilets: Phys. Rev. **A12** (1975) 748
8. G. Soff, P. Schlüter, B. Müller, W. Greiner: Phys. Rev. Lett. **48** (1982) 1465
9. W. Greiner: Opening remarks in: "Quantum Electrodynamics of Strong Fields", NATO-ASI at Lahnstein, ed. by W. Greiner, Plenum, N.Y. (1983)
10. B. Müller, J. Rafelski, W. Greiner: Phys. Lett. **47B** (1973) 5
11. G. Soff, B. Müller, W. Greiner: Phys. Rev. Lett. **40** (1978) 540
12. J. Bang, J. H. Hansteen: K. Dan. Vidensk. Selsk. Mat. Fys. Medd. **31** (1959) No 13
13. B. Müller, G. Soff, W. Greiner, V. Ceausescu: Z. Phys. **A285** (1978) 27
14. B. Müller, J. Reinhardt, W. Greiner, G. Soff: Z. Phys. **A311** (1983) 151
15. F. Bosch: Z. Phys. **A296** (1980) 11
16. D. Liesen et al.: Phys. Rev. Lett. **44** (1980) 983
17. W. Betz: Ph.D. thesis, University of Frankfurt (1980)
18. W. Greiner, B. Müller, and J. Rafelski: "Quantum Electrodynamics of Strong Fields", Springer, Berlin-Heidelberg (1985)
19. J. Reinhardt, B. Müller, and W. Greiner: Phys. Rev. **A24** (1981) 103
20. G. Soff, J. Reinhardt, B. Müller, and W. Greiner: Phys. Rev. Lett. **43** (1979) 1981
21. R. Anholt: Phys. Lett. **B88** (1979) 262

22. O. Graf, J. Reinhardt, B. Müller, W. Greiner, and G. Soff: Phys. Rev. Lett. **61** (1988) 2831

23. R. Krieg, E. Bozek, U. Gollerthan, E. Kankeleit, G. Klotz-Engmann, M. Krämer, U. Meyer, H. Oeschler, and P. Senger: Phys. Rev. **C34** (1986) 562

24. H. Backe, P. Senger, W. Bonin, E. Kankeleit, M. Krämer, R. Krieg, V. Metag, N. Trautmann, and J.B. Wilhelmy: Phys. Rev. Lett. **50** (1983) 1838

25. J. Rafelski, B. Müller, and W. Greiner: Z. Phys. **A285** (1978) 49

26. J. Reinhardt, U. Müller, B. Müller, and W. Greiner: Z. Phys. **A303** (1981) 173

27. U. Heinz, U. Müller, J. Reinhardt, B. Müller, and W. Greiner: J. Phys. **G11** (1985) L169

28. M. Clemente, E. Berdermann, P. Kienle, H. Tsertos, W. Wagner, C. Kozhuharov, F. Bosch, and W. Koenig: Phys. Lett. **B137** (1984) 41

29. U. Müller, G. Soff, T. de Reus, J. Reinhardt, B. Müller, and W. Greiner: Z. Phys. **A313** (1983) 263

30. P. Schlüter, T. de Reus, J. Reinhardt, B. Müller, and G. Soff: Z. Phys. **A314** (1983) 297

31. J. Schweppe, A. Gruppe, K. Bethge, H. Bokemeyer, T. Cowan, H. Folger, J.S. Greenberg, H. Grein, S. Ito, R. Schule, D. Schwalm, K.E. Stiebing, N. Trautmann, P. Vincent, and M. Waldschmidt: Phys. Rev. Lett. **51** (1983) 2261

32. T. Cowan, H. Backe, M. Begemann, K. Bethge, H. Bokemeyer, H. Folger, J.S. Greenberg, H. Grein, A. Gruppe, Y. Kido, M. Klüver, D. Schwalm, J. Schweppe, K.E. Stiebing, N. Trautmann, and P. Vincent: Phys. Rev. Lett. **54** (1985) 1761

33. "Physics of Strong Fields", ed. by W. Greiner, NATO Advanced Study Institute Series B, vol. **153**, Plenum, New York (1986)

34. J.S. Greenberg and P. Vincent: Heavy ion atomic physics - experimental, in: "Treatise on Heavy Ion Science", vol. **5**, ed. by D.A. Bromley, Plenum Press, New York (1985) p.141

35. A. Schäfer, J. Reinhardt, B. Müller, W. Greiner, G. Soff: J. Phys. G: Nucl. Phys. **11** (1985) L69-L74

36. T. Cowan, H. Backe, K. Bethge, H. Bokemeyer, H. Folger, J.S. Greenberg, K. Sakaguchi, D. Schwalm, J. Schweppe, K.E. Stiebing, P. Vincent: Phys. Rev. Lett. **56** (1986) 444

37. P. Salabura, H. Backe, K. Bethge, H. Bokemeyer, T.E. Cowan, H. Folger, J.S. Greenberg, K. Sakaguchi, D. Schwalm, J. Schweppe, K.E. Stiebing: Phys. Lett. **B245** (1990) 153

38. W. Koenig, E. Berdermann, F. Bosch, S. Huchler, P. Kienle, C. Kozhuharov, A. Schröter, S. Schuhbeck, H. Tsertos: Phys. Lett. **B218** (1989) 12

39. I. Koenig, E. Berdermann, F. Bosch, P. Kienle, W. Koenig, C. Kozhuharov, A. Schröter, H. Tsertos: Z. Phys. **A346** (1993) 153

40. Ch. Hofmann: Ph.D. thesis, University of Frankfurt (1994) and to be published

41. A.B. Balantekin, C. Bottcher, M. Strayer, and S.J. Lee: Phys. Rev. Lett. **55** (1985) 461

42. R.S. Van Dyck, P.B. Schwinberg, and H.G. Dehmelt in: "Atomic Physics 9", p. 53, ed. by R.S. Van Dyck and E.N. Fortson, World Scientific, Singapore (1984).

43. J. Reinhardt, A. Schäfer, B. Müller, and W. Greiner: Phys. Rev. **C33** (1986) 194

44. R. Barbieri and T.E.O. Ericson: Phys. Lett. **B 57** (1975) 270 ; U.E. Schröder: Mod. Phys. Lett. **A1** (1986) 157

45. A. Schäfer, J. Reinhardt, W. Greiner, and B. Müller: Mod. Phys. Lett. **A1** (1986) 1

46. A. Chodos and L.C.R. Wijewardhana: Phys. Rev. Lett. **56** (1986) 302

47. B. Müller and J. Rafelski: Phys. Rev. **D34** (1986) 2896
48. Y. Yamaguchi and H. Sato: Phys. Rev. **C35** (1987) 2156
49. A. Schäfer, B. Müller, and J. Reinhardt: Mod. Phys. Lett. **A2** (1987) 159
50. D. Carrier, A. Chodos, and L.C.R. Wijewardhana: Phys. Rev. **D34** (1986) 1332
51. A. Schäfer, J. Reinhardt, B. Müller, and W. Greiner: Z. Phys. **A324** (1986) 243
52. D. Neubauer, A. Schäfer, and W. Greiner: to be published
53. B. Müller and J. Reinhardt: Phys. Rev. Lett. **56** (1986) 2108
54. S. Barshay: Mod. Phys. Lett. **A1** (1986) 653
55. B. Müller in: "Intersections between Particle and Nuclear Physics", ed. by D.F. Geesaman, AIP Conf. Proceed. **150** (1986) 827
56. L.M. Krauss and F. Wilczek: Phys. Lett. **B173** (1986) 189
57. R.D. Peccei and H.R. Quinn: Phys. Rev. **D16** (1977) 1791;
    S. Weinberg: Phys. Rev. Lett. **40** (1978) 223;
    F. Wilczek: Phys. Rev. Lett. **40** (1978) 279
58. N.C. Mukhopadhyay and A. Zehnder: Phys. Rev. Lett. **56** (1986) 206
59. G. Mageras, P. Franzini, P.M. Tuts, S. Youssef, T. Zhao, J. Lee-Franzini, and R.D. Schamberger: Phys. Rev. Lett. **56** (1986) 2672
60. T. Bowcock et al. (CLEO collaboration): Phys. Rev. Lett. **56** (1986) 2676
61. H. Albrecht et al. (ARGUS collaboration): Phys. Lett. **B179** (1986) 403
62. A. Schäfer: Phys. Lett. **211B** (1988) 207
63. A. Konaka et al.: Phys. Rev. Lett. **57** (1986) 659
64. M. Davier, J. Jeanjean, and H. Nguyen Ngoc: Phys. Lett. **B180** (1986) 295
65. E.M. Riordan et al.: Phys. Rev. Lett. **59** (1987) 755
66. J.D. Bjorken et al.: Phys. Rev. **D38** (1988) 3375
67. D.J. Bechis, T.W. Dombeck, R.W. Ellsworth, E.V. Sager, P.H. Steinberg, L.J. Tieg, J.K. Joh, and R.L. Weitz: Phys. Rev. Lett. **42** (1979) 1511
68. C.N. Brown et al.: Phys. Rev. Lett. **57** (1986) 2101
69. H. Faissner et al.: Z. Phys. **C44** (1989) 557
70. J. Reinhardt, A. Scherdin, B. Müller, and W. Greiner: Z. Phys. **A327** (1987) 367
71. A. Scherdin, J. Reinhardt, W. Greiner, B. Müller: Rep. Prog. Phys **54** (1991) 1
72. J. van Klinken, W.J. Meiring, F.W.N. deBoer, S.J. Schaafsma, V.A. Wichers, S.Y. van der Werf, G.C.Th. Wierda, H.W. Wilschut, H. Bokemeyer: Phys. Lett. **B205** (1988) 223;
    H. Tsertos, C. Kozhuharov, P. Armbruster, P. Kienle, B. Krusche, K. Schreckenbach: Phys. Lett. **B207** (1988) 273;
    E. Lorenz, G. Mageras, U. Stiegler, I. Huszár: Phys. Lett. **B214** (1988) 10;
    H. Tsertos, C. Kozhuharov, P. Armbruster, P. Kienle, B. Krusche, K. Schreckenbach: Phys. Rev. **D40** (1989) 1397;
    A.L. Hallin, F.P. Calaprice, R.A. McPherson, E.R.J. Saettler: Phys. Rev. **D45** (1992) 3955;
    X.Y. Wu, P. Asoka-Kumar, J.S. Greenberg, S.D. Henderson, H. Huomo, K.G. Lynn, M.S. Lubell, J. McDonough, B.F. Phlips, M. Vehanen: Phys. Rev. Lett. **69** (1992) 1729
73. S.M. Judge, B. Krusche, K. Schreckenbach, H. Tsertos, P. Kienle: Phys. Rev. Lett. **65** (1990) 972;
    H. Tsertos, P. Kienle, S.M. Judge, K. Schreckenbach: Phys. Lett. **B266** (1991) 259
74. S.D. Henderson, P. Asoka-Kumar, J.S. Greenberg, K.G. Lynn, S. McCorkle, J. McDonough, B.F. Phlips, M. Weber: Phys. Rev. Lett. **69** (1992) 1733

75. E. Widmann, W. Bauer, S. Connell, K. Maier, J. Major, A. Seeger, H. Stoll, F. Bosch: Z. Phys. **A340** (1991) 209
76. J.J. Griffin: Phys. Rev. **C47** (1993) 351
77. L.S. Celenza, V.K. Mishra, C.M. Shakin, and K.F. Liu: Phys. Rev. Lett. **57** (1986) 55
78. D.G. Caldi and A. Chodos: Phys. Rev. **D36** (1987) 2876
79. Y.J. Ng and Y. Kikuchi: Phys. Rev. **D36** (1987) 2880
80. C.W. Wong: Phys. Rev. **D37** (1988) 3206
81. S. Barshay: Mod. Phys. Lett. **A7** (1992) 1843
82. E. Dagotto and H.W. Wyld: Phys. Lett. **B205** (1988) 73
83. R.D. Peccei, J. Solà, and C. Wetterich: Phys. Rev. **D37** (1988) 2492
84. J. Kogut, E. Dagotto, and A. Kocić: Phys. Rev. Lett. **60** (1988) 772
85. J.J. Griffin: J. Phys. Soc. Jpn. **58** (1989) 427;
    J.J. Griffin: Phys. Rev. Lett. **66** (1991) 1426;
    J.J. Griffin: Int. J. Mod. Phys. **A6** (1991) 1985
86. E.A. Hylleraas and A. Ore: Phys. Rev. **71** (1947) 493;
    Y.K. Ho: Phys. Rev. **A33** (1986) 3584
87. D.C. Ionescu, B. Müller, W. Greiner: J. Phys. **G15** (1989) L103
88. J.M. Bang, J.M. Hansteen, and L. Kocbach: J. Phys. **G13** (1987) L281
89. D.C. Ionescu, J. Reinhardt, B. Müller, W. Greiner, and G. Soff: J. Phys. **G14** (1988) L143
90. A.O. Barut and J. Kraus: Phys. Lett. **B59** (1975) 175;
    A.O. Barut and S. Komy: Fortschr. Phys. **33** (1985) 309
91. C.Y. Wong and R.L. Becker: Phys. Lett. **B182** (1986) 251
92. H. Dehnen, M. Shahin: Acta Phys. Pol. **B21** (1990) 309
93. K. Geiger, J. Reinhardt, B. Müller, and W. Greiner: Z. Phys. **A329** (1988) 77
94. C.W. Wong, C.-Y. Wong: Phys. Lett. **B301**, 1 (1993) and Nucl. Phys. **A562** (1993) 598
95. B.A. Arbuzov, E.E. Boos, V.I Savarin, S.A. Shicharin: Phys. Lett. **B240** (1990) 477;
    B.A. Arbuzov, S.A. Shichanin, E.E. Boos, V.I. Savrin: Mod. Phys. Lett. **5** (1990) 1441;
    B.A. Arbuzov, S.A. Shichanin, V.I. Savrin: Phys. Lett. **B275** (1992) 144
96. N.R. Walet, A.K. Klein, R.M. Dreizler: Phys. Rev. **D47** (1993) 844
97. J.R. Spence, J.P. Vary: Phys. Lett. **B254** (1991) 1 and Phys. Lett. **B271** (1991) 27
98. M. Horbatsch: Phys. Lett. **A161** (1992) 360
99. M. Grabiak, B. Müller, and W. Greiner: Ann. Phys. **185** (1988) 284
100. S. Schramm, B. Müller, J. Reinhardt, and W. Greiner: Mod. Phys. Lett. **A3** (1988) 783
101. S. Graf, S. Schramm, J. Reinhardt, B. Müller, W. Greiner: J. Phys. G **15** (1989) 1467
102. K. Johnson, C.B. Thorn: Phys. Rev. **D13** (1976) 1934
103. G. Soff, W. Greiner, W. Betz, B. Müller: Phys. Rev. **A20** (1979) 169
104. J. Reinhardt, B. Müller, W. Greiner: Phys. Rev. A **24** (1981) 103
105. N.K. Glendenning, T. Matsui: Phys. Rev. **D28** (1983) 2890
106. W. Koenig in: "Vacuum Structure in Intense Fields", ed. by H.M. Fried and B. Müller, NATO ASI Series **B255**, Plenum Press, New York (1991) 29
107. V.A. Novikov, L.B. Okun, M.A. Shifman, A.I. Vainshtein, M.B. Voloshin, V.I. Zakharov: Phys. Rep. **41** (1978) 1

108. B. Ehrnsperger: J. Reinhardt, A. Schäfer, W. Greiner: Z. Phys. **A346** (1993) 133 and 137
109. K. Danzmann, W.E. Meyerhof, E.C. Montenegro, E. Dillard, H.P. Hülskötter, N. Guardala, D.W. Spooner, B. Kotlinski, D. Cline, A. Kavka, C.B. Beausang, J. Burde, M.A. Delepranque, R.M. Diamond, R.J. McDonald, A.O. Macchiavelli, B.S. Rude, F.S. Stephens, J.D. Molitoris: Phys. Rev. Lett. **62** (1989) 2353
110. E. Stein, S. Graf, J. Reinhardt, A. Schäfer, W. Greiner: Z. Phys. **A340** (1991) 377
111. E. Stein, J. Reinhardt, A. Schäfer, W. Greiner: preprint UFTP 336/93

The Search for the Quark-Gluon Plasma

Herbert Ströbele

Fachbereich Physik, Johann Wofgang Goethe-Universität,
D-60486 Frankfurt am Main, Germany

1 Introduction

The notion of a Quark-Gluon Plasma is a challenge to cosmology, to theoretical and to experimental physics. A working hypothesis in cosmology and astrophysics is a Big Bang at the origin of our universe. It implies a simple correlation between lifetime (t), radius (R) and temperature (T) of the universe for the radiation dominated period [1]:

$$t \sim T^{-2} \sim R^2$$

A short lifetime corresponds to a small radius and a high temperature. Obviously any species of finite size particles can exist only, if the sum of their volumes does not exceed the volume of the universe. Hadrons, specifically, can be formed only after a lifetime of 10^{-5} seconds at a temperature of around 200 MeV. At earlier times only leptons and quarks populate the universe (together with the gauge bosons). Then all the strongly interacting particles may have formed a Quark-Gluon plasma. In such a scenario the plasma is a transient stage between the very high temperature regime with $T \sim 200$ GeV – the characteristics of which escape my imagination – and the current world made up of hadrons. The inverse transition from the hadronic to the quark phase may take place nowadays in the collapse of neutron stars and black holes, if gravitational forces compress nuclear matter to densities which exceed the density of hadrons. This means again that hadrons will overlap and consequently dissolve into quarks and gluons. The study of such a form of matter in the laboratory would be of high interest, because it will contribute to our understanding of the early universe, neutron stars and the way quarks and gluons are confined in the hadrons. This brings us to the challenge the notion of a Quark-Gluon Plasma presents to theoretical physics.

The theory of elementary particle physics has advanced enormously in the last 20 years. The theory of electroweak interactions is well in hand and quantum chromodynamics is the accepted theory of strong interactions. However, the latter can treat quantitatively only those interactions which involve large momentum transfers to or from a quark or a gluon. Large means momenta in excess of the equivalent binding energy in the hadrons. Under such conditions perturbative methods are adequate. Low momentum transfers, i.e. soft interactions, between colour carrying particles are affected by the confinement of quarks and gluons (which are also called partons) in hadrons. Therefore the scattering of one parton on another one is treated differently in these two domains. For high momentum transfers between asymptotically free partons, and thus high relative energies, perturbative QCD allows very precise calculations which are consistent with experimental results. The shortcomings of predictions can be quantitatively specified. For low momentum transfers, realized not necessarily at small relative energies(!), no well defined theory exists and, even worse, experiments are impossible, because there are no free partons. Low momentum transfers between partons are always a many-body problem which must be treated by theory in more or less phenomenological models. This dilemma could perhaps be cured by making a virtue of necessity: the strong interaction between many partons can be evaluated with numerical (Monte Carlo) methods using techniques from solid state physics, the complete lattice regularized QCD Lagrangian [2]. In this scheme the quarks and gluons form a thermal system the features of which can be calculated by considering the interactions between the partons only in finite regions of space, i.e. on the links between adjacent sites of the four dimensional lattice structure. One of the challenges of this approach is the huge computing power needed for lattice sizes which permit extrapolation to zero lattice spacing, especially in configurations with nonzero quark masses [3].

Experimentally the interaction between two partons at low relative energies cannot be studied because of confinement, or in other words, the superimposed interaction of the partons with the physical vacuum. Information about the interaction between free quarks may, however, be inferred from the features of the Quark-Gluon Plasma. Such a state of matter should be formed in experiments with ultra relativistic heavy ions. Its detection and subsequent characterization is the challenge of the experimental program with nuclear beams which is currently being persued at BNL and CERN.

With the advent of relativistic heavy ion beams in the year 1986 experimental physics got deeply involved into the search for the Quark-Gluon Plasma. It was clear from the very beginning that this search would be very difficult experimentally, because unprecedented high particle multiplicities in the final state of a central collision swamp the detectors. The collisions of high energy ^{16}O, ^{32}S and ^{197}Au ions with fixed targets can produce events with a very high number of particles in the final state. In fact central collisions of ^{16}O, ^{32}S and ^{197}Au with a heavy target nucleus (e.g. Au) obviously involve between 50 and 400 nucleons. In addition between 1 and 5 particles are produced per participating nucleon. The total number of particles in the final state of a central Au+Au collision can thus reach 2000 at 200 GeV/nucleon. The particle multiplicities represent an unprecedented challenge to the experiments.

Whatever the goals of the experiments are, from the detection of rare particles to the registration of as many as possible particles in the final state, the large number of neutral and charged particles asks for new detector technologies. This is only one of the problems of the experimental program. Even more serious is the lack of clear and well defined goals. So far there are only a few exotic signals, the observation of which would be accepted as an unequivocal evidence for the Quark-Gluon Plasma. More conventional predictions may look promising at the beginning, but it has turned out in the past that for almost each QGP signal an explanation in a hadronic picture can be found. Thus the experimental search for the QGP in the past may seem frustrating to some, because whenever an interesting (even predicted) signal (of the QGP) was observed, hadronic models were modified or newly developed, which explained the data. The lack of unique and at the same time quantitative predictions for the features of the QGP seems to leave the experiments in an impasse. One way out is an experimental program which studies nucleus-nucleus collisions in a systematic way as a function of energy and system size. From the resulting regularities we hope to extract the characteristic features of high energy density hadronic matter and, perhaps, identify signals of partonic origin.

This lecture will follow this line of arguments. The experimental results from nucleus-nucleus collisions at CERN energies (60 and 200 GeV/nucleon) are presented. The main emphasis is put on the description of the hadronic final state. The regularities of the nuclear stopping, the energy density, the production of light particles, the space and time configuration of their source, and the observables which are considered as signals of a QGP are described. For further and more detailed reading of the theoretical ideas and work in this field see the reports [4, 5] and the proceedings of the Quark Matter Conferences [6, 7].

2 General Features of Nucleus-Nucleus Collisions

The total inelastic cross section is one of the simple observables one may address in experiments with nuclear beams. Fig. 1 shows a compilation of data from the Dubna Synchrophasotron and the CERN SPS. The $\sqrt{\sigma_{prod}}$ is plotted as a function of the cube root of the nuclear radii [8]. It represents the cross section for all particle production processes. A linear dependence is observed independent of energy, which is expected for the collisions of spheres with radii given by $A^{1/3}$. The small deviations from the universal linear behavior can probably all be attributed to the experimental difficulty to define inelastic collisions. This result shows that collisions between opaque macroscopic objects are observed. If this is so, each nuclear collision can be characterized by its impact parameter. It is defined by the distance at which the the centers of the two objects pass each other, and it also determines an overlap region in both nuclei the sum of which is called the participant zone (cf. Fig. 2).

An observable measure of the impact parameter is the number of nucleons, which are not (directly) affected in the course of the collision. These nucleons (e.g. from the projectile in fixed target experiments) should continue their motion relatively undisturbed into the forward direction. A calorimetric detector covering a small forward cone will then detect mainly these spectator nucleons. The resulting signal is proportional to the sum of their kinetic energies. It is commonly used as a trigger to veto the uninteresting peripheral collision events. This detector is often called 'zero degree' or 'VETO' calorimeter. Distri-

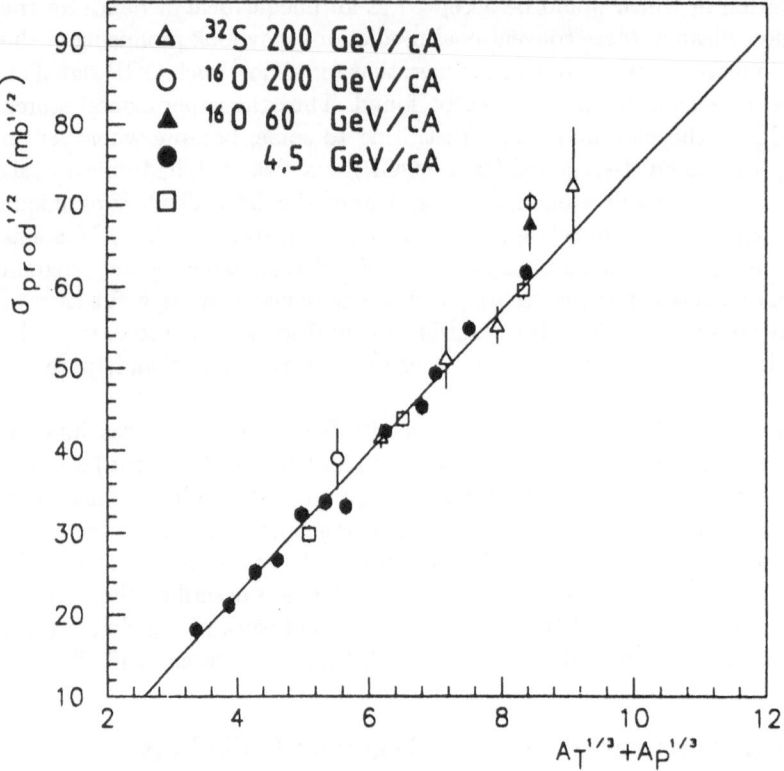

Fig. 1. Square root of the production cross section for 200 GeV/nucleon ^{32}S with Al, Fe, Cu, Ag, and Pb as a function of $A_P^{1/3} + A_T^{1/3}$. Data of lower energies and with an oxygen beam (squares) are also shown.

butions of the energy measured with such a device (E_{veto}) in various minimum bias nucleus-nucleus collisions are shown in Figs. 3 and 4. Dividing the energy scale by the corresponding beam energy per nucleon gives an estimate of the number of projectile spectator nucleons. The probability to observe a certain number is closely connected to the collision geometry. In a picture with straight line trajectories and well defined nuclear surfaces the number of spectator nucleons is directly correlated to the impact parameter. The distribution $d\sigma/dE_{\text{veto}}$ (with $E_{\text{veto}} \propto dN_S$, the number of projectile spectators) in Fig. 3 [9] exhibits a monotonic decrease of the cross section with centrality for the symmetric collision system S+S in contrast to what is observed in asymmetric systems for which the cross sections stay nearly constant over large ranges of energies. This behavior can again be explained by simple geometry. With increasing impact parameter larger portions of the projectile remain spectators. $d\sigma/dE_{\text{veto}}$ will stay constant, if $dE_{\text{veto}} \propto dN_S \approx b \cdot db$ as realized when a small nucleus overlaps by more than its radius with a much larger one. On the other hand, in symmetric systems the complement of the overlap volume (of the projectile) increases slower

Participant – Spectator Model

Before

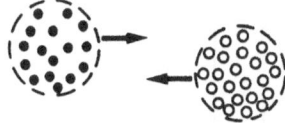

After

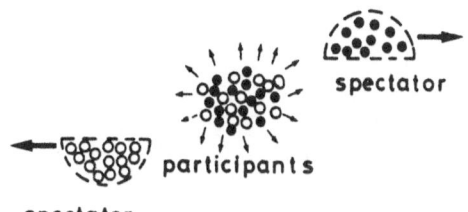

Fig. 2. Sketch of the participant spectator picture. The configuration of the nucleons in the nuclei are shown before and after the collision.

than $b \cdot db$. If the impact parameter becomes so small that the whole projectile dives into the target, ($A_{\text{projectile}} < A_{\text{target}}$) there should be no further change in E_{veto} provided that all nucleons in the overlap region interact. In this case the cross section $d\sigma/dE_{\text{veto}}$ should have a local maximum at low E_{veto}. Fig. 4 reveals that at 60 GeV/nucleon beam energy (Fig. 4a) the local maximum is most pronounced [10]. At the higher energy (200 GeV) there seems to be a finite probability for a projectile nucleon to pass through a gold nucleus. For the heavier projectile nucleus (^{32}S) the peak is even more suppressed. This first impression does not persist in a quantitative analysis. Integrating the cross section of the dive-in peak yields 1 barn at 60 GeV ($0 < E_{\text{veto}} < 210\,\text{GeV}$), 1 barn at 200 GeV ($0 < E_{\text{veto}} < 1000\,\text{GeV}$) and 0.77 barn for ^{32}S+Au ($0 < E_{\text{veto}} < 2000\,\text{GeV}$). The expectation from simple geometry for the dive-in cross section is 0.52 barn for ^{16}O+Au and 0.35 barn for ^{32}S+Au [9]. The integral under the peak corresponds always to twice the dive-in cross section. This regularity is impressive. The energy independence and the scaling with the projectile size are again supporting the simple geometrical picture we invoke to explain the pecularities observed in the distribution of the E_{veto} observable. The most probable E_{veto} (the peak position) corresponds to 1 spectator nucleon at 60 GeV/nucleon, two and three

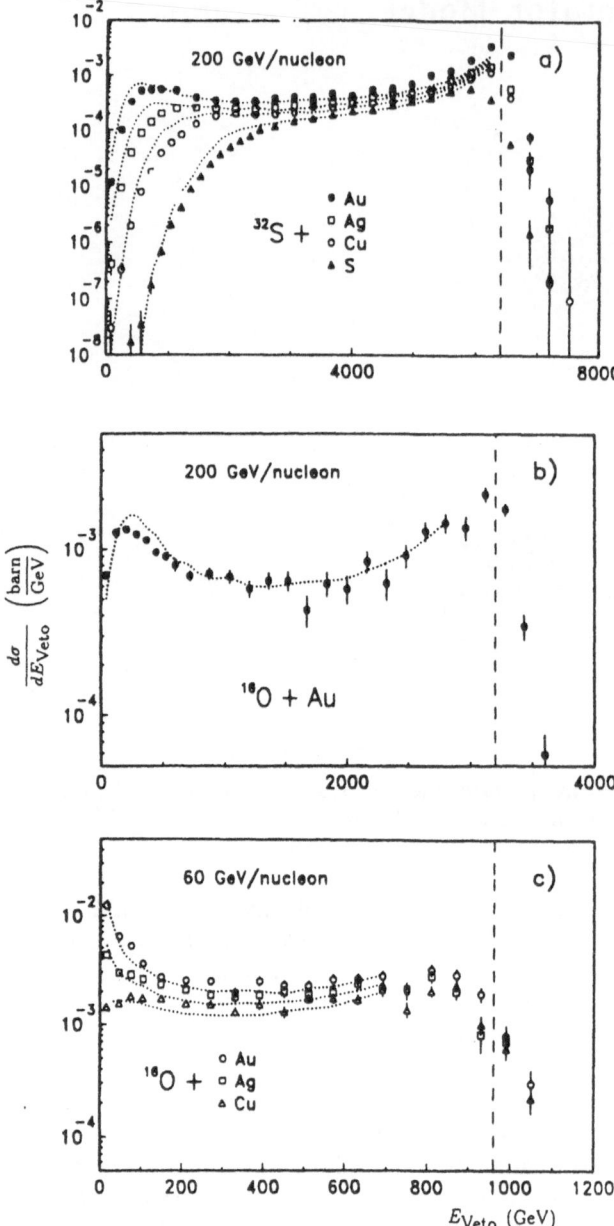

Fig. 3. Forward energy spectra in the Veto Calorimeter for a) S+Au, Ag, Cu, S at 200 GeV/nucleon b) O+Au at 200 GeV/nucleon and c) O+Au, Ag, Cu at 60 GeV/nucleon. The Fritiof spectra [12] are shown with dotted lines. Close to the beam energy (dashed vertical line) the spectra strongly depend on the interaction trigger condition.

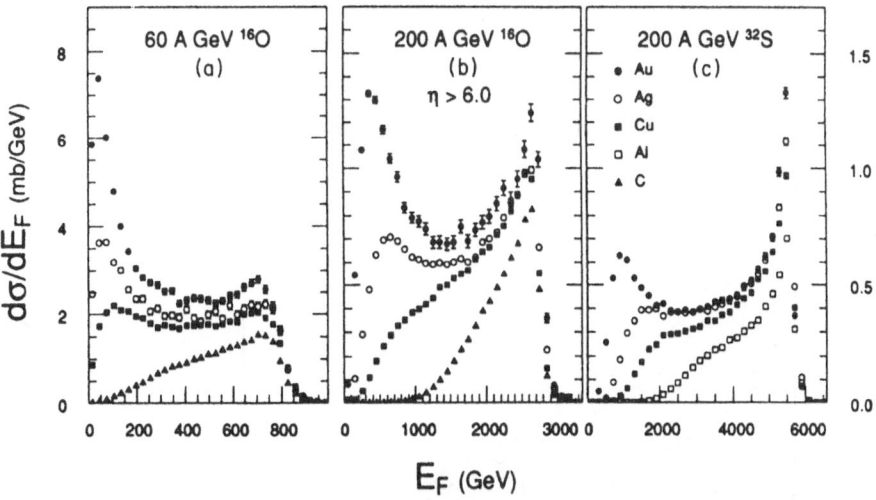

Fig. 4. Forward (same as Veto) energy spectra for heavy ion reactions induced by (a) 60 GeV/nucleon ^{16}O, (b) 200 GeV/nucleon ^{16}O and (c) 200 GeV/nucleon ^{32}S. The vertical scales in (b) and (c) are identical. Note the linear scale.

spectators for the Oxygen and Sulphur data at 200 GeV/nucleon. Part of this energy stems from forward going (participating) neutrons, K^0_{long} and photons from π^0 decays. The faster than linear increase of the dive-in energy with beam energy is probably due to the increase of the overall particle density. We conclude that the fraction of participant energy in central collisions observed at small forward angles varies from less than one unit (in terms of the corresponding energy of one spectator), via 1-2 units, to 3 units in ^{16}O+Au at 60 GeV, ^{16}O+Au at 200 GeV and ^{32}S+Au at 200 GeV/nucleon.

An observable complementary to E_{veto} is the transverse energy E_T. A large part of the incoming longitudinal energy will reappear as transverse energy. In Fig. 5 this anticorrelation between E_{veto} and E_T is demonstrated in the system O+Au at 200 GeV/nucleon [11]. The maximum transverse energy reaches 600 GeV in central S+Au collisions (not shown). For a realistic estimate of the energy density obtained in those violent collisions we consider the transverse energy per unit of rapidity as a function of the transverse energy in Fig. 6 [9]. According to Bjorken's picture [14] the transverse energy observed in one unit of rapidity can be linked to the energy density in the reaction volume. We obtain ≈ 3 GeV/fm^3, which is of the order of magnitude predicted by lattice QCD to be favorable for the formation of a Quark-Gluon Plasma.

It has been shown that $d\sigma/dE_T$ has a shape very similar to the multiplicity distribution [15]. Therefore we leave the discussion of the transverse energy and turn to the multiplicity distribution. Fig. 7 shows $d\sigma/dN^{\pm}$ for collisions of S-projectiles with various targets [16]. For heavy targets there is again a plateau region which underlines the inverse correspondence of E_{veto} and E_T (multiplicity). Up to 300 charged particles are observed in a single event, the highest

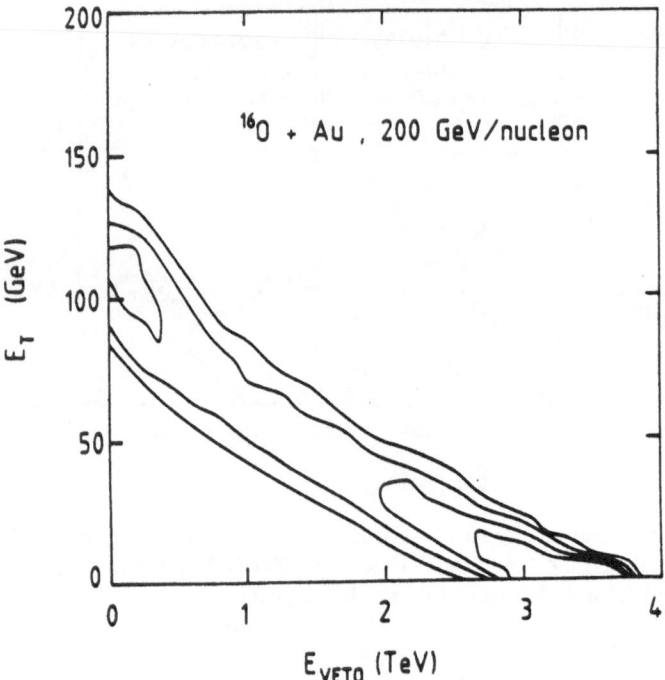

Fig. 5. Correlation between the transverse and veto (forward) energy in O+Au colli-sions. The contour lines give the event frequency.

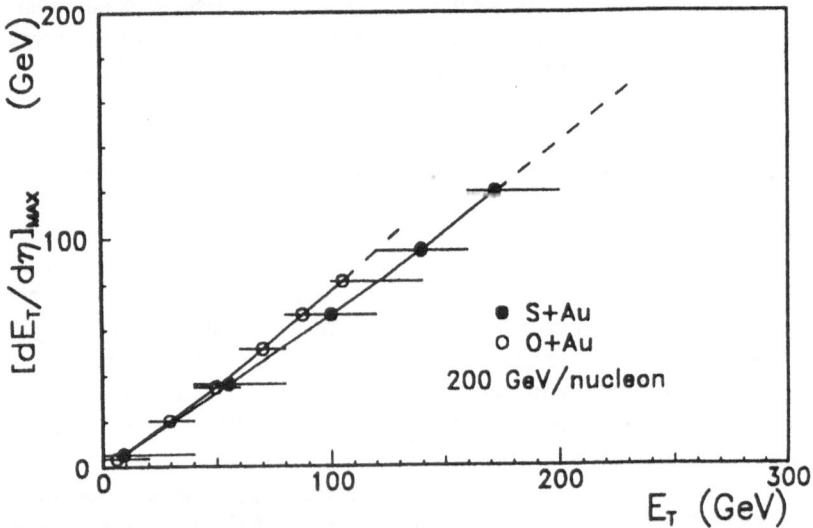

Fig. 6. Maximum pseudorapidity density $dE_T/d\eta$ for ^{32}S and ^{16}O on Au at 200 GeV/nucleon as function of the mean E_T in transverse energy bands indicated by the horizontal error bars. Dashed lines extrapolate to E_T values maximally reached for the reactions.

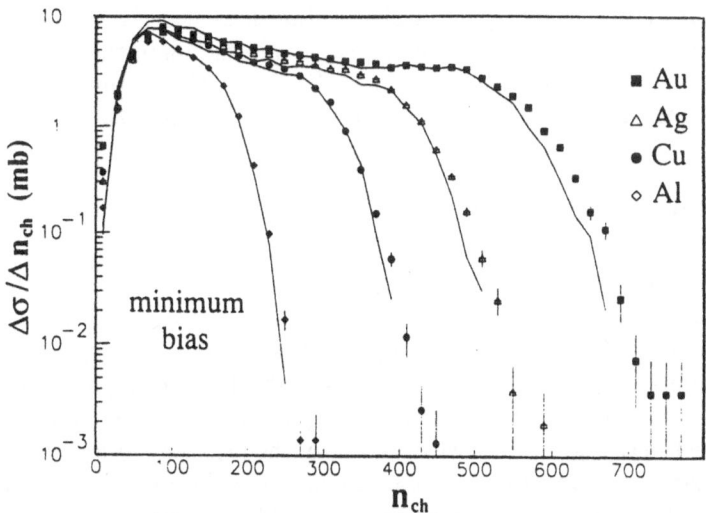

Fig. 7. Multiplicity distribution of charged particles in ^{32}S+Au, Ag, Cu, Al reactions at 200 GeV/nucleon. The rapidity interval covered is $-1.7 < \eta < 4.0$. Venus [13] results are indicated by the solid lines.

Table 1. Comparison of mean particle multiplicities in NN and AA collisions at 200 GeV/nucleon

particle type	N+N (p+p)	^{32}S+^{32}S (measured)	^{197}Au+^{197}Au (extrapolated:NNx200)
π	10	260	2000
nucleons	2	51	400
K_s^0	0.2	11	40
$K^+ + K^-$	0.4	22	80
$N, \overline{N}$ pairs	0.1	6	20
$\Sigma + \Lambda$	0.15	13	30
negatives	4	100	800
$\overline{\Sigma} + \overline{\Lambda}$	0.0.02	2.4	40

multiplicity observed so far in accelerator experiments. The shape of the multiplicity (and E_T) distributions can be reproduced by a convolution of p+A spectra by choosing an appropriate trigger for the p+A event sample. Thus the multiplicity (and E_T) distributions do not reveal any deviation from a 'normal' behavior.

So far our discussion of nucleus-nucleus collisions supports a simple geometrical picture with, however, hints for the realization of large energy densities, which are a prerequisite for the formation of the Quark-Gluon Plasma. Before turning to the characteristics of particle production in nucleus-nucleus collisions in the next section we anticipate a few results in Tab. 1 to give the reader an overview over average multiplicities and the particle composition in the average final state of central ^{32}S+S collisions at 200 GeV/nucleon. These numbers are compared to N+N interactions and extrapolated to Au+Au collisions.

3 Light Particle Production

At CERN energies pions are the most abundant particle species observed in the final state. The multiplicity of negatively charged particles $< n_- >$ is a good measure of the number of negatively charged pions, because the K$^-$ meson and anti-proton contributions to $< n_- >$ are below the 10% level. In high energy p+A experiments it was found [17] that $< n_- >$ is proportional to the number of participating nucleons. Thus secondary collisions of the incoming proton seem to be as effective for pion production as the first nucleon-nucleon interaction. A phenomenological picture of this finding was put forward by A. Bialas et al., as the wounded nucleon model [17]. As the name suggests the number of produced pions is proportional to the number of nucleons which have interacted at least once. It will be interesting to study this same correlation in A+A interactions. Fig. 8 reveals that in ^{32}S+S collisions this proportionality is also observed.

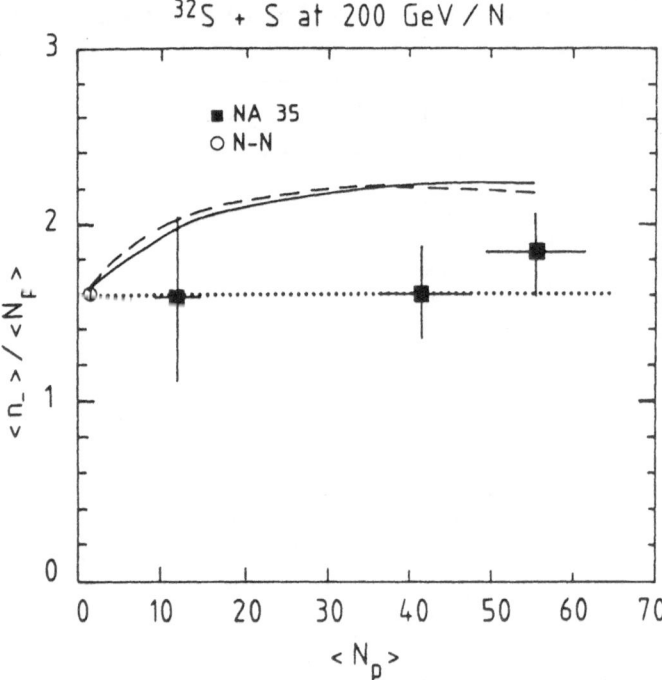

Fig. 8. Mean number of produced negatively charged particles per participating nucleon as function of the number of participating nucleons. The solid (dashed) line gives the result from Fritiof [12] (Venus [23]) model calculations.

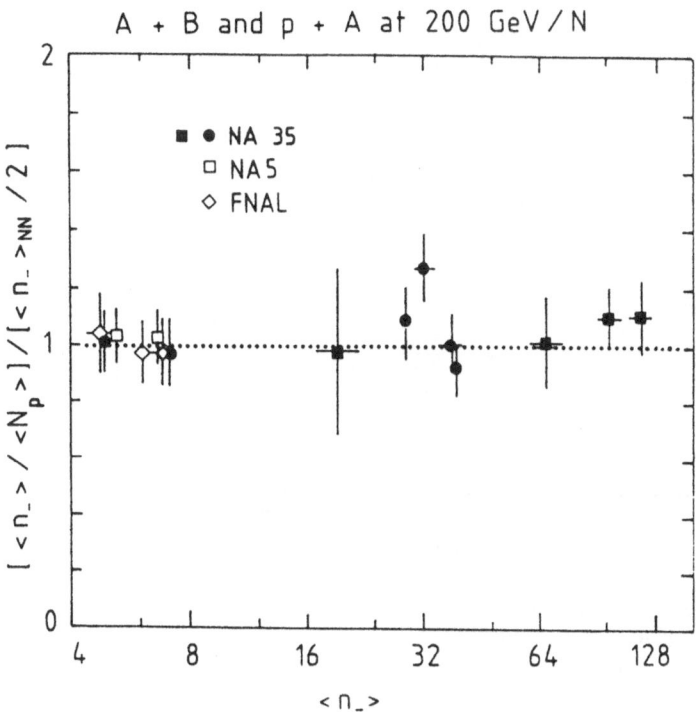

Fig. 9. Ratio of the mean number of produced negatively charged particles per participating nucleon for A+A and p+A interactions over nucleon+nucleon collisions as function of the number of negatively charged particles.

The interpretation of the experimental finding should take into account that pions can rescatter inelastically and may be absorbed by the surrounding baryons before leaving the reaction zone. It seems that these two effects together with the multiple collisions of the incident nucleons conspire to give the simple scaling behavior $< n_- > \propto N_{\text{participants}}$. Fig. 9 summarizes the situation at 200 GeV/nucleon [18]. Shown are double ratios $< n_- > / N_{\text{participants}}$ for A+A (p+A) to N+N collisions as function of $< n_- >$ which can be considered as a measure of the impact parameter. The double ratio is nearly constant and equal to one which indicates that the number of negative particles produced in any hadronic interaction is proportional to the number of participating nucleons independent of event multiplicity and thus impact parameter and collision system. The wounded nucleon model holds in p+A as well as in A+A interactions! Either no additional pions are produced in secondary collisions or absorption compensates for the additionally created pions. Overall, the multiplicity of n_- seems to be insensitive to rescattering effects.

Maybe the transverse momenta (p_T) of the pions 'feel' the multiple collisions. The corresponding measure could be the average p_T. The change of $< p_T >$ of negatively charged particles with centre-of-mass energy for p+p($\bar{p}$) and A+A

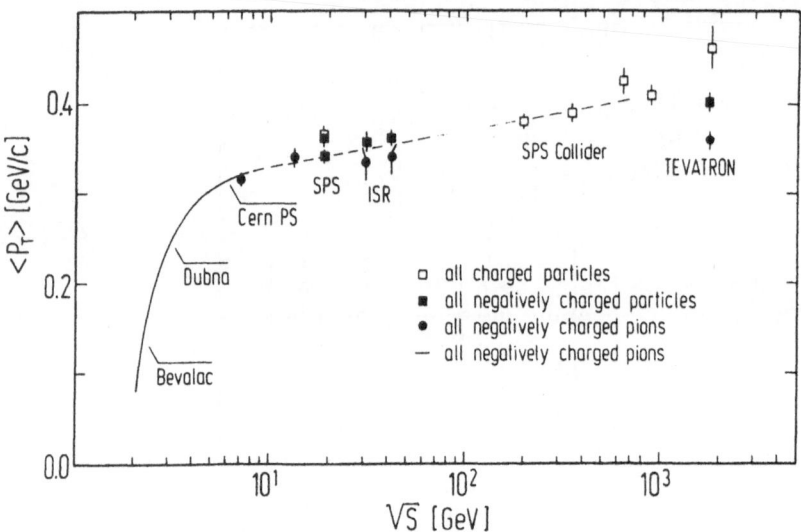

Fig. 10. Mean transverse momentum of produced particles as function of c.m. energy in p+p (p+$\overline{\text{p}}$) and nucleus-nucleus collisions.

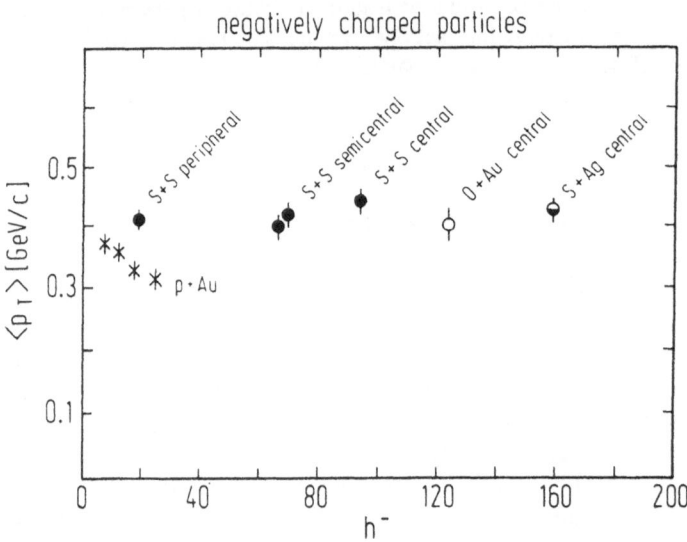

Fig. 11. Mean transverse momentum of negatively charged particles in nucleus-nucleus collisions at 200 GeV/nucleon as function of centrality (represented by the multiplicity of negatively charged particles).

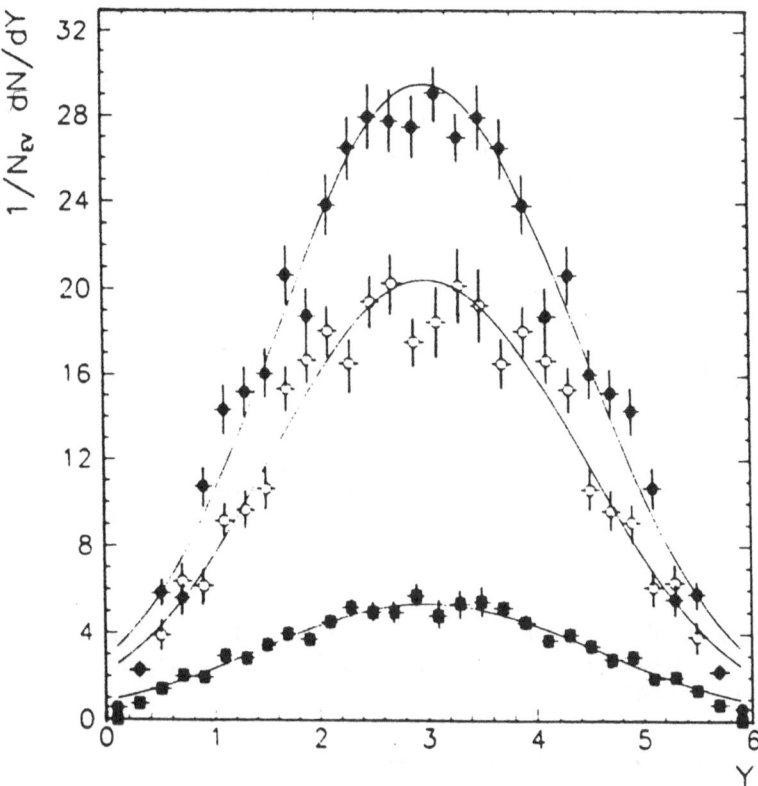

Fig. 12. Rapidity density distributions of negatively charged particles for peripheral (boxes), semiperipheral (open circles) and central (dots) ^{32}S+S collisions at 200 GeV/nucleon.

collisions is summarized in Fig. 10. $\sqrt{s}$ ranges from 2 GeV to 2000 GeV (per nucleon). At the highest (Tevatron, SPS collider) energies $< p_T >$ is subject to considerable uncertainty due to limited p_T coverage and the variation with energy in the mix of produced particles. All data follow a universal line which exhibits the increase of $< p_T >$ with available energy and its saturation at high energies at a value corresponding roughly to Hagedorn's limiting temperature [19]. There is no indication of differences between p+p and A+A collisions in this observable. In order to really pin down the existence or non-existence of such a difference we study $< p_T >$ as a function of centrality (multiplicity) in Fig. 11 . The results from p+p, p+A, ^{16}O+Au and ^{32}S+S collisions clearly demonstrate that multiple collisions, as expected to be present in central collisions, do not lead to an increase in $< p_T >$. (The only discernible variation is present in p+Au collisions in which high multiplicity events show a decrease in $< p_T >$. We understand this behavior as due to energy conservation.) One more observable is insensitive to multiple collisions of the incident nucleons in A+A collisions!

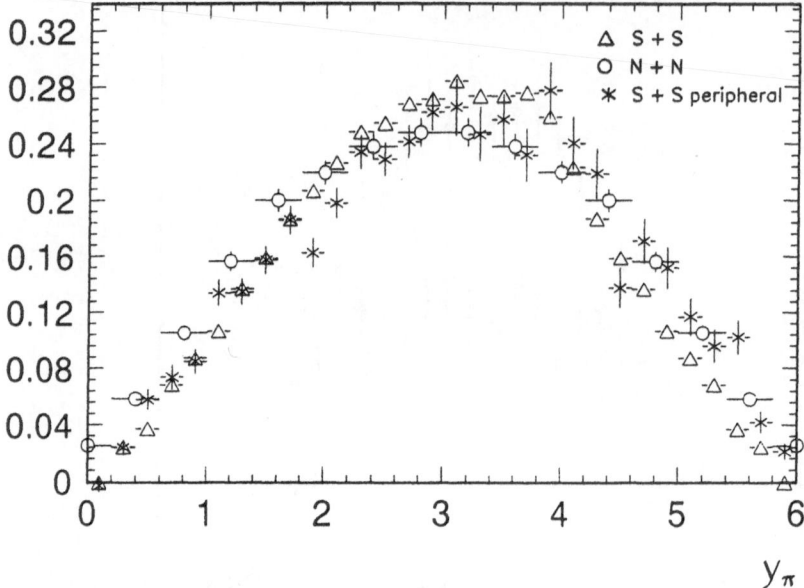

Fig. 13. Rapidity distributions of negatively charged particles for central and peripheral ^{32}S+S as well as nucleon-nucleon collisions at 200 GeV/nucleon. The distributions
are normalized to the same area.

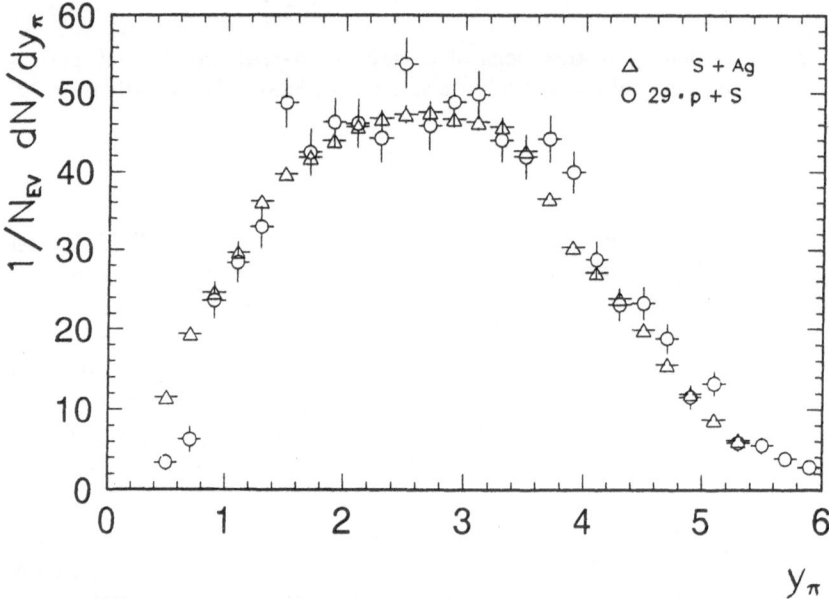

Fig. 14. Rapidity density distributions of negatively charged particles for ^{32}S+Ag (tri-
angles) and p+S (open circles) collisions at 200 GeV/ nucleon. The latter are scaled
by a factor of 29.

Before giving a tentative interpretation of this behavior we have to point out that the higher moments of the p_T distribution exhibit a change when going from p+p to A+A. However, this change is already present and even most pronounced in p+A collisions [20]. Therefore it doesn't seem to be important for the discussion of nucleus-nucleus collisions.

Let us consider for a moment the scenario that the pions do not feel multiple collisions. In this case production and absorption in secondary collisions must balance even differentially. A simple, although far reaching assumption would explain such a behavior: in collisions between both nucleons and nuclei the pions are in (or close to) thermal and chemical equilibrium (at a common temperature).

Moving on to the rapidity distribution of negatively charged particles which here are all treated as pions, we check again for multiple nucleon-nucleon scattering effects. The rapidity density distribution dn/dy is shown in Fig. 12 for peripheral, semi-peripheral and central ^{32}S+S collisions [21]. The yields of π^- change according to the expectations: more pions are produced in central collisions.

The other important feature of the distributions are that their widths are all twice as broad as expected for a sample of pions emitted from a single thermal source. This constraints the assumption of thermalisation to only a local equilibration with some ordered longitudinal motion. We will come back to this feature later in the discussion of two-pion correlations.

The comparison of peripheral and central events should reveal rescattering effects. In Fig. 13 the normalized dn_-/dy distributions obtained from N+N, peripheral ^{32}S+S and central ^{32}S+S collisions are compared (the latter stem from a newly analyzed high statistics event sample [22]). There are only minor differences and thus no significant rescattering effects seen. Similar data exist for ^{32}S+Ag and and p+S interactions. These latter two systems have roughly the same ratio of participating nucleons in the projectile and target and should have very similar shapes of the dn_-/dy distributions, if rescattering effects are neglected. Fig. 14 confirms this expectation. In addition it provides a convenient small system reference. p+S interactions reproduce central ^{32}S+Ag collisions, if scaled by a factor of 29. The scale factor depends on the applied trigger selections. We will use this scale factor later in the comparison of heavy particle production ratios. Only minor differences in the shapes of rapidity density distributions between small and large systems are found (in symmetric as well as in asymmetric collisions). We conclude that the number of created negatively charged particles per nucleon and their rapidity and $<p_T>$ distributions are either not sensitive to multiple collisions, or that there are $\underline{no}$ multiple collisions. This question will be addressed in the next setion.

4 Participant Baryon Distributions

In nucleon-nucleon interactions the incident particles scatter once and experience a reduction in longitudinal motion and a boost in transverse direction. The change of (centre-of-mass) energy of both incident particles goes into particle

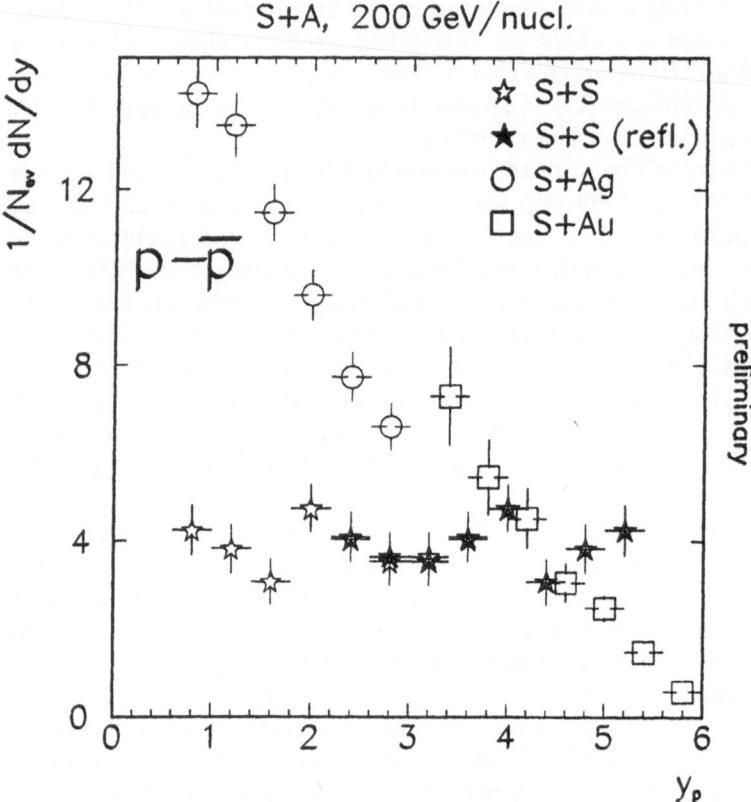

Fig. 15. Rapidity density distribution of participant protons in S+S, S+Ag, and S+Au collisions.

production. In nucleus-nucleus collisions the incident nucleons will in general scatter more than once. The smaller the impact parameter, i.e. the larger the overlap between the two nuclei, the more collisions per nucleon are expected. This correlation is more pronounced in heavy projectiles. For example in central ^{32}S+S collisions 2.8 collisions per nucleon are expected in a geometrical picture with straight line trajectories. Nucleons which scatter more than once will experience a higher loss in longitudinal momentum and thus also in energy. The energy available per participating nucleon for particle production should be higher in central than in peripheral nucleus-nucleus collisions. This picture seems to fail for the lightest hadrons, the pions, as demonstrated in the preceeding section. We examine in the following whether the influence of multiple collisions of nucleons is seen in the rapidity and transverse momentum distributions of the participating nucleons. The comparison of nucleon-nucleon or peripheral nucleus-nucleus data with central nucleus-nucleus collisions should show, for example, higher average rapidity shifts, higher mean p_Ts and higher energy losses of the nucleons when multiple collisions are present. For the determination of the energy

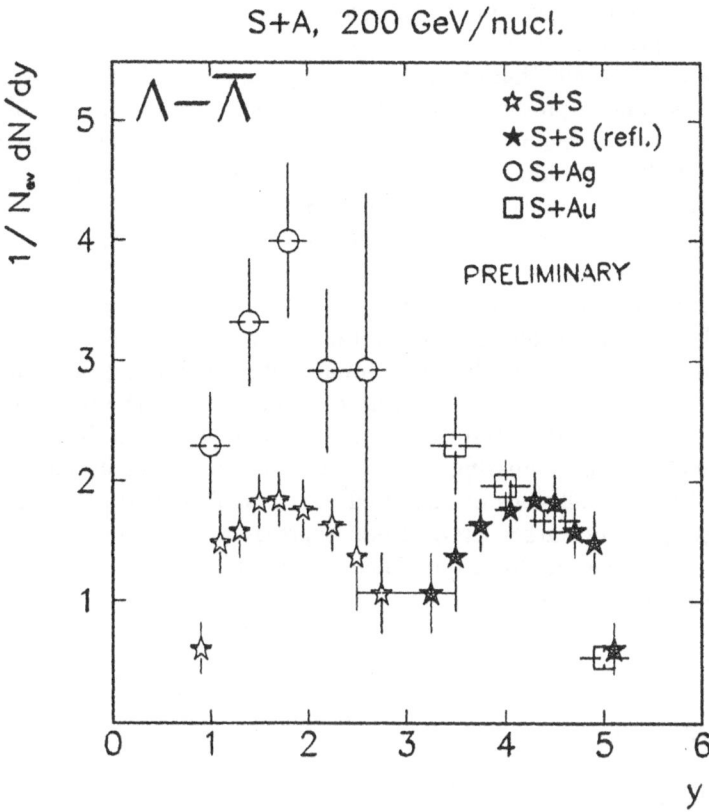

Fig. 16. Rapidity density distribution of participant hyperons ($\Lambda - \overline{\Lambda}$) in S+S, S+Ag, and S+Au collisions.

loss the centre-of-mass of the colliding system has to be known. In symmetric projectile-target configurations this comes for free. We therefore consider first ^{32}S+S collisions. The experiment NA35 at the CERN SPS has measured the spectra of participant protons by subtracting negative from positive particle distributions [24]. In Fig. 15 we compare their rapidity distributions in central and peripheral collisions. The data points in the target hemisphere ($0.2 < y < 3$) have been measured, and the points in the projectile hemisphere ($3 < y < 5.8$) are obtained from reflection symmetry. The peripheral collision data are very similar to results from p+p experiments and exhibit clearly the diffractive peak near target rapidity and a strong dip in rapidity density around midrapidity. In contrast, central collisions lead to an almost flat rapidity distribution. Multiple collisions reduce the longitudinal motion of the incident nucleons! The mean rapidity shift which is ≈ 1.0 in p+p interactions increases to a value of 1.5 in central ^{32}S+S collisions. Part of the stopped energy should go into their transverse motion and, indeed, the $< p_T >$ of participant protons increases from 450 MeV/c in p+p to 620 MeV/c in central A+A collisions [24]. Before the energy loss of the nucleons can be evaluated we have to consider hyperon production. At SPS energies strangeness production proceeds predominantly via the so called associated production which means that one incoming nucleon is transformed into a hyperon with the associated production of a K-meson. The relevant observable is the difference between the number of hyperons and anti-hyperons represented by $\Lambda - \overline{\Lambda}$ in Fig. 16. This rapidity distribution has a similar shape as the one of the nucleons except near target rapidity where the production of hyperons is suppressed due to energy conservation. Combining the energy loss of all baryons (protons, lambdas with a correction factor of 1.6 for unseen hyperons) we obtain an average energy loss of the incoming nucleons of 6.5 GeV which is $\approx$2.5 times larger than calculated for p+p collisions. The total energy per collision available for particle production is given by the total number of participating nucleons multiplied by the average energy loss. The sum of the integral of the distributions in Fig. 15 and Fig. 16 with a factor of 1.6 for unseen hyperons on the latter yields 54 for the number of participating nucleons. Tab. 2 summarizes the energy balance for central ^{32}S+S collisions as measured in the NA35 experiment. The bulk of the energy goes into meson production. Approximately 10% appear in heavy particles like nucleons and hyperons. Their production characteristics will be discussed later in this lecture.

Table 2. Energy balance in ^{32}S+^{32}S collisions

energy loss of the (54) nucleons	326 $\pm$ 20
energy in the produced particles	
π	212 $\pm$ 20 GeV
K,$\overline{K}$	47 $\pm$ 5 GeV
Hyperons, from pair production	9 $\pm$ 1 GeV
η, only from π^0 excess	6 $\pm$ 1 GeV
N,$\overline{N}$ only pair production	27 $\pm$ 12 GeV
sum	301 $\pm$ 30 GeV

So far we have characterized the final state of heavy ion collisions by the distribution of nucleons and pions in the final state. Before turning to the Quark-Gluon Plasma signatures we try to learn about the space time picture of such reactions by studying two-pion correlations in the next section.

5 Two-Pion Correlations

Correlations between two or more particles can be due to forces between the particles or the result of quantum statistical effects. An example for the former is the Coulomb force. The latter are also known as the Pauli Principle and the Bose Einstein enhancement. These are relevant only for identical particles which have a reduced, respectively enhanced, probability to be found in the same 6-dimensional phase space cell. In general, particle correlations are measured in momentum space for an ensemble of equivalent but independent events or particle pairs. It is intuitively clear that the correlation observed in momentum space will be weak, if the particles are far away in configuration space. Let us specify the meaning of "weak" for the case of the Coulomb force: two particles with small relative momenta will experience a small (large) change in relative momentum, if they are at large (small) distance in configuration space; in the limit of infinite distance no change in relative momentum will occur. Thus "weak" here means a small change in momentum difference.

As a practical example for the construction of a correlation we consider the distribution of the momentum differences between all combinations of two particles from a large ensemble. The gross features of its shape are given by the square of the single particle density distribution in momentum space. We are not interested in this trivial type of correlation. Dynamical and statistical effects can be extracted by taking the ratio of the true distribution of momentum differences to a distribution which contains only the correlation due to the single particle phase space distribution. The latter is obtained, e.g., by pairing only particles from different events and the former by pairing particles from the same event. This ratio as a function of the momentum difference represents the relevant correlation function. Its shape is affected by the average distance between the two particles which form the pair. To be more precise: the distribution of distances between the particles of all pairs in the ensemble considered is reflected in the shape of the correlation function with large (small) distances affecting small (large) momentum differences.

We turn now to the correlation between pairs of identical pions. The laws of (Bose-Einstein) statistics tell us that two such pions have an enhanced probability to be found in the same 6-dimensional phase space cell given by $\Delta\mathbf{p}\cdot\Delta\mathbf{r} \sim \hbar^3$. Comparing two ensembles of pairs coming from sources with $\Delta\mathbf{r}$ small and large (on the average) will result in correlation functions showing an enhancement in a large and small range of $\Delta\mathbf{p}$ (momentum differences) respectively. This inverse correspondence holds true also for each component of the momentum difference and distance. Fig. 17 shows an example of a correlation function as obtained from central S+Au collisions at 200 GeV/nucleon. The variable is the Lorentz invariant momentum difference $Q_{\text{inv}} = \sqrt{-(p_1 - p_2)^2}$ with p_i being the four-vector of particle i. The Bose Einstein enhancement is clearly seen at low Q_{inv}. For a more detailed view of what can be measured we consider all six degrees of freedom spanned by the 3-momenta of two pions $(\mathbf{p}_1, \mathbf{p}_2)$. Three momentum differences are chosen in the following way: $\Delta p_{\text{long}} = Q_{\text{long}}$, in the direction of the incident beam; $\Delta p_{\text{T(out)}} = Q_{\text{out}}$, the component of the transverse momentum difference vector projected onto the sum vector $(\mathbf{p}_{1\text{T}} + \mathbf{p}_{2\text{T}})$; $\Delta p_{\text{T(side)}} = Q_{\text{side}}$, the remaining orthogonal component of the transverse momentum difference. The corresponding size parameters are R_{long}, R_{side}, and R_{out}. The special role of R_{long} will be discussed below. The meanings of R_{out} and R_{side} are illustrated in Fig. 18 . They stand for a measure of the depth of the source and its transverse size, respectively. Details of the evaluation of the radius parameters and their interpretation can be found in reference [25]. With 3 out of 6 degrees exhausted we consider the remaining three, which are the components of the vector sum $p_1 + p_2$. One degree of freedom, the azimuthal angle, should be integrated over because of the rotational symmetry with respect to the beam axis. The two remaining degrees of freedom are chosen to be the rapidity (y) and the transverse momentum (p_T) of the pair. With these categories it is natural to study R_{long}, R_{side}, and R_{out} as function of rapidity and p_T. An experimental difficulty should be mentioned here: the influence of non genuine pion pairs, in which at least one particle does not originate from the main interaction point but instead from a decay or a secondary interaction, is not corrected for in the results presented below. The corresponding distortion will cause only minor systematic errors on the radius parameters, but will reduce the λ-parameter, which is introduced explicitly for this type of effects [27].

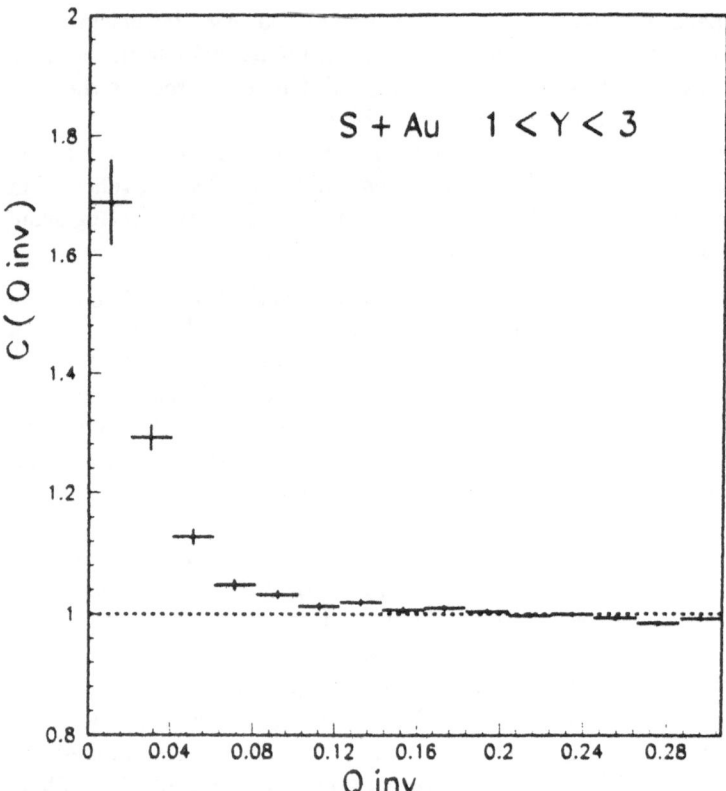

Fig. 17. Two-pion correlation as a function of the invariant momentum difference.

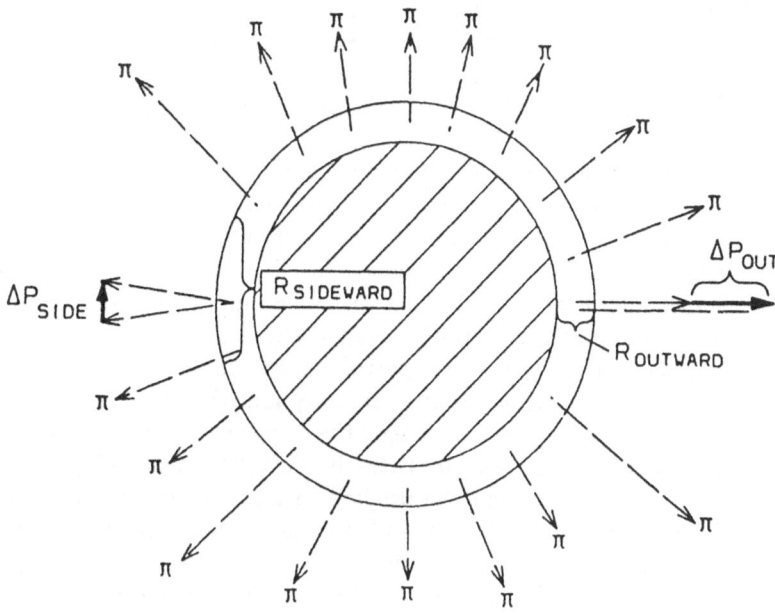

Fig. 18. Illustration of the meaning of the outward (R_{out}) and sideward (R_{side}) radii.

Different scenarios for the shape and time evolution of the pion emitting source will be discussed next. The simplest one is a static spherical source which emits all pions instantly and with a momentum distribution which could be thermal or of any other simple shape. In this case an observer will see the same shape of the correlation function in Q_{out}, Q_{side} and Q_{long}, if the momentum measurements are done in the Lorentz frame of the source. In the case of relative motion (e.g. in the longitudinal direction) the source will appear Lorentz contracted (i.e. R_{long} will be smaller than R_{out} and R_{side}). The correlation functions will all be independent of where in phase space the momentum measurements are done. Thus the shape of the correlation functions will look alike in all intervals of rapidity and p_{T}.

Another extreme scenario is a static source in the same configuration but with continuous emission of pions. In this case (and if in addition the measurement time is short enough) R_{out} will be larger than R_{side}, because pions emitted radially and at different times may lead to pairs of pions in phase cells which are outside of the original source volume. (In fact both scenarios could be described more appropriately by a spherical set of radius parameters.) We will not pursue the picture of a static source any further, because it is not consistent with the rapidity distributions of produced particles as stated in Sect. 3. They are roughly two times broader than expected for thermal emission, whereas in the transverse directions the momentum distributions resemble closely a thermal distribution. There seems to be a preference for longitudinal motion, for which several mechanisms have been proposed, one being Bjorken's longitudinal scaling expansion [14] and, another, two or more sources having longitudinal velocities, which depend on their relative distance in configuration space. The longitudinal motion of the observed pions would then be the superposition of a thermal component and the motion of several different pion sources (or one expanding pion source).

In both pictures a strong correlation between rapidity and the longitudinal coordinate in configuration space is established. In the expansion regime particles emitted from the centre of the source will be observed at midrapidity ($y = 0$) and particles emitted from a certain distance from the centre will be shifted by a corresponding Δy. (The thermal motion will wash out this 1-to-1 relation somewhat.) In the scenario of a longitudinal expanding source the correlation between R_{long} and Q_{long} is lost, because particle pairs emitted from distant locations in the source will never be near in rapidity and thus never at small Q_{long}! The correlation in Q_{long} measures instead the duration of expansion from the formation of the source until the particles become free (freezout!). This can be qualitatively understood from the original 'definition': the probability to observe two identical bosons in the same phase space cell is enhanced. The corresponding longitudinal phase space cell dimension is given by $\hbar = \Delta x \cdot Q_{\text{long}}$. In the longitudinal expansion scheme Δx (e.g. the distance to the origin) will increase with time according to $\Delta x = \tau \cdot v_{\text{expansion}}$. On the other hand $v_{\text{expansion}}$ is correlated with the longitudinal momentum difference, thus with Q_{long}. With this correlation the uncertainty relation reads $\hbar = \tau \cdot f(Q_{\text{long}}) \cdot Q_{\text{long}}$. With Δx eliminated from the expression, the measurement of the Bose enhancement as a function of Q_{long} will give information on τ only. These arguments hold, if the observer measures pairs which have small longitudinal momenta (rapidities) in his c.m. frame. As soon as one looks at parts of the source which are far away they move away too, and thus the scale in configuration space is Lorentz contracted, the same way as in the expanding universe the light from distant stars is blue shifted. The clear prediction is that $\tau = \Delta x \cdot /v_{\text{expansion}}$ decreases with increasing rapidity.

These qualitative arguments can be easily checked by studying the dependence of R_{long} on rapidity. The complete theory ([28, 29, 30, 31]) predicts:

$$R_{\text{long}} = \sqrt{2T/m_T} \cdot \tau \cdot \cosh^{-1}(y - y_0)$$

Figs. 19a,b show the experimental results obtained from central S+S and S+Ag collisions. The data points represented by squares are derived from the TPC detector; they nicely extend the rapidity coverage beyond midrapidity and show good agreement with the corresponding value below midrapidity in the symmetric S+S system. For R_{long} the $\cosh^{-1}$ dependence is obvious. Using the experimental values T=180 MeV and m_T =200 MeV and R_{long} at $y = y_0$ yields $\tau \simeq 5$ fm. Before turning to the results on R_{side} and R_{out} a word of caution is neccessary: the experimental evidence for a longitudinal expanding source is quite solid and in accordance with the shape of the single particle rapidity distribution. The quantitative result on τ, however, is subject to many uncertainties and, thus, has a large (50%) systematical error.

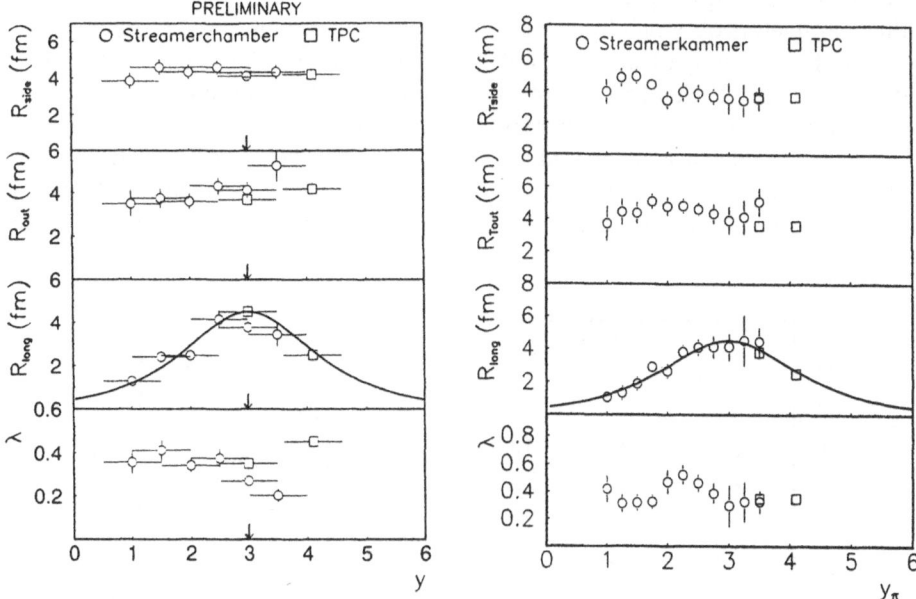

Fig. 19. Radius parameters R_{side}, R_{out}, and R_{long} in units of fm as function of rapidity in central S+Ag (left) and S+S (right) collisions at 200 GeV/nucleon. All pions in each rapidity interval are combined to form the pairs which are used to calculate the correlation functions.

We turn now to the other radius parameters which are presented in Fig. 19 disregarding, however, the λ-parameter (see above). No significant rapidity dependence of R_{out} and R_{side} are observed. The difference between R_{out} and R_{side} being small we don't observe an effect which was predicted in references [32, 33] for the emission of pions from a long lived source (like the Quark-Gluon-Plasma).

A systematic overview of transverse source sizes of different collision systems is presented in Fig. 20 [34]. On the abszissa various projectile/target combinations are grouped at arbitrary distances; they are ordered, however, according to the geometrical transverse size of of the interaction volume (see below). R_{side} seems to increase monotonically with the transverse size of the system. Any statement about absolute magnitudes or comparisons with sizes of nuclei are dangerous. The transverse source size parameter (R_{side} which is computed as the rms-width of a Gaussian density distribution) should be compared to the mean transverse radius of the interaction volume as defined by the initial conditions. More precisely, the relevant size has to be deduced from the distribution of the pion production points which may be identified with the loci of the nucleon-nucleon interaction points. The density distribution of these points is given by the product of the nucleon density distributions in the overlapping parts of the two colliding nuclei. The projection of the resulting density distributions onto a plane perpendicular to the beam direction, or rather their rms, provides the quantity to which R_{side} must be compared. It turns out (see Fig. 20) that the radius parameters obtained from the two-pion correlation analysis is consistently a factor of $1.5 - 2$ larger than expected from the initial distribution of the pion production points. If this first impression is confirmed by experiments with better statistics and smaller systematical errors and/or larger projectiles and higher energies one could consider the scaling factor a sort of normalization of the method of determining the radius parameter. An increase of this parameter with energy or pion density could be an indication of transverse expansion before the pions decouple.

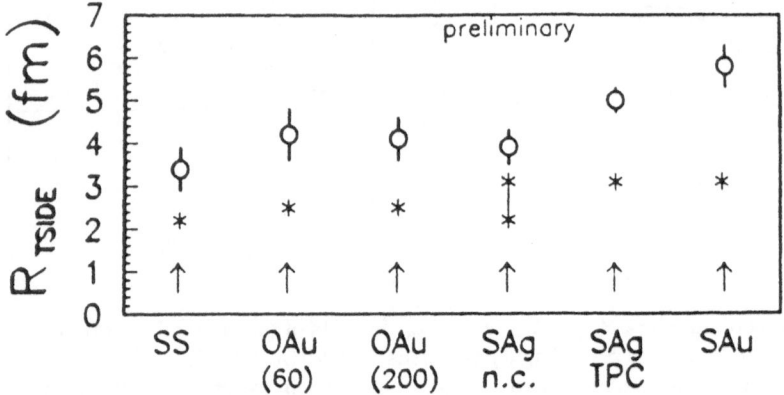

Fig. 20. Radius parameters R_{side} as obtained in various nucleus-nucleus collisions at 60 and 200 GeV/nucleon. The '*' points represent effective transverse sizes of the incident nuclear densities.

So far the rapidity dependence of the source size parameters have been looked at. What do we expect for their variation with transverse momentum? In the case of the longitudinal dimension in the longitudinal expanding scenario the R_{long} dependence on m_T as given by the formula presented earlier provides a concrete prediction. R_{long} should decrease with inreasing p_T as $1./\sqrt{m_T}$. We demonstrate this sort of dependence by showing the correlation functions of Q_{long} for low and high transverse momentum intervals (Fig. 21) as seen in central S+Au collisions. The difference between the correlation functions is striking. Clearly the high p_T pairs give a smaller R_{long} and thus a smaller apparent decoupling time. More detail of the p_T dependence is given in Fig. 22. Here R_{long} values obtained from central S+S, S+Ag, and S+Au collisions are plotted as a function of p_T. The solid lines indicate the $1./\sqrt{m_T}$ function. Again good agreement with the longitudinal expansion scenario is apparent.

Going back to Fig. 21 we consider the p_T dependence of R_{side} and R_{out}. Here the statistical errors are too large to draw any conclusions. The fits to the correlation functions yield smaller radii at large transverse momenta, but a simple overlay of the measured correlations in the two momentum intervals confirms that the differences are not significant. A decrease of the transverse radii with transverse momentum would be indicative of transverse expansion. More

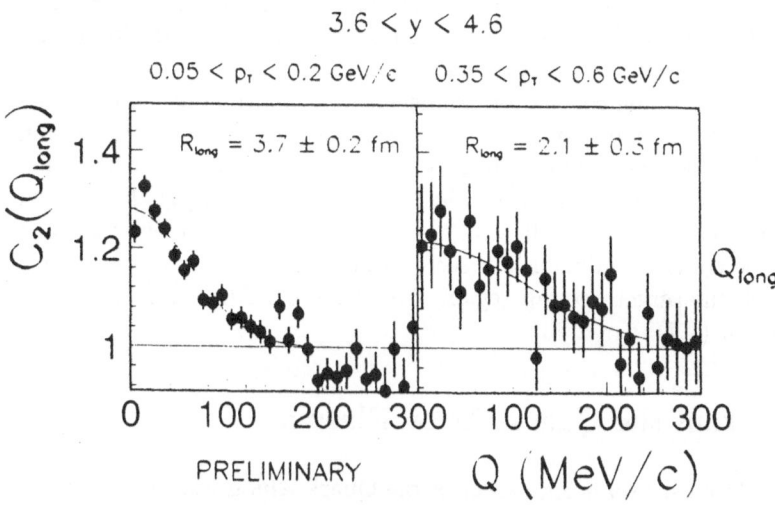

Fig. 21. Two pion correlation functions as obtained in two different transverse momentum intervals in ^{32}S+Au collisions at 200 GeV/nucleon.

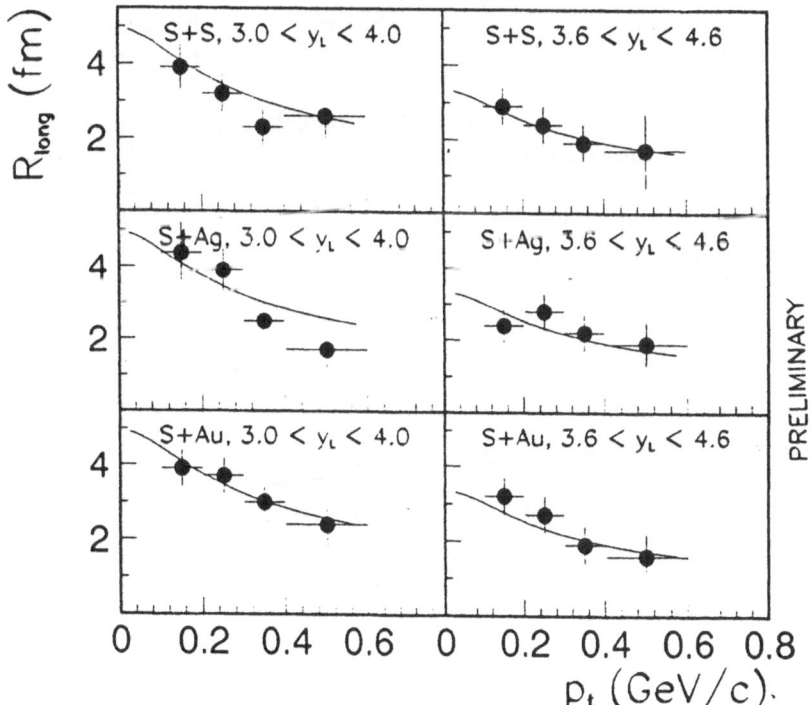

Fig. 22. Radius parameters R_{long} as function of transverse momentum in central ^{32}S+^{32}S, ^{32}S+Ag and ^{32}S+Au collisions.

high statistics data are needed before the existence of transverse expansion can be claimed. To this end NA35 has embarked on the analysis of 15000 Streamer Chamber events. Results should be available in 1994.

In summary two pion correlation studies provide information on the dimension of the particle emitting source. The size parameters are determined as function of rapidity and transverse momentum of the pairs. Subject to the uncertainties of various model assumptions the results strongly suggest that the pion source is expanding longitudinally and that the decoupling time is of the order of 5 fm/c. The transverse dimensions seem to be systematically larger than the effective projectile-target overlap region. The corresponding factor is approximately constant (1.7).

6 Signals from the Quark-Gluon Plasma

The potentially most promising aspect of the Quark-Gluon Plasma is the prospect to study parton interactions at low momentum transfer. These soft collisions will dominate the plasma dynamics. The hope is to learn about them from the bulk properties of the plasma. Before embarking on this problem one has not only to demonstrate the existence of the plasma but also to establish methods with which it can be formed under controlled conditions. In this section we discuss the most commonly quoted signals some combination of which may, once they are found, constitute sufficient evidence for the existence of the Quark-Gluon Plasma (QGP).

There are at least two ways to categorize the signals: there are those which are sensitive to the phase transition from the QGP to hadrons (1) and those which 'feel' the plasma state itself rather than the phase transition (2). If the signals are studied as a function of centre-of-mass energy the resulting excitation function may show the onset of the plasma formation. In the following we will first list examples for both categories. The suppression of J/Ψ mesons and the production of strange particles will be discussed in more detail afterwards. Before doing so we want to come back to the difficulty of establishing reliable signals for the formation of the Quark-Gluon Plasma: the search for a specific signal implies knowledge or at least an estimate of the signal strength relativ to the background. Quantitative predictions must come from theoretical considerations. Except for lattice QCD calculations all theoretical approaches rely on phenomenological models. Their validity depends on the many assumptions which can be subject to more or less severe criticism. Thus single signals are not good signals and we reiterate at this point that the most promising approach is the study of a complete set of observables which describe the final state. They will contain perhaps so many positive signals that only the model with the right ingredients can describe them correctly.

A simple and common signal of a phase transition (category 1) can be recorded in each kitchen. The measurement of the temperature in a heated water pot will exhibit an increase with time until 100°C is reached. Then the temperature stays constant for a while and will rise again once all the water has evaporated. Instead of time one could also use the energy deposited in the system. In particle collisions the temperature can be represented by the mean transverse momentum of produced particles and the energy density by the beam energy or the multiplicity of produced particles. Averaging would be done either per event or event ensembles. The dependence of $< p_T >$ on c.m.-energy and on n_- are shown in Fig. 10 and Fig. 11. $< p_T >$ as a function of c.m.- energy exhibits, after a steep rise, a saturation which is not followed by a second rise. The characteristic feature of a phase transition is not visible. At fixed beam energy the energy density depends on the multiplicity of produced particles represented by n_-. It in turn is a function of projectile size and impact parameter. No variation with n_- is seen in Fig. 11 . The picture of a pure hadronic fireball at low beam energy (or small collisions systems), a mixed phase at intermediate energies and a formation of the QGP for the highest energies is not suitable.

Another phenomenon characteristic of a phase transition is the occurrence of strong fluctuations. They have been searched for, but no unexpected high fluctuations have been found so far.

Many signals from the plasma phase itself have been proposed (category 2). We list some of them here with a few characterizing remarks:

– Thermal photons should be emitted more abundantly from a plasma than from a hadron gas. They originate from parton-parton annihilation and bremsstrahlung [35]. A large background from $\pi^0 \Rightarrow \gamma\gamma$ and other electromagnetic decays dilute a possible signal. Experimental hints for direct (thermal) photons have been reported by the CERN experiment WA80 [36].

– Strangelets are baryons made up by more than 3 quarks and a large number of strange quarks. They may be formed in a baryon rich QGP, if the decay or hadronization of the plasma proceeds predominantly by emission of $q - \bar{q}$ pairs. In this process the abundant u- and d-quarks will combine with relatively high probability with $\bar{s}$ quarks leaving s-quarks behind. In the end, the initial number of baryons, which is a conserved quantity, will contain a large amount of strangeness. Combinations of several strange quarks with the normal u- and d-quarks can lead to stable heavy hadrons with unsual charge [37]. Experimental searches for such particles are underway (see Ref. [60]).

– Jet quenching is one of the QGP signals specific to future experiments at heavy ion colliders. From hadron-hadron collisions we know that the cross section for hard parton scattering, which becomes apparent in jets, increases with energy. At RHIC and LHC energies they can be studied even in the high multiplicity events produced in the collision of nuclei. In the QGP these parton jets may propagate differently than in the physical vacuum or a hadronic gas [38].

– The restauration of chiral symetry is expected to occur in a QGP. If hadronization of the plasma is a relatively slow process the hadron masses may be affected differently in a pure hadronic fireball and a QGP. Therefore, the study of resonances with lifetimes so short that they decay inside the reaction volume may help to differentiate between the two scenarios [39].

6.1 Heavy Particle Production

Heavy particles are promising probes for studying the Quark-Gluon Plasma. Several different mechanisms for their creation have to be considered. In the normal hadronic world the production of heavy particles is suppressed with respect to the light ones, because their creation costs more energy. This simple suppression effect may be modified, if the particles are born in a QGP or in the course of its decay. It is intuitively clear that this has to do with the partonic structure of the hadrons. On strong interaction time scales flavour is a conserved quantum number which means that the quarks are always created in pairs. The modification of the suppression of heavy relative to light quarks is governed by the mass of the $q\bar{q}$-pairs: (i) If the mean kinetic energy (temperature) of the partons in the plasma is of the same order of magnitude as the mass of the $q\bar{q}$-pair, then the creation of these quarks should occur frequently. In current A+A experiments at the AGS and the CERN SPS temperatures of 100-200 MeV are found. Thus the strange quark pairs ($m \approx 200$ MeV) are affected. Based on this argument an enhanced production of strangeness was proposed as a signal for QGP formation [40]. It will be discussed in the second part of this section. In the third part the production of antibaryons will be addressed. (ii) Very heavy quarks like $c\bar{c}$ will be created exclusively in the first generation of N+N interactions, which provide the highest c.m.-energies. The number of these very heavy quarks will not be affected by the presence of a QGP, whereas the mix of charmed hadrons in the final state may be modified. Qualitatively this is easily understood for the J/Ψ meson: if in a free nucleon-nucleon interaction a bound $c\bar{c}$ state is produced, it will appear as a J/Ψ hadron in the final state; if, however, such a $c\bar{c}$ pair has to live for a while in a QGP , its binding forces are screened and the two quarks are free to move away from each other. The consequences of this effect are discussed in the following.

With the number of $c\bar{c}$ pairs given by the number of hard first collisions the most sensitive final state observable would be the ratio R of J/Ψ to all other charmed hadrons in the final state. In interactions which pass through the QGP phase the multiplicity of J/Ψ mesons should be suppressed relative to the multiplicity of D mesons. Experimentally it is very difficult to measure the corresponding inclusive cross sections. Only the detection of the J/Ψ decays into two muons has been mastered in such a way that systematic studies are possible. As originally proposed in [41] the yield of J/Ψ mesons relative to the (Drell-Yang) background was measured by the NA38 collaboration at CERN as a function of centrality in A+A collisions [42]. Fig. 23 clearly demonstrates a relative suppression of J/Ψ mesons in central ^{32}S+U central collisions (compared to less centrally selected events). This is the classical situation of a well founded prediction followed by the experimental confirmation. Does this finding constitute proof of the existence of the Quark-Gluon Plasma? Unfortunately the answer is NO, simply because the above argument is a neccessary but not sufficient condition! In fact a similar effect is expected for a scenario of high energy density hadronic matter. If the $c\bar{c}$ density is small, the rate of absorption of the J/Ψ particle in hadronic interactions will be larger than the rate of creation, resulting in a net J/Ψ suppression. So far the experimental results do not allow to distinguish between the hadronic and partonic scenario, but they suggest that in central A+A collisions the J/Ψ mesons traverse a region of high energy density. Efforts are ongoing to use the transverse momentum dependence of the J/Ψ suppression for decisive experiments. Details of charm production in A+A interactions may be looked up in [5]. All arguments made for charm production hold also for b-quarks. Their higher mass and thus production threshold makes them the subject of future studies at the heavy ion colliders RHIC and LHC.

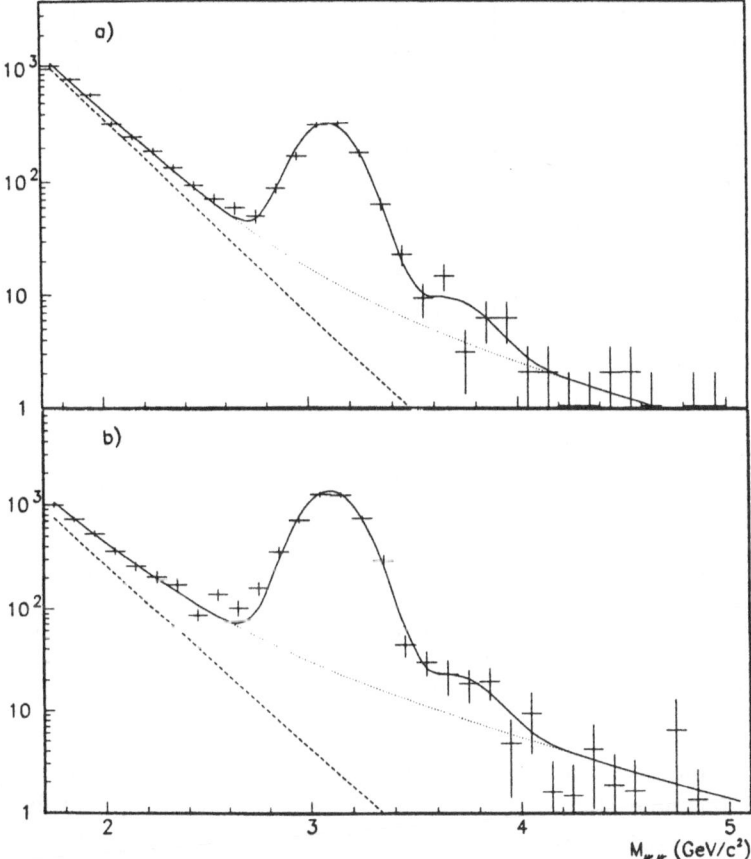

Fig. 23. Mass spectra of muon pairs for high(a) and low(b) E_T ^{32}S+U interactions as measured by NA38. The background is represented by the dashed line. The dotted line represents the sum of background and continuum.

Strange quarks can be found in mesons (kaons) and baryons (hyperons). Their production requires the preceding creation of a s$\bar{s}$ pair. These two strange (and antistrange) quarks may stay together forming a ϕ, they both may pick up an u- or d- (anti-)quark to form two K mesons, or each may collect two fellow

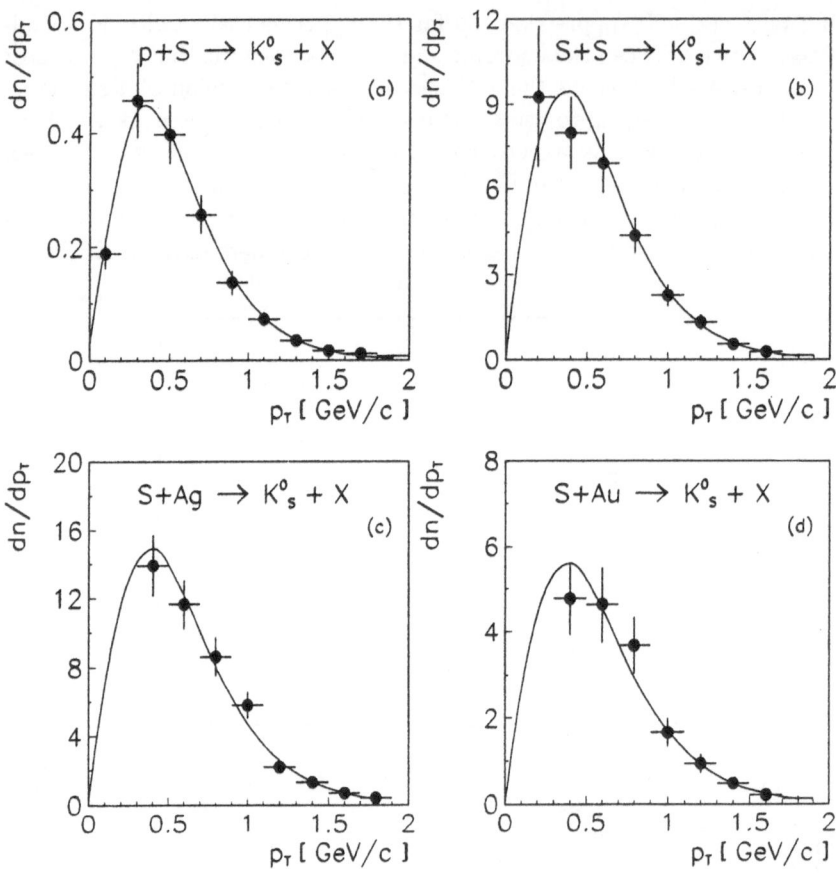

Fig. 24. Transverse momentum distributions for K_s^0 particles produced in p+S (a) interactions and central ^{32}S+S (b), ^{32}S+Ag (c) and ^{32}S+Au (d) collisions at 200 GeV/nucleon. The vertical scale is given in $(GeV/c)^{-1}$. For rapidity ranges see reference [46].

quarks to form (anti-)hyperons. In contrast to the very heavy flavours charm and bottom the number of strange quarks is not given by the number of first generation nucleon-nucleon interactions in A+A collisisons. Subsequent interactions between hadrons (or partons in the QGP) may produce additional strange quark pairs. The strangeness abundance in the final state will obviously be a complicated function of many production and, perhaps less frequent, absorption processes. In such a situation one usually makes a virtue out of neccessity and goes to the limit of complete equilibration. In our case one would, e.g., assume that the collision of two nuclei leads to a hadronic fireball which lives long enough to completely equilibrate. Then the relative particle abundances can be calculated from their masses provided the temperature is known. The resulting ratio R of strange to nonstrange quarks in the final state happens to be larger by about a factor of two than in binary hadron-hadron collisions at the same energy. This

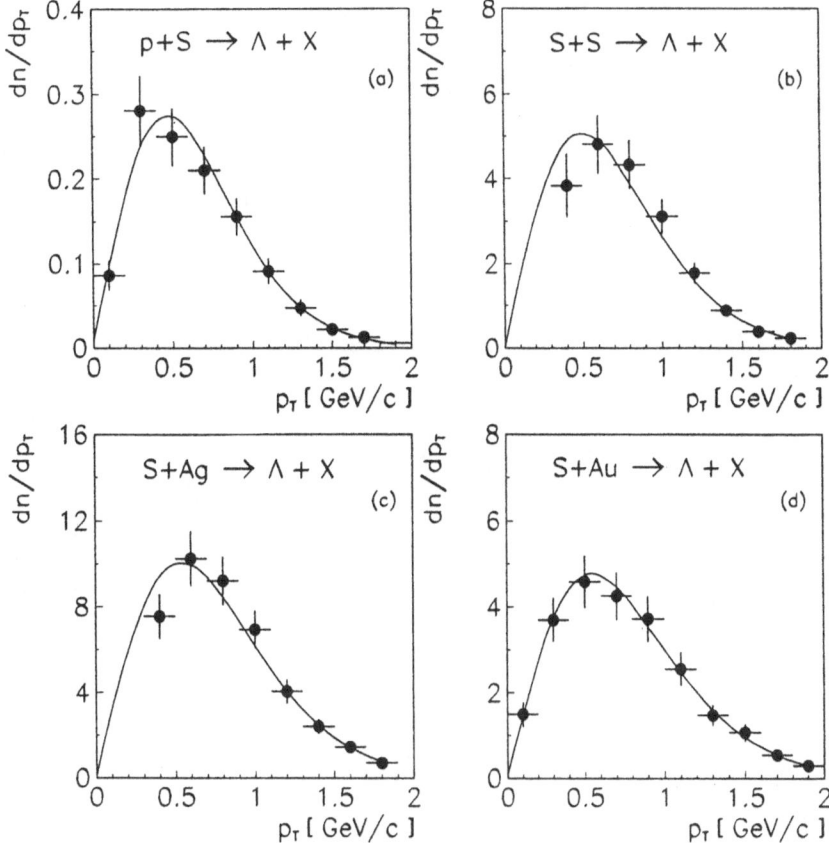

Fig. 25. Same as Fig. 24 but for Λ particles.

is so for the QGP and the hadronic fireball. The double ratio R_{A+A}/R_{N+N} is normally used to quantify the strangeness enhancement. So far we have learned that an equilibrated system will yield enhancements of a factor of two. The question is whether it is possible to reach (chemical!) equilibrium. There are many theoretical arguments that the equilibration time in a hadronic fireball is larger than 20-30 fm/c [43] which is too long for its anticipated lifetime. The results from the previous section suggest lifetimes of 5 fm/c. If this is so, the hadronic fireball scenario will produce a strangeness enhancement of only 20% to 30%. The QGP, on the other hand, will equilibrate much faster due to the large number of particles and their small mass. Here theory predicts equilibration times below 10 fm/c [44]. This argument is the principle basis for calling the strangeness enhancement a signal for the Quark-Gluon Plasma. The experimental challenge is to measure the multiplicity of strange and nonstrange particles in A+A interactions, determine the ratio of strange to nonstrange quarks and compare it to the corresponding number in nucleon-nucleon collisions. Since we are looking for a factor of two we don't need to bother at this point about contributions at

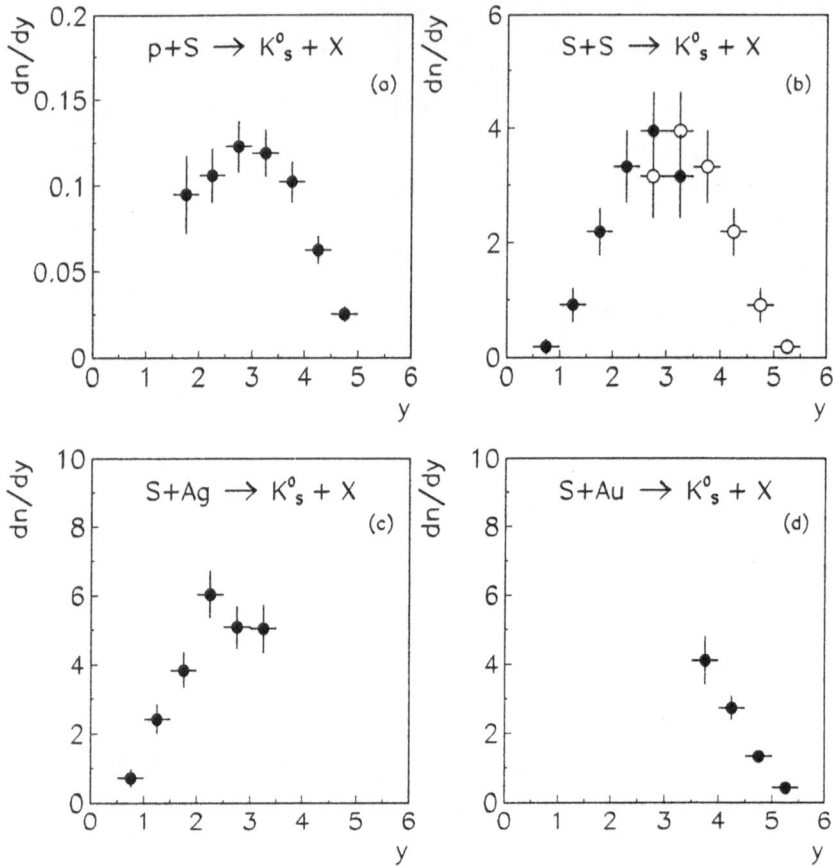

Fig. 26. Rapidity distributions for K_s^0 particles produced in p+S (a) interactions and central ^{32}S+S (b), ^{32}S+Ag (c) and ^{32}S+Au (d) collisions at 200 GeV/nucleon. The data are extrapolated to the full p_T range. The open circles show the distributions for central ^{32}S+S collisions reflected at $y_{cm} = 3.0$.

the few percent level. This means that the determination of the total number of strange quarks requires the measurement of the hyperon and kaon multiplicity only. They are obtained from the charged decays of neutral particles. This topology comprises $K_s^0 \Rightarrow \pi^+\pi^-$, $\Lambda \Rightarrow \pi p$ and $\Sigma^0 \Rightarrow \Lambda\gamma$ decays. The extrapolation from K_s^0 to all kaons by a factor of 4 is straightforward for isospin symmetric systems. The extrapolation from $\Lambda + \Sigma^0$ to all hyperons is more involved and subject to systematic errors as long as the multiplicities of charged and multistrange hyperons have not been determined experimentally. Unfortunately strangeness production in nucleon-nucleon interactions is not measured very accurately (cf. Ref. [45]). Either these measurements will have to be repeated or peripheral nucleus-nucleus collisions will have to be used as reference.

The CERN experiment NA35 has measured Λ and K_s^0 decays in regions of phase space which are large enough to allow safe extrapolations to 4π [46]. As

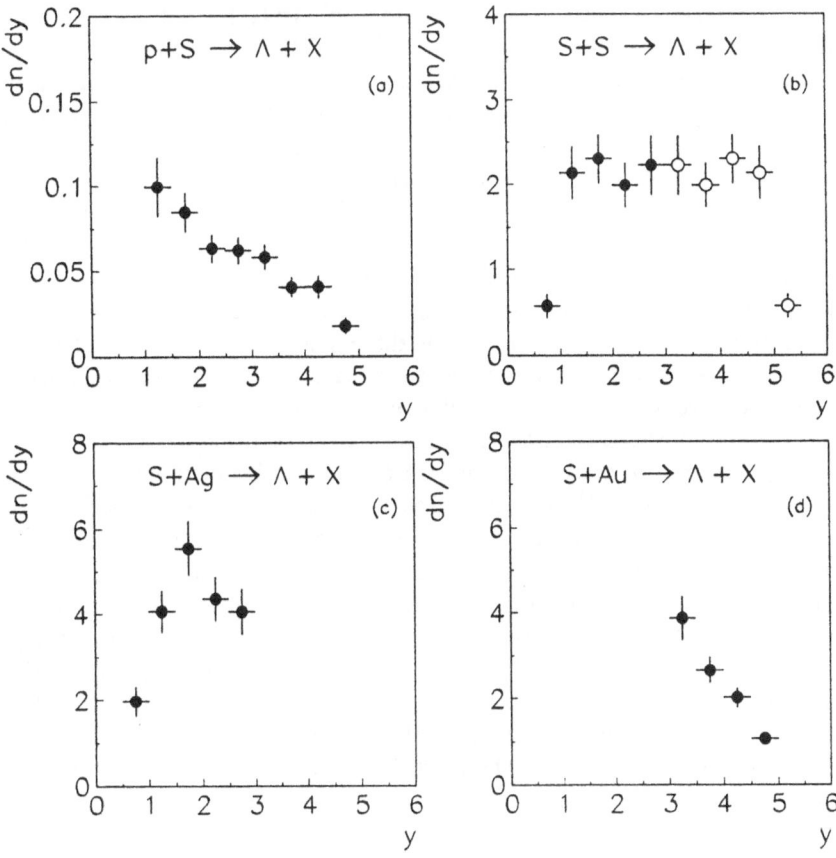

Fig. 27. Same as Fig. 26 but for Λ particles.

examples we present in Figs. 24 and 25 the transverse momentum distributions of K_s^0 and Λ particles in various central A+A collisions. For comparison the calculations from a thermal model are shown as solid lines. The corresponding temperatures are all in the 200 MeV range. In Figs. 26 and 27 the rapidity density distributions are shown. Since Λ and K_s^0 detection beyond rapidity 3 was performed only for the reaction S+Au the extrapolation of S+Ag data was obtained from averaging S+Au and S+S data, the latter being obtained below $y = 3$ but reflected to the forward hemisphere, which is a valid procedure in symmetric systems. Tab. 3 summarizes the measured particle multiplicities in central S+S and S+Ag collisions. The qoted multiplicities for p+p and nucleon-nucleon interactions at 200 GeV/nucleon are taken from Ref. [45]. Clearly the strange particle over pion yields are up by almost a factor of 2 with respect to what was measured in N+N interactions. It is important to note that the shape of the Λ rapidity distribution in central A+A collisions is similar to the one in N+N interactions. This indicates that the production is strong in regions of phase space where it is substantial already in N+N interactions. The conjecture of a special strangeness emitting fireball at midrapidity [47] is not supported by the NA35 data. Other experiments at the CERN SPS have reported a similar strangeness enhancement. Data are available on the K/π ratio [48], the production of ϕ mesons [49], and high p_T hyperons [50]. More recent results from NA36 [51] also find the strangeness enhancement. In addition this collaboration reports a high yield of Λs at midrapidity which is not confirmed by other experiments. In summary all experiments at 200 GeV/nucleon consistently find that strangeness production in Sulphur induced central collisions is enhanced by approximately a factor of two with respect to nonstrange particles. Here again we have a confirmation of a predicted signal. Unfortunately this finding is by itself again not conclusive. One important difference between A+A and N+N interactions are reinteractions which should contribute to strangeness production. Above we have mentioned that equilibrium hadronic processes are unlikely to be the source of the a strangeness enhancement of a factor of two. However, non equilibrium processes like interactions of the incident nucleons with fireball hadrons may boost the creation of s$\bar{\text{s}}$ pairs. Quantitative comparisons of hadronic and partonic predictions need detailed model calculations which have a phenomenological base and are therefore subject to uncertainties. To cut a long story short we find that the microscopic models, which reproduce more or less the results on strangeness production in p+p and A+A collisions at high energy all invoke some non standard string or hadronic feature. These comprise the formation of colour ropes as the result of overlapping strings in RQMD [52], the modification of the strangeness content in the sea quarks in the DPM [53], and the connection of double strings to the leading quark in VENUS [54]. These partonic effects can be considered the onset or precursors of a Quark-Gluon state.

Table 3. Mean total multiplicities of negative hadrons, Λ, $\overline{\Lambda}$ and K_s^0 produced in p+S, central ^{32}S+S and central ^{32}S+Ag collisions at 200 GeV per nucleon

Reaction	$\langle h^- \rangle$	$\langle \Lambda \rangle$	$\langle \overline{\Lambda} \rangle$	$\langle K_s^0 \rangle$
p+p	2.85±0.03	0.096±0.015	0.013±0.005	0.17±0.01
N+N	3.22±0.06	0.096±0.015	0.013±0.005	0.20±0.03
p+S	5.7±0.2	0.28±0.03	0.049±0.006	0.38±0.05
^{32}S+S	95±5	9.4±1.0	2.2±0.4	10.5±1.7
^{32}S+Ag	160±8	15.2±1.2	2.6±0.3	15.5±1.5

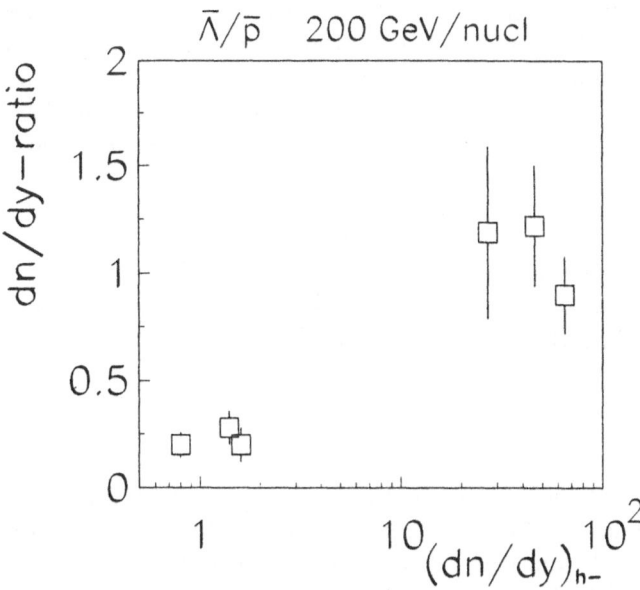

Fig. 28. Ratio of $\overline{\Lambda}/\overline{p}$ around midrapidity as function of negative particle multiplicity density in various hadronic systems. The 3 highest points are from central ^{32}S+A collisions and the lowest from p+p, p+S and p+Au interactions [55]

So far we have left aside the antihyperons, mainly because they should be considered together with the antinucleons. The first remarkable aspect of antihyperon production is the difficulty the microscopic models have to reproduce the experimental yields. Very energetic intermediate states are needed to boost antihyperon production. Such sources should obviously also enhance antinucleon production. Astonishingly enough the ratio $\overline{\Lambda}/\overline{p}$ does not stay constant when comparing N+N with A+A interactions, but rather shows a sharp increase (Fig. 28) [55]. Going back to the concept of an equilibrated system one finds the same ratio $\overline{\Lambda}/\overline{p}$ (≈ 1) in both the hadronic and partonic worlds. For the former we have stated already that chemical equilibration is hard to imagine. As non equilibrium processes a high absorption of $\overline{p}$ but not $\overline{\Lambda}$ can be the reason. This is the picture evolving from RQMD calculations [52]. However, experimental information on $\overline{\Lambda}$ absorption is scarce and this prediction thus subject to large uncertainties. Needless to say that chemical equilibration is a reasonable configuration, if the Quark-Gluon Plasma has been formed in the course of the A+A collisions.

The role of multistrange hyperons (Ξ and Ω) is both encouraging and a problem for the subject of strangeness enhancement. The ratio of the yields of these hyperons to those of their antiparticles are predicted to be more sensitive to the differences between partonic and hadronic scenarios (for example the heavy $\overline{\Xi}$ would need a very long time to approach its equilibrium abundance in a hadronic fireball). On the other hand these multistrange hyperons decay into Λs and $\overline{\Lambda}$s. Their decay products are difficult to separate from primary hyperons. Corrections based on so far unknown (thus estimated) inclusive cross sections for the production of $\Xi\overline{\Xi}$ and $\Omega\overline{\Omega}$ have to be done. With this in mind we conclude that the experimental results on antibaryons should be considered preliminary and subject to large systematical errors. Nevertheless their yields are very important observables.

It is important to study the signals of a strangeness enhancement as a function of c.m.-energy. At the AGS ($\sqrt{s}$=5 GeV as compared to $\sqrt{s}$=20 GeV at the CERN SPS) a similar strangeness enhancement has been found [56]. Its interpretation, however, is different, because microscopic models reproduce strangeness production without invoking the special effects mentioned above.

7 Summary

The search for the Quark-Gluon Plasma is well underway. The playground is the field of ultrarelativistic heavy ion collisions. Experiments have demonstrated that the final state of such collisions can be studied in detail in spite of the large number of produced particles per collision event. In particular the transverse energy, the charged particle multiplicity and the transverse momenta of the produced particles have been measured. Results from different experiments consistently find a high energy density state of (hadronic) matter. Such a state is one of the prerequisites for the formation of a Quark-Gluon Plasma.

For ensembles of central collision events the yields of different particle species present in the final state were determined. Comparison with particle production in nucleon-nucleon interactions shows that relatively few pions but many strange particles are observed. This strangeness enhancement was predicted for hadrons coming from the QGP. It is also seen at lower energies (AGS). Here microscopic models can explain the strange particle surplus with standard hadronic or string interactions. Only detailed studies of the strangeness enhancement especially in multistrange hyperons as function of system size and energy, together with comparisons with model calculations may lead to an unequivocal interpretation of the experimental data on particle production in central nucleus-nucleus collisions at CERN energies.

The detection of rare processes like the decay of the J/Ψ into muons, the decay of mesonic resonances in two leptons and the creation of single photons is difficult but feasible. In the QGP a $c\overline{c}$ pair will have a reduced probability to form a J/Ψ due to colour screening. Experimentally such a suppression was found. This again is only a neccessary but not sufficient condition for plasma formation, since the J/Ψ can undergo inelastic scattering processes in dense hadronic matter in the course of which the two charmed quarks will 'disappear' into D-mesons. So far the experimental results are interesting and promising but not yet conclusive. In this situation one may ask with which probability and to what extent the QGP may be created in ultrarelativistic heavy ion collisions. Maybe the phenomenon we are searching for is a rare event. A reason for the lack of clear evidence could be that the signals become observable only if a large

fraction of the reaction zone turns into the plasma state. This means that the
search has to use more sensitive methods. Testing for several signals at the same
time, i.e. in the same event, allows to check for correlations between the signals.
Such event-by-event analysis will become possible in forthcoming second (and
third) generation experiments. Due to the size of the colliding nuclei and the high
energy the particle yields in the final state will be so high that particle production
characteristics and even source size measurements will be possible on an event-
by-event basis. The large experiments aim at a complete detection of the final
state including particle identification. At CERN the SPS experiment NA49 is
designed to achieve this task. Details of the RHIC and LHC experiments can be
found in Refs. [57, 58, 59]. A short survey for reference purposes is available as
a booklet [60].

There is an experimental programme to search for the Quark-Gluon Plasma
which extends well into the next decade. Our understanding of strong interactions and the structure of the hadrons strongly suggests that the QGP is real.
We end with an intriguing question: what would be the consequence of not finding the Quark-Gluon Plasma in the next ten years? Would we have to change
our picture of the hadron structure?

References

1. K.A. Olive et al.: Nucl. Phys. **A418** (1984) 289c
2. F. Karsch: "Simulating the Quark-Gluon Plasma on the Lattice", CERN-TH-
   5498/89
3. F.R. Brown et al.: Phys. Rev. Lett. **65** (1990) 2491
4. A. Adami and G.E. Brown: Phys. Rep. **234** (1993)
5. C.P. Singh: Phys. Rep. **236** (1993) 147
6. T.C. Awes, F.E. Obenshain, F. Plasil, M.R. Strayerand, and C.Y. Wong (eds.):
   Nucl. Phys. **A525** (1992)
7. E. Stenlund, H.-A. Gustafsson, A. Oskarsson and I. Otterlund (eds.): Nucl. Phys.
   A566 (1994)
8. E.Andersen et al.: Phys. Lett. **B220** (1989)328
9. J. Bächler et al. (NA35 Coll.): Z. Phys. **C52** (1991) 239
10. R. Albrecht et al. (WA80 Coll.): Phys. Rev. **C44** (1992) 2736
11. W. Heck et al. (NA35 Coll.): Z. Phys. **C38** (88) 19c
12. B. Anderson, G. Gustafson, B. Nilsson-Almquist: Nucl. Phys. **B281** (1978) 289
13. K. Werner and P. Koch: Phys. Lett. **B242** (1990) 251
14. J.D. Bjorken: Phys. Rev. **D27** (1983) 140
15. J. Stachel: "Relativistic Heavy Ion Physics at CERN and BNL", Ann. Rev. Nucl.
    Part. Sci. **42** (1992)
16. R. Albrecht et al. (WA80 Coll.): Z. Phys. **C55** (1992) 539
17. A. Bialas, M. Blezsynski, and W. Czyz: Nucl. Phys. **B111** (1976) 461
18. J. Bächler et al. (NA35 Coll.): Z. Phys. **C51** (1991) 157
19. R. Hagedorn: Suppl. Nuovo Cim. **3** (1965) 147
20. J. Simon-Gillo: Nucl. Phys. **A566** (1994) 175c
21. H. Rothard: PhD thesis Frankfurt 1990
22. D. Röhrich (NA35 Coll.): Nucl. Phys. **A566** (1994) 35c

23. K. Werner: Phys. Lett. **B219** (1988) 211
24. J. Bächler et al. (NA35 Coll.): Phys. Rev. Lett. **72** (1994) 1419
25. M. Plümer, S. Raha and R.M. Weiner (eds.): CAMP Workshop Marburg 1990, World Scientific (1991) ISBN 981-02-0331-4
26. D.Ferenc: PhD Thesis, Rudjer Bošković Institute, Zagreb, 1992
27. M. Deutschmann et al.: Nucl. Phys. **B204** (1982) 333
28. K. Kolehmainen and M. Gyulassy: Phys. Lett. **B180** (1986) 203
29. Y. Hama and S. Padula: Phys. Rev. **D37** (1988) 3237
30. A. Makhlin and Yu.M. Sinyukov: Z. Phys. **C39** (1988) 69
31. B. Lorstad and Yu.M. Sinyukov: Phys. Lett. **B265** (1991) 159
32. G.F. Bertsch and G.E Brown: Phys. Rev. **C40** (1989) 1830
33. S. Pratt: Phys. Rev. **D33** (1986) 1314
34. D. Ferenc et al.: Nucl. Phys. **A544** (1992) 531c
35. E. Shuryak: Phys. Lett. **B78** (1978) 150;
    S. Raha and B. Sinha: Int. J. Mod. Phys. **A6** (1991) 517
36. R. Santo et al. (WA80 Coll.): Nucl. Phys. **A566** (1993) 66c
37. H. Liu and G. Shaw: Phys. Rev. **D30** (1984) 1137;
    C. Greiner et al.: Phys. Rev. Lett **58** (1987) 1825 and Phys. Rev. **D44** (1991) 3517
38. M. Gyulassy and M. Plümer: Phys. Lett **B243** (1990) 432
39. R. Pisarski: Phys. Lett. **B110** (1982) 155
40. J. Rafelski and B. Müller: Phys. Rev. Lett. **48** (1982) 1066;
    J. Rafelski: Phys. Rep. **88** (1982) 331
41. T. Matsui and H. Satz: Phys. Lett. **B178** (1986) 416
42. C. Baglin et al.: Phys. Lett. **B220** (1989) 471;
    C. Baglin et al.: Phys. Lett. **B255** (1991) 459
43. P. Koch: Progr. Part. Nucl. Phys. **26** (1991) 253;
    H.C. Eggers and J. Rafelski: Int. J. Mod. Phys. **A6** (1991) 1067
44. M.H. Thoma: Phys. Lett. **B269** (1991) 144;
    G. Baym et al.: Phys. Rev. Lett. **64** (1990) 1867;
    E. Shuryak: Phys. Rev. Lett. **68** (1992) 3270
45. M. Gazdzicki and O. Hansen: Nucl. Phys **A528** (1991) 754
46. T. Alber et al. (NA35 Coll.): accepted for publ. in Z. Phys. C (June 1994)
47. J. Rafelski et al.: Phys. Lett. **B294** (1992) 131
48. M.A. Mazoni et al. (Helios Coll.): Nucl. Phys. **A566** (1994) 95c
49. R. Ferreira et al. (NA38 Coll.): Nucl. Phys. **A 544** (1992) 497c
50. S. Abatzis et al. (WA85 Coll.): Nucl. Phys. **A566** (1994) 225c;
    S. Abatzis et al. (WA85 Coll.): Phys. Lett. **B270** (1991) 123
51. E. Andersen et al. (NA36 Coll.): Phys. Lett. **B294** (1992) 127
52. H. Sorge et al.: Phys. Lett. **B289** (1992) 6
53. J. Ranft et al.: Phys. Lett. **B320** (1994) 346
54. K. Werner: Phys. Rep. **232** (1993) 87
55. The A+A data are from NA35 and very preliminary. The figure is for illustration only and should not be used as a reference.
56. T. Abbot et. al. (E802 Coll.): Phys. Rev. Lett. **64** (1990) 847;
    S.E. Eiseman et al. (E810 Coll.): Phys. Lett. **B297** (1992) 44
57. J. Harris et al. (Star Coll.): Nucl. Phys. **A566** (1994) 277c
58. S. Nagamiya: Nucl. Phys. **A566** (1994) 287c
59. J. Schukraft et al.: Nucl. Phys. **A566** (1994) 311c

60. Copies of the booklet are available at a price of 5 USD from: The secretary of Cosmic and Subatomic Physics, University of Lund, Solvegatan 14, S-22362 Lund, Sweden

Physics at LHC

Michel Della Negra

PPE-Division, CERN,
CH-1211 Geneve 23, Switzerland

Abstract: The Large Hadron Collider (LHC) at CERN will allow to address the most crucial questions of today's particle physics. In these lectures I will review the physics case for the LHC (Lecture 1), the design and performance of the two approved proton-proton experiments (Lecture 2) and the potential of the LHC to study CP violation effects in beauty decays (Lecture 3). I briefly summarize below the content of my lectures.

1 The Physics Case for the LHC (Lecture 1)

The Standard Model (SM) of elementary particles is a triumph of the last 20 years of particle physics. It is a quantum field gauge theory. The fields of matter are the quark and lepton fields. The quanta are fundamental pointlike fermions: the quarks and the leptons regrouped into 3 families. They interact through three elementary forces, weak, electromagnetic and strong, represented by gauge fields, whose quanta are the gauge bosons: W^+, W^-, Z^0, photon, and 8 gluons. The photon and the gluons are massless because the symmetries of the electromagnetic and strong interactions are exact. The W and Z bosons have masses of order 100 GeV.

Hence the gauge symmetry of the weak interactions is broken. In the minimal standard model the mechanism responsible for this symmetry breaking is called the Higgs mechanism. The Higgs field is a scalar field that is modifying the symmetry properties of the vacuum. By interacting with the Higgs field the W and Z become massive.

The quantum of the Higgs field is the Higgs particle. Its mass is not predicted in the standard model, but once its mass is fixed, its coupling to all fundamental particles of the theory are determined. There is no experimental evidence in favor of the Higgs sector of the SM. Direct searches for the Higgs at LEP show that its mass is greater than 62.5 GeV. Unitarity in the combined scattering amplitudes WW, ZZ, HH and HZ imposes that its mass is less than 1 TeV. An accelerator able to probe WW (or ZZ) scattering up to about 1 TeV in the center of mass will necessarily elucidate the Higgs mechanism of the minimal standard model.

This is the main motivation to build the LHC. It is a proton-proton collider reaching 14 TeV in the center of mass. Study of WW scattering up to 1 TeV should be possible with such a machine. The Ws are radiated by the quarks contained inside the colliding protons.

The three gauge coupling constants associated with the three forces are running with energy. Their evolution depends on the number of particles in the theory. If the SM remains valid up to very high energies, one predicts that the 3 coupling constants become approximately equal at about 10^{15} GeV (Grand Unification Theory, GUT) called the GUT scale. Precise measurements at LEP show that in fact unification cannot occur unless there exists a new family of particles with masses around 1 TeV. This is one of the motivations for Supersymmetry or SUSY. Another reason in favor of SUSY is given by the evolution of the Higgs mass. If there is no new physics between the weak scale (300 GeV) and the GUT scale (10^{15} GeV) the Higgs mass will suffer from quadratic divergences because of loop diagrams involving bosons and fermions. One elegant and radical solution to this problem is to introduce exact BOSON/FERMION symmetry. This is SUSY! SUSY partners enter in the loops with identical couplings but opposite signs and the cancellation of the divergences is exact if bosons and fermions have identical masses. In SUSY each fermion (spin 1/2) has a superpartner with spin 0. Likewise each gauge boson (spin 1) has a superpartner which is a fermion (spin 1/2). Clearly supersymmetric bosons (squarks and sleptons) cannot have the same mass as the known fermions otherwise SUSY would have been discovered already. Stabilization of the Higgs mass up to the GUT scale requires M(SUSY) less than or equal 1 TeV in agreement with the constraint on the evolution of the coupling constants.

In the minimal supersymmetric extension of the Standard Model (MSSM) 5 fundamental Higgs particles are predicted: 2 charged and 3 neutrals.

In summary a zoo of new particles is likely to be discovered in the TeV range: fundamental Higgses, SUSY particles, new heavy gauge bosons W' and Z' as predicted by GUT. If no fundamental Higgs exists, the study of WW, WZ and ZZ scattering in the TeV range should reveal that their interactions become strong, with possibly resonances in the s channel in full analogy with pion-pion scattering. For example rho-like resonances could be discovered in the WZ channel.

With an energy of 14 TeV and a design luminosity of 10^{34} cm^{-2}s^{-1} the LHC should allow to have a first look at the 1 TeV particle spectroscopy.

2 The LHC Proton-Proton Experiments (Lecture 2)

Standard model Higgs hunting at LHC requires high luminosity. At the design luminosity of 10^{34}cm^{-2}s^{-1} the interaction rate of inelastic proton-proton collisions is 10^9 events/s. This has experimental implications in the design of the detectors. The detectors and the electronics have to be radiation tolerant. The crossing rate is 40 MHz (one crossing every 25 ns). The average number of inelastic collisions per crossing (so called minimum bias events) is about 20. To reduce the channel occupancy and pile-up effects the detector cells have to be very small and the shaping time of the signals has to be of order 25 ns. Typical LHC detectors will have more than 10^4 calorimeter channels and more than 10^6 tracking channels. The number of events of interest to study electro-weak symmetry breaking phenomena is small. Therefore an efficient trigger selection has to be applied on-line to reduce the rate of recorded events from 10^9 Hz to 100 Hz. Pipelines of at least 3 microsecond depth will be needed to guarantee deadtime-less operation.

The Higgs production cross-section can be computed in the SM with little theoretical uncertainty if we assume that the top mass is 174 GeV as suggested by the recent results from CDF at FNAL. The dominant mechanism is gluon-gluon fusion via a top loop. Only at about 1 TeV the process of W fusion : qq→ qqH becomes comparable in cross-section. The main decay mode for Higgs discovery at LHC is H→ ZZ→ 4 leptons (muons or electrons). The signature is very clean and the backgrounds are small. This mode allows to explore the mass range from 130 GeV to 800 GeV. Below 130 GeV the only practicable decay mode is H→ γγ. In this low mass range the Higgs is very narrow and the signal over background ratio is entirely dominated by the experimental resolution. This calls for a very precise electromagnetic calorimeter. Above 800 GeV and up to about 1000 GeV one has to rely on more difficult channels with higher branching ratios such as H→ ZZ→ 2 neutrinos +2 muons or electrons, or H→ ZZ→ 2 leptons +2 jets. The proposed detectors for LHC have been optimized for SM Higgs detection as a bench mark requirement. The requirements for Higgs detection turn out to be compatible with the requirements for a general purpose detector, which should also allow the study of heavy flavour physics (top and bottom decays) as well as the search for SUSY particles (squarks and gluinos).

The main requirements are:

– Muon and electron detection over a rapidity range [-3, 3]. This is needed for the goldplated channel H→ 4 muons or electrons.

– Photon detection over the same range with good energy resolution and measurement of the photon direction. This is needed for the difficult channel: H→ γγ below 130 GeV.

– Jet detection up to very forward rapidities of about 5. This is needed to ensure good hermeticity for missing transverse energy measurements.

– Capability of measuring missing transverse energy. This is needed in particular to detect gluinos or squarks as expected in the MSSM.

– Tau's are useful for SUSY Higgs detection.

– K^0 and secondary vertex reconstruction are fundamental for CP violation studies in bottom decays.

Out of three letters of intent two general purpose detectors have been selected by the LHC experimental committee:

ATLAS: **A** Toroidal **LHC** Apparatu**S** (Fig. 1)

CMS: **C**ompact **M**uon **S**olenoid (Fig. 2)

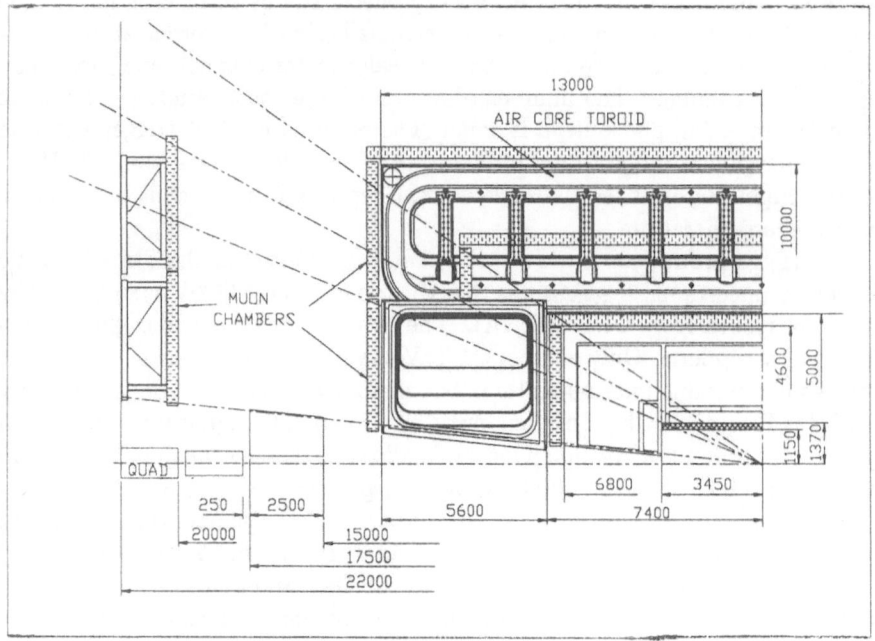

Fig. 1. The ATLAS detector

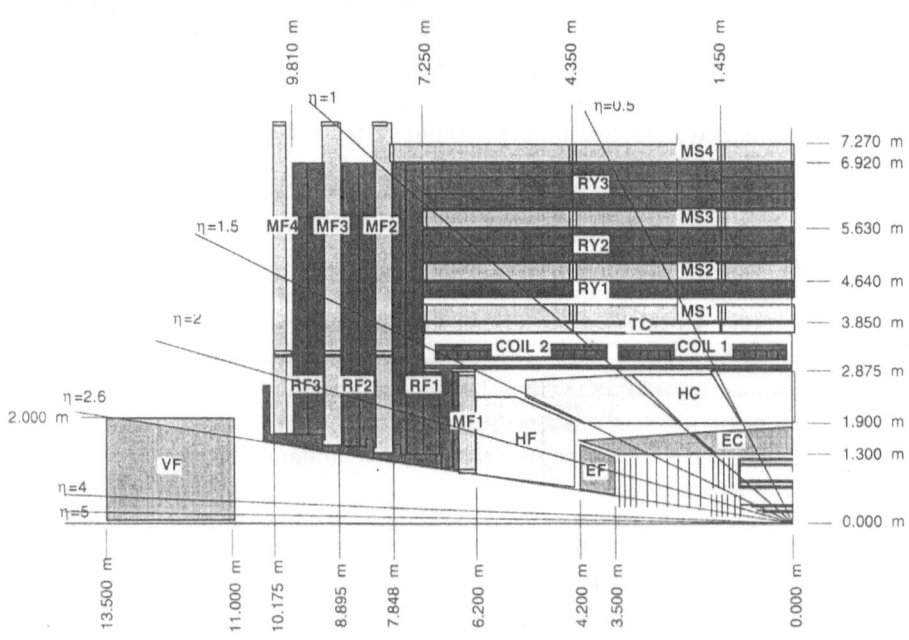

Fig. 2. The CMS detector

The two detectors are complementary in their technological choices. In case of claim of discovery, they will allow the necessary cross-checks. The main difference between the two detectors is in the choice of the magnet for muon measurements.

With an air core toroidal magnet providing an average magnetic field of 0.6 Tesla over a radius range [5m, 10m] ATLAS will measure muons after the calorimeters in a stand alone mode independent of the inner tracking. They aim at a precise sagitta measurement in the air with a systematic error of 45 micrometers. This leads to a momentum masurement precision of about 7% for muons of transverse momenta 1 TeV over the rapidity range [-2.5, 2.5].

With a 15 m long solenoid of inner radius about 3 m delivering a strong magnetic field of 4 Tesla CMS will measure the sagitta of muon trajectories in the plane transverse to the beam over the full radius of the solenoid starting at the vertex. CMS aims at a systematic error on the sagitta measurements of 100 micrometers leading to a momentum precision error for transverse momenta of 1 TeV of 4% up to rapidity of 1.5 (the acceptance of the solenoid) degrading to about 15% at rapidity of 2.4 (maximum acceptance of the muon system). The choice of a strong magnetic field of 4 Tesla leads to a more compact design for the CMS detector. The overall dimensions of CMS are 20 m in length and 15 m in diameter. The corresponding figures for ATLAS are 40 m in length and 20 m in diameter.

The whole calorimetry of CMS sits inside the solenoid. Hence there is no interference between the coil and the calorimetry. CMS aims at a very precise electromagnetic calorimeter working in 4 Tesla. The baseline solution is a lead/scintillator sandwich read by orthogonal wave length shifter fibers (so called Shashlik calorimeter). A more ambitious solution exists consisting of some 100,000 towers of CeF_3 crystals. Cost is a limiting factor. With the crystal calorimeter the Higgs could be discovered in the two-photon mode in the mass range [90 GeV, 150 GeV] for an integrated luminosity of $2 \times 10^4 pb^{-1}$ (5 st. dev.).

The baseline choice of ATLAS for the electromagnetic calorimeter is a sampling calorimeter lead/liquid argon. Large prototypes with projective geometry exist for the barrel and the end caps based on the 'accordion' technics. They have demonstrated an homogeneous response over a large area equivalent to 16 degree in azimuth and 0.65 units of rapidity of better than 1% (rms of 0.6%). The toroid does not generate any magnetic field in the inner tracking cavity. A small solenoid of radius 1.15 m delivering a magnetic field of 2 Tesla has been added in front of the electromagnetic calorimeter. With the coil in front and a preshower to correct for the energy lost in the coil and the vacuum vessel ATLAS expects to discover the Higgs in the two-photon mode in the mass range [90 GeV, 150 GeV] for an integrated luminosity of $10^5 pb^{-1}$ (5 st. dev.).

The inner tracking detector of ATLAS combines high precision discrete layers for momentum measurements with continuous tracking for pattern recognition. The precision layers consist of silicon pixels and microstrips in the barrel and MicroStrip Gas Chambers (MSGC) in the forward regions. The continuous tracking is achieved by 4 mm diameter straw tubes combined with foil radiators for transition radiation measurements. With a magnetic field of 2 Tesla and a radius of 1.15 m the inner tracking detector of ATLAS allows to determine the sign of charged particles at $P_t = 500$ GeV up to rapidities of 2.2.

The inner tracking of CMS consists of only precision layers made either of silicon microstrips in the inner layers or MSGCs in the outer layers. In the barrel the average number of precision measurements is 12. With a magnetic field of 4 Tesla and a radius of 1.3 m the inner tracking detector of CMS allows to determine the sign of charged particles at $P_t = 1.5$ TeV up to rapidities of 2.2.

In both detectors hermeticity of the hadron calorimeter is achieved by adding very forward calorimeters to detect jets up to rapidities of 5. These very forward calorimeters have to be extremely radiation resistant. Different solutions are still under study in CMS and ATLAS.

Both detectors should allow the discovery of gluinos and squarks in the 1 TeV mass range. The signature would be an excess of events with large missing transverse energy ($E_t > 500$ GeV) accompanied by at least 3 jets with $E_t > 200$ GeV.

In summary the question about the nature of electro-weak symmetry breaking can most probably be answered by CMS and ATLAS at LHC. Direct searches can be made for most of the particles associated with the mechanism.

High luminosities and severe experimental conditions pose unique challenges to the machine and to the detectors. Present design of CMS and ATLAS should be able to cope with the difficult experimental environment. The challenge can be accepted with some confidence.

3 B Physics and CP Violation (Lecture 3)

The weak transition amplitude of a down quark i to a top quark j has a V-A structure and is proportional to the Cabibbo-Kobayashi-Maskawa (CKM) matrix element V_{ij}. The CKM matrix is unitary and phase freedom leaves 4 parameters: 3 mixing angles and one phase. This phase is the source of CP violation in the SM. If V_{ij} would be real there would be maximal P and C violation, but CP would be conserved. CP violation requires the CKM matrix to be complex. In the Wolfenstein parameterization the 4 real parameters describing the CKM matrix are λ, A, ρ and η. These 4 parameters are parameters of the SM equally important as M(Z) or M(top) for example.

In the SM CP violation effects (η different from zero) are expected in B decays, which can be tested at LHC. The unitarity of the CKM matrix leads to 6 orthogonality conditions, which eventually should be all tested. One particular orthogonality condition has a good chance to be tested experimentally in B decays, this is the db orthogonality. It leads to the so-called db unitarity triangle in the complex ρ, η plane.

The three angles of the db triangle are called α, β and γ. Our present knowledge of the CKM matrix comes from experimental results on strange and bottom semileptonic decays, the ratio charmless to charm B decays, CP violation in K^0 decays and B_d^0 mixing. The parameters ρ and η are poorly known. The best fit to all data gives $\sin 2\beta > 0.15$ at 90% C.L. This implies that η is different from zero and that large CP violation effects should be present in B decays. Significant improvement on the knowledge of the CKM matrix requires measurement of B_s^0 mixing and of CP violation in B decays. Both measurements can be done at LHC with much better precision than in the planned B factories at SLAC, Cornell and KEK.

To measure B_s^0 oscillations at LHC it is necessary to reconstruct primary and secondary vertices for exclusive B_s^0 decays such as $B_s^0 \to D_s^- \pi^+$. Time integrated asymmetries are too small to be measured. Direct measurement of oscillations is necessary. Hence one has to reconstruct the proper time of the decay, i.e., the distance d between the primary and the secondary vertex and the momentum $P(\mathrm{B})$ of the B particle. Typical required resolutions on d are 1 mm in the forward regions ($P(\mathrm{B}) = 300$ GeV in average) and 0.05 mm in the central region ($P(\mathrm{B}) = 10$ GeV after trigger) to reach x_s (the oscillation parameter) values of 20. This calls for precise vertex detectors. One dedicated B experiment is planned at LHC, which should be able to measure x_s up to the maximum expected value of 40. CMS and ATLAS will also be able to measure B_s^0 oscillations with a sensitivity depending on their final choice for the vertex detector.

Measurement of CP violation effects in B^0 decays induced by mixing gives access in principle to the 3 angles α, β and γ. Measurement of B^0 decays into CP eigenstates are required. For example the decay $B_d^0 \to J/\Psi\, K_s^0$ gives access to the angle β, the decay $B_d^0 \to \pi^+\pi^-$ to the angle α. In practice the precise determination of the angle depends on the assumption that a single amplitude contributes to the decay. Because of Penguin diagrams having different weak phases than the tree diagram for the decay $B_d^0 \to \pi^+\pi^-$, the determination of α, for example, will be affected by theoretical uncertainties. This is not the case for β in the decay $B_d^0 \to J/\Psi\, K_s^0$ for which the determination of β is safe to $< 1\%$. This channel is therefore taken as a benchmark for experiments to be able to observe CP violation in B decays. The expected asymmetry in this channel is large ($> 8\%$) and the final state is easy to identify. Taking into account all branching ratios and efficiency of tagging, one concludes that at least 10^9 $b\bar{b}$ pairs should be produced in order to have the few 100 reconstructed events necessary to have a significant measurement on the asymmetry. In this respect hadron colliders have a net advantage over e^+e^- B factories as for example LHC can produce 3×10^{12} $b\bar{b}$ pairs per year at 10^{33} luminosity (1/10 of nominal luminosity), when B factories can only produce 10^7 $b\bar{b}$ pairs per year. As an example ATLAS and CMS expect to measure $\sin 2\beta$ with a statistical error of < 0.05 in one LHC year at 10^{33}, whilst the SLAC B factory will need few years to reach an error of 0.1!

In summary, study of CP violation effects in B decays is an essential part of the program to measure accurately the parameters of the SM. Only one parameter is responsible of CP violation: η. Many precision tests are therefore possible. In particular it is important to measure all three angles. Two generations of experiments will be in charge of this program. Before 2002 e^+e^- B factories and Hera -B will make first investigations on CP violation effects in B decays. After 2002 precision measurements will come from LHC from CMS, ATLAS and the dedicated B experiment still to be approved.

References

1. I.J.R. Aitchison and A.J.G. Hey: "Gauge Theories in Particle Physics", Adam Hilger, Bristol (1982)
2. J.F. Gunion, H.E. Haber, G.L. Kane and S. Dawson: "The Higgs Hunter's Guide", UCD 89-4
3. G. Altarelli: "The Standard Electroweak Theory and its Experimental Tests", CERN TH 6867/93; and in "Substructures of Matter as Revealed with Electroweak Probes" (Proceedings of the 32. Int. Univ. Wochen für Kern- u. Teilchenphysik, Schladming 1993), ed. by L. Mathelitsch and W. Plessas, Lecture Notes in Physics **426**, Springer-Verlag Heidelberg (1994)
4. T.J. Virdee: "Prospects in Hadron Collider Physics", Rapporteur's talk at EPS Marseille (1993);
   CMS Letter of Intent, CERN/LHCC 92-3;
   ATLAS Letter of Intent, CERN/LHCC 92-4;
   Large Hadron Collider Workshop, CERN 90-10, 3 volumes, Aachen (1990)
5. E.D. Commins and P.H. Bucksbaum: "Weak Interactions of Leptons and Quarks", Cambridge University Press (1983)
6. A. Ali: "Standard Model of Flavour Mixing and CP Violation. Status Report and Perspectives", CERN TH 7123/93
7. P.F. Harrison: "CP Violation at Future Colliders", Rapporteur's talk at EPS Marseille (1993)

Seminars

Several participants contributed interesting seminar talks fitting to the general theme of the School. It was not possible to include them here, but we list the seminar speakers (alphabetically ordered) and the topics covered by them.

V.V. Burov (JINR Dubna):
Investigations of fundamental matter at JINR

G. Duckeck (CERN):
LEP physics - latest results from OPAL

L. Jenkovszky (Acad. Sci. Ukraine):
Bubble free energy in a first-order phase transition

G. Kozlov (JINR Dubna):
Does an exotic light scalar X-boson really exist?

B. Krishnan (TU Wien):
Deconfinement phase-transition in a fluctuating space-time geometry

A. Moroz (Czech. Acad. Sci. Prague):
Instability of matter in 2+1 dimensions

H. Rosu (Univ. Guanajuato):
Möbius function and Hawking radiation

W. Sakuler (TU Wien):
Topological fluctuations in hadronic and quark matter

D. Schwarz (TU Wien):
Thermal field-theoretic approach to cosmological perturbations

A.V. Sidorov (JINR Dubna):
Neumann-Wigner resonances in a relativistic system of two particles

Lecture Notes in Physics

New Series m: Monographs